AF361644

# Statistics in Hydrology

中國統計水文
STAHY
CSH-CNC-IAHS

# Statistics in Hydrology

Editors

**Yuanfang Chen**
**Dong Wang**
**Dedi Liu**
**Binquan Li**
**Ashish Sharma**

MDPI • Basel • Beijing • Wuhan • Barcelona • Belgrade • Manchester • Tokyo • Cluj • Tianjin

*Editors*

Yuanfang Chen
Hohai University
China

Dong Wang
Nanjing University
China

Dedi Liu
Wuhan University
China

Binquan Li
Hohai University
China

Ashish Sharma
University of New South
Wales
Australia

*Editorial Office*
MDPI
St. Alban-Anlage 66
4052 Basel, Switzerland

This is a reprint of articles from the Special Issue published online in the open access journal *Water* (ISSN 2073-4441) (available at: https://www.mdpi.com/journal/water/special_issues/Statistics_Hydrology).

For citation purposes, cite each article independently as indicated on the article page online and as indicated below:

LastName, A.A.; LastName, B.B.; LastName, C.C. Article Title. *Journal Name* **Year**, *Volume Number*, Page Range.

ISBN 978-3-0365-4321-5 (Hbk)
ISBN 978-3-0365-4322-2 (PDF)

# Contents

# About the Editors

**Yuanfang Chen**

Yuanfang Chen is a professor at the College of Hydrology and Water Resources in Hohai University, China. His primary expertise is in the areas of hydrology and water resources engineering with emphasis on stochastic hydrology, statistical methods for river flow long-middle term prediction, hydrologic design and water resources management. He has been President of Commission on Statistical Hydrology, Chinese National Commission for the International Association of Hydrological Sciences (CSH-CNC-IAHS) since 2013, and a Vice President of ICSH/IAHS since 2015. He has worked as a key researcher in a number of sponsored research projects, and received several professional and educational awards, including two state educational prizes of second-class and five provincial science and technology advanced prizes. He obtained a national honor for his excellent contribution to science and technology from the State Council of China in 2012.

**Dong Wang**

Dong Wang is a professor of hydrology and water resources at Nanjing University. Much of his research has focused on stochastic hydrology, fuzzy and AI method in hydrology, and water resources management. He has served as the Vice President of Commission on Statistical Hydrology, Chinese National Commission for the International Association of Hydrological Sciences (CSH-CNC-IAHS) since 2020.

**Dedi Liu**

Dedi Liu is a professor of hydrology and water resources at Wuhan University, China. His primary expertise is in the areas of hydrological modeling, hydrological frequency analysis and water resources allocation. He has served as the President of the Youth League under the China Africa Water Association (YLCAWA), the Vice President of CSH-CNC-IAHS, and the Deputy Director of Youth Working Committee of Water Resources Commission, China Society of Natural Resources. He has authored or co-authored over 100 papers and received the Youth Project of National Talent Program and four other provincial science and technology awards.

**Binquan Li**

Binquan Li, Prof. Dr., obtained his Doctor of Engineering degree in 2014, Hohai University, China, and joined the College of Hydrology and Water Resources, Hohai University as a faculty member after graduation. His main research interests include watershed hydrological modeling, coupled hydrology and hydrodynamics simulation, hydrological uncertainty, cryosphere hydrological cycle, etc. Binquan has served as the Secretary-General of CSH-CNC-IAHS since 2020.

**Ashish Sharma**

Ashish Sharma is a Professor of Civil and Environmental Engineering at the University of New South Wales, Sydney, Australia. He is an engineering hydrologist interested in problems involving hydrological uncertainty. Much of his research has focused on the impact of climate change and variability on hydrological practice, along with applications related to remote sensing, formulating stochastic approaches, developing hydrological models, and the two big hydrology bread-and-butter problems: design flood estimation and water resources management. Ashish has served as the

Past-President of the International Commission of Hydrologic Sciences (IAHS) Commission on Statistical Hydrology (STAHY). He has served on the Australian Research Council's College of Experts twice, and the Technical Committee overseeing the development of the Australian Rainfall and Runoff Design Flood Estimation guidelines (ARR2016).

*water*

MDPI

*Editorial*

# Statistics in Hydrology

Yuanfang Chen [1], Dong Wang [2], Dedi Liu [3], Binquan Li [4,5,*] and Ashish Sharma [6]

[1] College of Hydrology and Water Resources, Hohai University, Nanjing 210098, China; chenyuanfang@hhu.edu.cn
[2] Department of Hydrosciences, School of Earth Sciences and Engineering, State Key Laboratory of Pollution Control and Resource Reuse, Nanjing University, Nanjing 210046, China; wangdong@nju.edu.cn
[3] State Key Laboratory of Water Resources and Hydropower Engineering Science, Wuhan University, Wuhan 430072, China; dediliu@whu.edu.cn
[4] State Key Laboratory of Simulation and Regulation of Water Cycle in River Basin, China Institute of Water Resources and Hydropower Research, Beijing 100038, China
[5] State Key Laboratory of Hydrology-Water Resources and Hydraulic Engineering, Hohai University, Nanjing 210098, China
[6] School of Civil and Environmental Engineering, University of New South Wales, Sydney, NSW 2052, Australia; a.sharma@unsw.edu.au
* Correspondence: libinquan@hhu.edu.cn

**Citation:** Chen, Y.; Wang, D.; Liu, D.; Li, B.; Sharma, A. Statistics in Hydrology. *Water* **2022**, *14*, 1571. https://doi.org/10.3390/w14101571

Received: 9 May 2022
Accepted: 11 May 2022
Published: 13 May 2022

## 1. Introduction

Statistical methods have a long history in the analysis of hydrological data for designing, planning, infilling, forecasting, and specifying better models to assess scenarios of land use and climate change in catchments. The effectiveness of statistical descriptions of hydrological processes reflects the enormous complexity of hydrological systems, which makes a purely deterministic description ineffective. Many different statistically oriented methodologies are used in hydrological studies, with multiple examples in the literature where the statistical aspect of the work is incomplete, unreasonable, or even unacceptable.

On 19–20 October 2019, the 10th International Workshop on Statistical Hydrology (STAHY 2019), which was organized by the International Commission on Statistical Hydrology, International Association of Hydrological Sciences (ICSH-IAHS), took place in Nanjing, China. A total of 132 participants from 13 countries including Australia, Belgium, Canada, China, Germany, Italy, Poland, Spain, Switzerland, the United States, etc., registered to participate in the conference, and more than 300 graduate students and young scholars from universities in Nanjing also attended the conference. In addition, 28 early career scholars from eight countries participated in the first Early Career Course (ECC) of ICSH held in Nanjing. The ECC was held in advance of the STAHY 2019 workshop for one day (October 18). Multiple new academic innovations and achievements were inspired during and after the conference. The authors of the presentations proposed to publish their research results in the Special Issue of "Statistics in Hydrology". Thus, the main purpose of this Special Issue is to share the latest research in the field of statistics in hydrology with reference to the discussions held during STAHY 2019.

## 2. Overview of the Contributions

The call for papers was announced in April 2021, and after a rigorous peer-review process, a total of 11 papers have been published [1–11]. To gain a better insight into the essence of the Special Issue, a brief overview of these papers is shown below.

Statistical methods provide effective tools for the analysis of changes in hydrometeorological variables and extreme hydrological/meteorological events, as well as the correlation between different variables. Ahemaitihali and Dong [1] analyzed the spatiotemporal characteristics and driving forces of flash floods in the Altay Prefecture, China. They examined the kernel density, standard deviational ellipse, and spatial gravity center of flash floods and analyzed the temporal and spatial variations. Several statistical methods including multiple linear regression, principal component analysis and random forest were used to analyze the driving force of flash floods. This study uses a variety of statistical methods to provide insights into the spatial–temporal dynamics of flash floods and a reference case for similar studies. Dong et al. [2] applied multiresolution analysis and continuous wavelet analysis approaches to evaluate the influences of precipitation and river level fluctuations on groundwater. Their results showed that the wavelet technique was more effective than spectral analysis in detecting the correlations among precipitation, river level, and aquifer level in the temporal domain. Lang et al. [3] explored the characteristics of extreme precipitation events over the Henan province, China using the latest ERA5 dataset. Their results showed that precipitable water, wind, and relative humidity were the most common drivers for extreme precipitation events over the Henan province for the 1981–2021 period. This study provides insights for further flood estimations and forecasts.

The forecasting and prediction of hydrological elements such as precipitation and runoff is one of the main topics in the hydrological field. The statistical method is central to this. He et al. [4] developed statistical forecasting models for summer rainfall and streamflow over the Yangtze River valley using a relatively small number of samples. The logistic regression was used to make probability forecasts for the binary classification, and then three testing procedures, i.e., predictability assessment (PA), model output statistics (MOS), and the reanalysis-based (RAN) approach, were explored. Their results showed that the RAN approach was better at generating exceedance probability forecasts than MOS, as RAN can utilize more samples. The findings provide a useful reference for these statistical methods in long-term hydrological forecasts. Data-driven machine learning (ML) methods are also widely used in hydrology. Under the stacking ensemble learning framework, Gu et al. [5] used four ML models, namely k-nearest neighbors (KNN), extreme gradient boosting (XGB), support vector regression (SVR), and artificial neural networks (ANN), for monthly rainfall prediction. Their results showed that the stacking ensemble learning method can synthesize the advantages of each single ML model to produce a good simulation accuracy in their study area (Taihu Basin, China).

Uncertainty in hydrological forecasting has always existed. Statistical methods are an effective means to quantitatively deal with this uncertainty. Romero-Cuellar et al. [6] developed an extension of the Model Conditional Processor (MCP), which merged clusters with Gaussian mixture models to offer an alternative solution to manage heteroscedastic errors. Case studies indicated that this new post-processer had significant potential in generating more reliable, sharper, and more accurate monthly streamflow predictions than the MCP and MCP using a truncated normal distribution, especially in dry catchments.

Statistical methods are highly important in engineering hydrological design, such as sample processing, model fitting, and statistical parameter estimation in frequency analysis. The Special Issue published five papers on this topic [7–11]. Among them, the first two papers [7,8] are mainly about model selection and parameter estimation in frequency analysis under stationary conditions, and the last three papers [9–11] mainly focus on frequency analysis under nonstationary conditions caused by changing environment. Shao et al. [7] revised the method for the regional frequency analysis of extreme precipitation using regional L-moments methods. A Monte Carlo (MC) simulation was conducted to determine the appropriate probability distribution. A case study in Jiangsu province, China shows that the frequency estimations based on this revised regional frequency analysis are in good agreement with observations. This study provides a new perspective in regional frequency analysis. Song et al. [8] used the maximum likelihood estimation (MLE) method to estimate the parameters and confidence intervals of quantiles of the four-parameter exponential gamma (FPEG) distribution. The FPEG distribution was then applied to precipitation data of the Weihe watershed in China. The results showed that FPEG distribution is a good candidate for modeling annual precipitation data. Considering that the use of the FPEG distribution has received only limited attention from the hydrological community, this finding may provide guidance for estimating design values of random variables in other parts of the world.

Due to the changing environment, hydrological design such as frequency analysis faces nonstationary difficulties. Zeng et al. [9] used Bayesian nonstationary time-varying moment models to investigate the annual maximum flood peak (AMFP) risk and return period. Two climate covariates were defined to exhibit a significant positive correlation with AMFP. The results indicated that the climate-informed model demonstrated the best performance on extreme flood frequency analysis as well as sufficiently explaining the variability of extreme flood risk. This study provides a good alternative to extreme flood frequency analysis under the context of climate change. Li et al. [10] found that the 1-day annual maximum flood volume (AMFV) exhibits a significant correlation with AMFP in a case study at Longmen Reservoir (China). Moreover, they developed a copula-based bivariate nonstationary flood frequency analysis to investigate environmental effects on the dependence of flood peak and volume. The results showed that the design floods estimated by bivariate nonstationary joint distribution would increase largely compared with the ones estimated in a univariate nonstationary context. On this subject, Li and Qin [11] introduced the mechanism-based reconstruction (Me-RS) method into this topic and demonstrated a case study on the calculation of design annual runoff under nonstationary conditions in the Jialu River Basin, China. In the Me-RS framework, the nonstationary hydrological series was transformed into stationary series for forthcoming hydrological design calculation. The results showed that the Me-RS method not only had theoretical support, but also its obtained design values were consistent with the actual condition and had much smaller uncertainty. Thus, the method provides an effective tool for annual runoff frequency analysis under nonstationary conditions. The findings are very useful because the statistical characteristics of many rivers around the world exhibit complex nonstationary changes.

## 3. Conclusions

The above 11 papers contribute to the increasing interest in the studies of hydrometeorological changes, hydrological prediction and uncertainty analysis, and engineering hydrological design under stationary/nonstationary conditions. The Guest Editors hope that readers will be inspired by this Special Issue and further innovate in the field of statistical hydrology. In particular, the era of "big data" has arrived, which will bring new development opportunities to statistical hydrology.

**Funding:** This research was funded by the National Natural Science Foundation of China, grant number 41877147, and the Open Research Fund of State Key Laboratory of Simulation and Regulation of Water Cycle in River Basin (China Institute of Water Resources and Hydropower Research), grant number IWHR-SKL-202013.

**Acknowledgments:** The authors of this paper, who served as the Guest Editors of this Special Issue, wish to thank the journal editors, all authors submitting papers, and the referees who contributed to revise and improve the 11 published papers. In addition, we would like to thank Xia Wu of Hohai University and Shanika Zhu, section managing editor of this journal for their contributions to communication and coordination in the organization of this Special Issue.

**Conflicts of Interest:** The authors declare no conflict of interest.

## References

1. Ahemaitihali, A.; Dong, Z. Spatiotemporal Characteristics Analysis and Driving Forces Assessment of Flash Floods in Altay. *Water* **2022**, *14*, 331. [CrossRef]
2. Dong, L.; Guo, Y.; Tang, W.; Xu, W.; Fan, Z. Statistical Evaluation of the Influences of Precipitation and River Level Fluctuations on Groundwater in Yoshino River Basin, Japan. *Water* **2022**, *14*, 625. [CrossRef]
3. Lang, Y.; Jiang, Z.; Wu, X. Investigating the Linkage between Extreme Rainstorms and Concurrent Synoptic Features: A Case Study in Henan, Central China. *Water* **2022**, *14*, 1065. [CrossRef]
4. He, R.; Chen, Y.; Huang, Q.; Wang, W.; Li, G. Forecasting Summer Rainfall and Streamflow over the Yangtze River Valley Using Western Pacific Subtropical High Feature. *Water* **2021**, *13*, 2580. [CrossRef]
5. Gu, J.; Liu, S.; Zhou, Z.; Chalov, S.R.; Zhuang, Q. A Stacking Ensemble Learning Model for Monthly Rainfall Prediction in the Taihu Basin, China. *Water* **2022**, *14*, 492. [CrossRef]

6.   Romero Cuellar, J.; Gastulo-Tapia, C.J.; Hernández-López, M.R.; Prieto Sierra, C.; Francés, F. Towards an Extension of the Model Conditional Processor: Predictive Uncertainty Quantification of Monthly Streamflow via Gaussian Mixture Models and Clusters. *Water* **2022**, *14*, 1261. [CrossRef]
7.   Shao, Y.; Zhao, J.; Xu, J.; Fu, A.; Wu, J. Revision of Frequency Estimates of Extreme Precipitation Based on the Annual Maximum Series in the Jiangsu Province in China. *Water* **2021**, *13*, 1832. [CrossRef]
8.   Song, S.; Kang, Y.; Song, X.; Singh, V.P. MLE-Based Parameter Estimation for Four-Parameter Exponential Gamma Distribution and Asymptotic Variance of Its Quantiles. *Water* **2021**, *13*, 2092. [CrossRef]
9.   Zeng, H.; Huang, J.; Li, Z.; Yu, W.; Zhou, H. Nonstationary Bayesian Modeling of Extreme Flood Risk and Return Period Affected by Climate Variables for Xiangjiang River Basin, in South-Central China. *Water* **2022**, *14*, 66. [CrossRef]
10.  Li, Q.; Zeng, H.; Liu, P.; Li, Z.; Yu, W.; Zhou, H. Bivariate Nonstationary Extreme Flood Risk Estimation Using Mixture Distribution and Copula Function for the Longmen Reservoir, North China. *Water* **2022**, *14*, 604. [CrossRef]
11.  Li, S.; Qin, Y. Frequency Analysis of the Nonstationary Annual Runoff Series Using the Mechanism-Based Reconstruction Method. *Water* **2022**, *14*, 76. [CrossRef]

*Article*

# Spatiotemporal Characteristics Analysis and Driving Forces Assessment of Flash Floods in Altay

Abudumanan Ahemaitihali [1,2] and Zuoji Dong [1,*]

1   School of Economics and Management, University of Chinese Academy of Sciences, Beijing 100190, China; abudoumanan19@mails.ucas.ac.cn
2   Altay Regional Committee of the Communist Youth League, Altay 836500, China
*   Correspondence: dongzuoji@ucas.ac.cn

**Abstract:** Flash floods are devastating natural disasters worldwide. Understanding their spatiotemporal distributions and driving factors is essential for identifying high risk areas and predicting hydrological conditions. In this study, several methods were used to analyze the changing patterns and driving factors of flash floods in the Altay region. Results indicate that the number of flash floods each year increased in 1980–2015, with two sudden change points (1996 and 2008), and April, June, and July presented the highest frequency of events. Habahe and Jeminay were known to have high flash flood incidences; however, currently, Altay City, Fuhai, Fuyun, and Qinghe are most affected. In terms of driving force analysis, precipitation and altitude performance have a key impact on flash flood occurrence in this settlement compared to other subregions, with a high percentage increase in the mean squared error value of 39, 37, 37, 37, and 33 for 10 min precipitation in a 20-year return period, elevation, 60 min precipitation in a 20-year return period, 6 h precipitation in a 20-year return period, and 24 h precipitation in a 20-year return period, respectively. The study results provide insights into spatial–temporal dynamics of flash floods and a scientific basis for policymakers to set improvement targets in specific areas.

**Keywords:** flash flood; spatiotemporal change; driving factor; Altay

**Citation:** Ahemaitihali, A.; Dong, Z. Spatiotemporal Characteristics Analysis and Driving Forces Assessment of Flash Floods in Altay. *Water* **2022**, *14*, 331. https://doi.org/10.3390/w14030331

Academic Editors: Yuanfang Chen, Dong Wang, Dedi Liu, Binquan Li and Ashish Sharma

Received: 30 November 2021
Accepted: 21 January 2022
Published: 24 January 2022

**Publisher's Note:** MDPI stays neutral with regard to jurisdictional claims in published maps and institutional affiliations.

## 1. Introduction

Flash floods are disaster events with high peak flows and short response times in mountainous watersheds of tens to numerous square kilometers, usually triggered by heavy rainfall [1,2]. Globally, flash floods are among the most devastating natural disasters, often resulting in loss of life and significant economic damage [3]. In America, for example, flash floods ranked first among the causes of death, with approximately 100 deaths every year [4]. Additionally, between 1950 and 2006, 40% of flood related deaths in Europe were from flash floods [5], while this proportion exceeded 80% in southern Europe [6]. China is a vast country with a complex ecological and geographical environment, which is also influenced by heavy precipitation, human activities and other factors [7]. China has suffered the most serious flash floods worldwide [8], and the incidence of these events shows an upward trend [9]. From a regional point of view, affected by the complex terrain, the distribution of heavy rainfall presents an obvious spatial difference pattern. Indeed, according to the latest IPCC report, under a global warming scenario of 1.5 °C, flash floods will be more frequent and more violent in Asia [10]. Meanwhile, the Altay ecosystem is extremely sensitive to climate change and human activity owing to its vulnerability and geographical conditions. From this perspective, it is crucial to bridge the knowledge gap between the spatial and temporal patterns of flash floods and the driving patterns in the Altay region.

In recent decades, most relative studies have illustrated that flash floods are related to a combination of spatial and temporal factors [11–14]. Currently, research on flash floods has concentrated on three aspects: (1) the assessment of flash flood risk [15,16], (2) flash flood mechanisms [17,18], and (3) spatiotemporal distribution and influencing factors [19,20]. The risk assessment of flash floods is mainly to identify high and low risk areas at the local or national scale [15,21,22], which are mainly quantified by three methods: the scenario simulation method [23,24], a historical data based method [25,26], and an index based system method [27,28]. Additionally, the complexity of flash flood formation has led many of these studies to be conducted only in typical watersheds [29]. In general, when researching flash flood disasters, it is necessary to make a comprehensive record of past events. According to this theory, it is very meaningful to analyze the temporal and spatial components of flash flood disasters and explore the driving factors behind them. Concerning spatial–temporal distribution and influencing factors, numerous studies have used kernel density estimation, spatial autocorrelation, spatial gravity center migration, and standard deviational ellipse to discover the temporal and spatial specialties of flash floods [8,30]. Numerous methods have been applied to conduct the driving force analysis of flash floods (i.e., principal component analysis, geographical detector, multiple linear regression, and random forest), each with its advantages. On this basis, precipitation, terrain, normalized difference vegetation index (NDVI), and human activities, among others, have been selected by many scholars as influencing factors [31]. This type of study generally selects a country (or province) as the study object.

Despite extensive research into the mechanisms and drivers of flash floods, problems remain. Firstly, some studies indicate that land use has changed dramatically due to urban expansion and increased human activity [20], and there is an urgent need for research that starts with land-use factors for driving force analysis. Second, certain static factors respond to dynamic factors, while ignoring the spatiotemporal aspects of dynamic factors, which may lead to inaccurate and objective results. Current studies on the influences of flash floods on urban expansion and human activities and intensification of rainfall are needed, although some studies have explained the response mechanisms between human activities and flash floods. Moreover, some previous studies were performed to detect only the interaction between two drivers [31,32], but they did not adequately reflect the rate of contribution of each driver in the different models. Finally, the current flood prediction research is mostly focused on areas with high flooding [33,34], while ignoring arid areas such as Altay. However, the past flash flood disasters in Altay also caused great harm to local people and economic losses. In particular, with the influence of the Altay Mountains and the Gurbantunggut Desert, the geological features of Altay show uniqueness and climate diversity; thus, there is an urgent need to explore and analyze the spatial-temporal distribution and driving factors of flash floods in Altay.

The purpose of this research is to analyze the changing patterns of historical flash floods and explore the driving factors behind flash floods in the Altay region. The primary objectives were: (1) discovering the spatial–temporal variability of flash floods based on the M-K test, kernel density estimation, standard deviational ellipse, and spatial gravity center model; and (2) analyzing the drivers of flash flooding in four land-use type subdivisions using four methods. The research results can provide scientific reference and decision support for the development of reasonable disaster prevention measures and effective flood risk management.

## 2. Materials and Methods

### 2.1. Study Area

The Altay Prefecture is located in the heartland of Eurasia, spanning over $45°0000''–49°1045''$ N and $85°3136''–91°0423''$ E. The area is in northwest China, bordered by Mongolia, Russia, and Kazakhstan. The Altay Prefecture is composed of seven counties and covers $1.18 \times 10^5$ km$^2$. Precipitation and temperature in the Altay Prefecture vary considerably. Owing to the blockage of the Arctic and Atlantic monsoons by the Altay Mountains, high precipitation and low temperatures occur in the northern mountains. Meanwhile, the southern plain has a dry climate and scarce precipitation because of the effect of the Gurbantunggut Desert. A flash flood is a violent surface water runoff event, which is usually caused by precipitation in a small watershed. According to the flash floods inventory database provided by National Mountain Flood Disaster Investigation Project (NMFDIP), the spatial distribution of flash floods in 1980–2015 is plotted in Figure 1. The position of flash flood points is defined as the central location of the historical flash flood ditch.

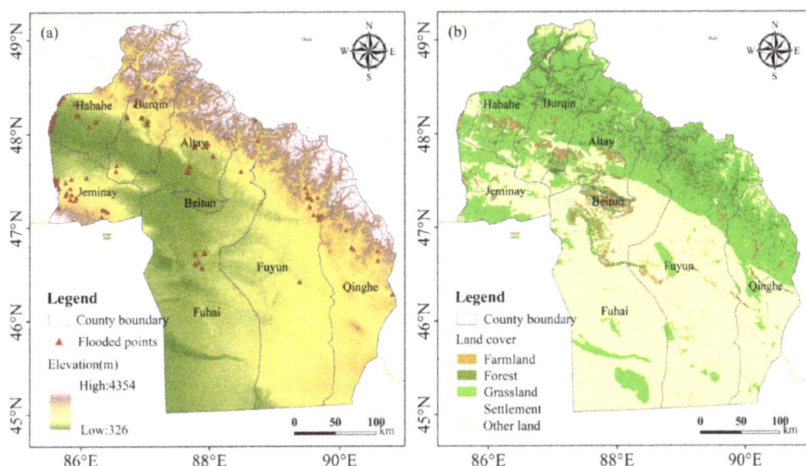


**Figure 1.** The study area: (**a**) the flash floods inventory map in Altay; (**b**) landcover types in Altay.

Affected by hydrothermal conditions, climate change, and human activities, the land cover in 2020 is mainly cropland (1.91%), forest (4.92%), grassland (11.39%), meadow (10.94%), and desert (61.84%). Altay is one of the key pastoral areas in China, with rich natural grassland resources accounting for nearly 15% of the Xinjiang Uygur Autonomous Region, where flash floods are rapidly increasing, posing a serious threat to people's lives and property.

*2.2. Dataset*

Flash floods can be caused by a variety of factors, such as heavy rainfall, dam and levee breaches, landslides, and urbanization [35], but heavy rainfall is generally considered to be the most common cause, and in this study, heavy rainfall is the factor of flash floods. Previous studies have shown that factors such as precipitation, topography, and human activities are closely related to the occurrence of flash floods [11,36]. Therefore, the scientific data used in this research are mainly divided into two types: (1) Flash floods inventory database, which is provided by NMFDIP, which was launched by the Ministry of Water Resources of China and the Ministry of Finance of China (MWRCMFC) in 2013. Collected by field surveying, the dataset records the occurrence time, longitude, latitude, and other attributes of historical flash floods. It is worth mentioning that not every cause of flash floods is mentioned in this database, however, among the causes mentioned in this study, the main factor of flash floods in Altay is heavy rainfall. This database had been strictly inspected, and widely used in a large number of studies [37–39]; and (2) The cover raster data and the driving force factor, which is consistent with previous studies in selecting driving factors, mainly including precipitation factors, representing topography and human activities data. Precipitation data is grid data with four indicators, which are the 6 h rainstorm (H06_20), 24 h rainstorm (H24_20), 10 min rainstorm (M10_20), and 60 min rainstorm (M60_20) in the 20-year return period. This dataset belongs to the flood inventory dataset mentioned above and is also widely used. Data representing topography are digital elevation model (DEM) raster data and its derivatives (slope raster data (SLP) and topographic relief raster data (TR)). NDVI, gross domestic product (GDP), and population density (PD) data were used to represent human activities.

According to the description of the above data, we determine that the number of historical flash flood points in Altay is 210, from the flash flood database. Precipitation raster data of 1 km × 1 km is derived from the inverse distance weighted interpolation of the precipitation factor grid. We calculate the TR and SLP based on DEM data using the focal statistical method. The NDVI raster data is the multi-year average value of MODIS NDVI products (MOD13A1) from 2000 to 2015, calculated from the Google Earth Engine platform. The corresponding historical PD and GDP raster data are obtained from the Resource and Environment Science and Data Center of the Chinese Academy of Sciences (RESDC). More descriptions of the datasets are represented in Table 1 and shown in Figure 2.

**Table 1.** Data source and description.

| Type | Factors | Spatial Resolution | Temporal Resolution | Description |
|---|---|---|---|---|
| Basic data | Flash floods | Vector data | 1980–2015 | China, National Mountain Flood Disaster Investigation Project [31]. |
| | Landcover | 1 km × 1 km | 2010 | RESDC. |
| Driving force factor data | Precipitation factors | Vector data | 2015 | China, National mountain flash flood disaster survey and evaluation data. |
| | DEM | 90 m × 90 m | 2010 | Geospatial Data Cloud of China. |
| | NDVI | 1 km × 1 km | 2000–2015 | Google Earth Engine. |
| | Population density | 1 km × 1 km | 2000, 2005, 2010 and 2015 | RESDC. |
| | GDP | 1 km × 1 km | 1995, 2000, 2005, 2010 and 2015 | RESDC. |

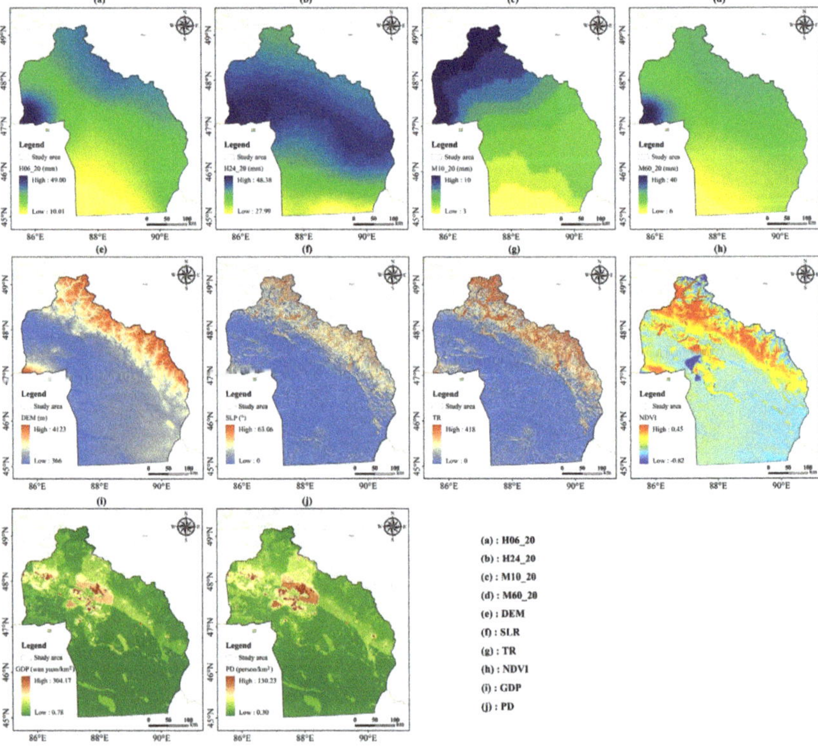


**Figure 2.** Data source plotting and visualization.

## 3. Methods

*3.1. Spatiotemporal Analysis Method*

### 3.1.1. Temporal Analysis

Proposed by Kendall in 1979, The Mann–Kendall (M-K) test is a mature method, which is widely used in the studies of hydrology, meteorology, and natural disasters [37,38]. Thus, the M-K test method was used in this study to recognize the mutation point and the trend in the temporal variation of flash floods from 1980 to 2015. Specifically, annual is used as the time scale, and the entire Altay region is used as the spatial scale, and the mutation point and the changing trends were detected. The time node when the occurrence of flash floods changed significantly can be revealed by mutation point. The M-K test can obtain two results for UF and UB. The UF expresses the trend of flash floods over time in a normal time series. The UB is the statistical sequence obtained from the inverse time series. When UF is greater than 0, the occurrence of flash floods shows an upward trend. Conversely, is a downward trend. If the value of UF exceeded the critical line ($p$ = 0.05), it indicates that the rising or falling trend was significant. Draw UF and UB into a curve, and the intersection of the two lines is the mutation point in the time series [30]. Flash flood detailed information can be found in a previous study [39].

### 3.1.2. Spatial Analysis

(1)  Estimation of the kernel density, KDE

Here, the kernel density estimation technique is used to create a representation of flash floods, for the kernel function allows this estimate to be considered as the average of the effects of the kernel function, which is centered on the flash flood location and evaluated at each point. Using kernel density estimation, we calculated the spatial intensity of flash floods as follows:

$$\lambda_h(P) = \frac{1}{nh} \sum_{i=1}^{n} m\left(\frac{P - P_i}{h}\right) \tag{1}$$

where $\lambda_h(P)$ represents the probable spatial intensity of flash floods, and $m(.)$ is the kernel function, which is an invariant function but not necessarily a positive function, $P_1, \dots, P_n$ represents the location of $n$ of the observed flash floods, and $h$ is the bandwidth, which determines the semidiameter of the circle centered at $P$. The spatial density of flash floods was calculated for kernel density estimation using ArcGIS 10.6 [40].

(2)  Standard deviational ellipse, SDE

The standard deviation ellipse is always used as a general purpose GIS tool for measuring binary distribution features. The tool is typically used to depict trends in the spatial distribution of elements by summarizing the direction and dispersion of elements. When the flash flood data are presented as points, the direction and trends were generally determined by SDE method. Based on ArcGIS10.6, the SDE is mainly influenced by three main factors, namely, mean location, the concentration or dispersion of features, and direction. Additionally, the SDE can be expressed as one to three standard deviations, in this study, one standard deviation was used [41].

(3)  Spatial gravity center model, SGCM

The spatial center of gravity model facilitates the study of the spatiotemporal migration of elements by analyzing their center of gravity trajectories. Here, the distribution of factors in two-dimensional space and their evolutionary characteristics are revealed directly and precisely with this model [42]. On this basis, we calculated the coordinates of the center of gravity and the average annual precipitation for flash floods [43]. The coordinates of the center of gravity of flash floods are as follows:

$$G_f(x) = \frac{\sum_1^n f_i(x)}{n}, \ G_f(y) = \frac{\sum_1^n f_i(y)}{n} \tag{2}$$

where $G_f(x)$ and $G_f(y)$ denote the center of gravity coordinates of annual flash floods, $n$ denotes the number of annual flash floods, while $f_i(x)$ and $f_i(y)$ denote the geometric coordinates of the $i$th flash flood ($i$ = 1, 2, . . . , $n$) [44]. In addition, the center of gravity coordinates of the average annual precipitation was as follows:

$$G(x) = \frac{\sum_{i=1}^{n} x_i \times m_i}{\sum_{i=1}^{n} m_i}, \ G(y) = \frac{\sum_{i=1}^{n} y_i \times m_i}{\sum_{i=1}^{n} m_i} \tag{3}$$

where $G(x)$ and $G(y)$ denote the annual center of gravity coordinates of this element (annual average precipitation), ($x_i$, $y_i$) denotes the geometric coordinates of the $i$th weather station for annual average precipitation, and $m_i$ is the attribute value of the $i$th weather station [42].

*3.2. Analysis of Influencing Factors*

In this section, we will discuss the influencing factors of flash floods based on land use cover change data. Since all historical flash floods occurred in grassland, settlement, farmland, and forest areas, we chose these four types of land cover. Due to the vast area of the Altay region, 1000 random points in each subregion were selected as the sample points. Then, we could take these point value data to explore the influence mechanism between mountain flash floods and driving forces factors under different land use. The correlation and interaction between influencing factors and kernel density, the weight of influencing factors, and their contribution to the flash flood will be analyzed and quantified. Finally, 210 flood points and non-flood point data were selected to analyze the driving factors of disaster points using a random forest.

### 3.2.1. Correlation Coefficient Calculation

The Pearson correlation coefficient [45] is a classical method to measure the linear correlation of $x$ and $y$. This coefficient is the product of the covariance of the two variables divided by their standard deviation, essentially making it is a standardized measurement of covariance such that the result is always between $-1$ and 1. This statistic is denoted as:

$$r_{xy} = \frac{\sum_{i=1}^{n} (x_i - \overline{x})(y_i - \overline{y})}{\sqrt{\sum_{i=1}^{n} (x_i - \overline{x})^2} \sqrt{\sum_{i=1}^{n} (y_i - \overline{y})^2}} \tag{4}$$

where $r_{xy}$ is the correlation coefficient of the two sets of data $x$ and $y$. $x_i$ and $y_i$ represent the corresponding values indexed in $i$, respectively. Here, $\overline{x}$ and $\overline{y}$ are the mean of each list data; n is the size of the samples. The $r_{xy}$ values and corresponding correlation levels are listed in Table 2 [46].

**Table 2.** Classification of the range level of Pearson correlation coefficient.

| $r_{xy}$ Value | Relevance |
|---|---|
| $r_{xy} = 0$ | no association or no correlation |
| $0 < \left| r_{xy} \right| < 0.25$ | very weak correlation |
| $0.25 < \left| r_{xy} \right| < 0.5$ | weak correlation |
| $0.5 < \left| r_{xy} \right| < 0.75$ | strong correlation |
| $0.75 < \left| r_{xy} \right| < 1$ | very strong correlation |
| $\left| r_{xy} \right| = 1$ | perfect correlation |

### 3.2.2. Multiple Linear Regression, MLR

A phenomenon is often related to multiple factors; therefore, multiple linear regression statistical methods are involved. The goal of multiple linear regression is to establish a linear relationship model between the explanation (independent variable) and response (dependent variable). The MLR equation is as follows:

$$y_i = \alpha_0 + \alpha_1 x_{i1} + \alpha_2 x_{i2} + \ldots \alpha_p x_{ip} + \varepsilon \ (i = 1, 2, \ldots n) \tag{5}$$

where $y_i$ represents the dependent variable; $x_i$ represents the explanatory variable. $\alpha_0$ is the constant term; $\alpha_p$ is the slope coefficient for each explanatory variable, and $\varepsilon$ is the residual term of the MLR.

### 3.2.3. Principal Component Analysis, PCA

The essence of PCA [45] is to find the most important aspect of the data through orthogonal transformation and use the most important aspect to replace the original data. For a set of data that may have a linear correlation between different dimensions, PCA can transform this set of data into data that are linearly independent of each dimension through orthogonal transformation. It is a dimensionality reduction method of unsupervised learning. It only needs eigenvalue decomposition to compress and denoise data. PCA mainly has the following three advantages: it only needs to measure the amount of information by variance, not that affected by factors other than the data set; the orthogonal between the principal components can eliminate the mutual influence factors between the original data components; and the calculation model is considered to be simple, and the main operation is eigenvalue decomposition, which is easy to implement.

### 3.2.4. "Random Forest", RF

Random forest, as the name suggests, is to build a forest randomly. There are many decision trees but there is not a correlation between each decision tree in the random forest. Random forest is a method that uses multiple classification trees to distinguish and classify data. While classifying the data, it can also give the importance score of each variable (gene), and evaluate how each variable is the role played in classification. The following technique was implemented in the R package "random forest" [47].

It has many advantages: it is not easy to fall into overfitting and has good antinoise ability; it can handle relatively high dimensional data without feature selection and has strong adaptability to data sets. It can handle discrete data, process continuous data, and does not need to be standardized. In the training process, the mutual influence between features can be detected, and the implementation is relatively simple.

SHAP, whose name comes from SHapley Additive ExPlanation, originated from cooperative game theory. It is inspired by cooperative game theory to construct an additive explanatory model, and all features are regarded as contributions. For each prediction sample, the model produced a prediction value, and the SHAP value was the value assigned to each feature in the sample. The SHAP value is given by the following equation:

$$y_i = y_{base} + f(x_{i1}) + f(x_{i2}) + \ldots + f(x_{ij}) \tag{6}$$

where $j$ is the number of features; $x_{ij}$ is the jth feature of the ith sample; $y_i$ is the predicted value of the model for the sample; $y_{base}$ is the baseline for the entire model; $f(x_{ij})$ is the SHAP value of $x_{ij}$. If $f(x_{ij}) > 0$, it means that the feature improves the predicted value and has a positive effect. On the contrary, it indicates that the feature reduces the predicted value, which has a negative effect. Compared with the traditional feature importance calculation method, the SHAP value can reflect the influence of the characteristics in each sample, while also showing a positive and negative effect.

The technical flowchart of this study is drawn based on the data and methods previously presented (Figure 3). First, based on the mountain torrent data, the MK test is used to analyze the time scale, and the KDE, SDE, and SGCM are used to analyze the space scale, and the results of the time–space analysis are obtained. Second, the kernel density estimation results and the mountain torrent driving force factor data are combined, and the Pearson coefficient calculation, PCA, and MLR are used to obtain the interaction and contribution rate of the influence factors; the RF and SHAP value are used to show how each variable affects the mechanism of torrents.

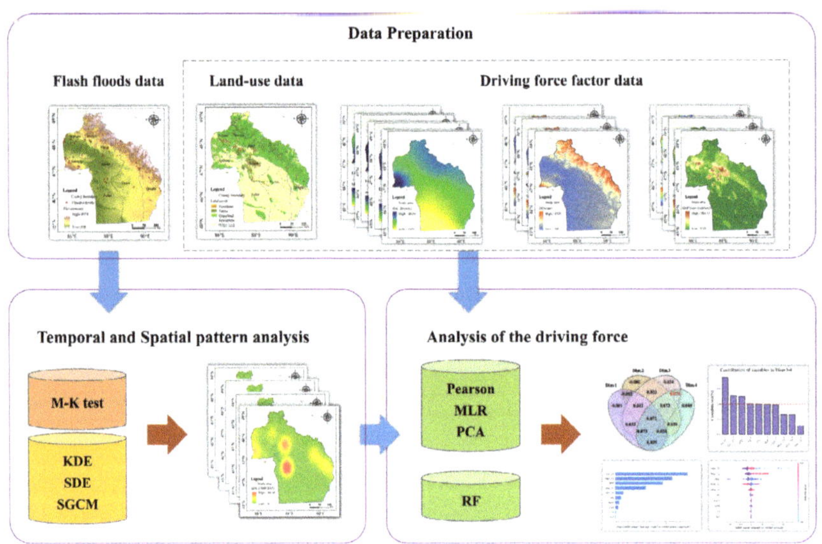

**Figure 3.** Flowchart of this study.

## 4. Results

### 4.1. Temporal Pattern of Flash Floods

As shown in Figure 4, the variation characteristics of flash floods in the time series from 1980 to 2015 were expressed in period scale, yearly scale, and monthly scale. As shown in Figure 4b, the intersection of UF and UB occurred in 2008, indicating that 2008 was the mutation point of a flash flood. In addition, the trend of UF indicated an almost constant increase in flash floods after 1996. Thus, 1996 is also an important time point. According to the result of the M-K test in Figure 4b, the occurrence of flash floods in the yearly scale was divided into three time periods, which are 1980–1996, 1997–2008, and 2009–2015, when the number of flash floods was 18, 78, and 112. Figure 4a indicated that few flash floods occurred during the first period, except in 1995. In the second period, the number of flash floods increased, especially in 1997, 1998, and 2002. The number of flash floods in the third period continued to increase, peaking at 56 in 2013. The results of the M-K test indicated that the year 2008 is the mutation point (the intersection point of UF and UB), and it indicated that the number of flash floods steadily increased after 2008, which was consistent with the statistical results. In addition, the value of UF steadily increased after 1996, indicating that the frequency of flash floods rose after 1996. Figure 4c shows that the most flash floods occurred in July, accounting for 37.5%. April and June also had many flash flood records. No flash floods occurred in January, February, October, November, and December.

### 4.2. Spatial Pattern of the Flash Floods

The kernel density analysis results are presented in Figure 5. In the 1980–1996 period, Habahe (13 flash floods) was the area with the most frequent occurrence of flash floods. Altay City had two flash floods, and other districts had very few flash floods (Figure 5a). In the period from 1997–2008, the high frequency of flash floods was concentrated in Habahe (44 flash floods) and Jeminay (12 flash floods), and Burqin had 5 flash floods. In addition, the frequency of flash floods has increased significantly in the period from 2007 to 2015. The number of flash floods in Habahe (9 flash floods) and Jeminay (11 flash floods) decreased. A high frequency of flash floods occurred in Altay City (24 points) and Fuhai (30 points). Fuyun (14 points) and Qinghe (22 points) showed an increase (Figure 5c). In general, the area with a high incidence of flash floods was concentrated early on in Habahe and more recently in Altay City.

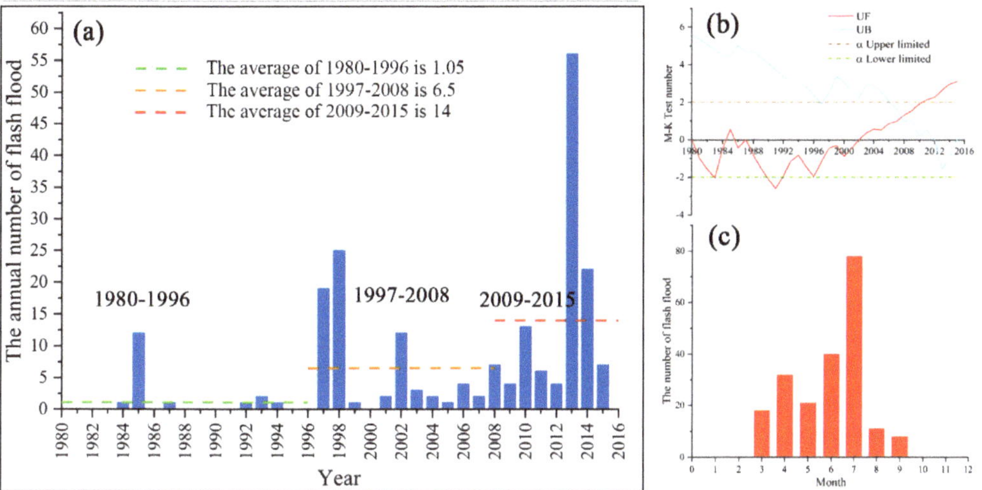

**Figure 4.** Characteristics of temporal variation of the flash floods: (**a**) the histogram of the number of flash floods per year and the average of the periods; (**b**) flash flood mutation and trend analysis result by using the M-K test; (**c**) characteristics of monthly flash floods.

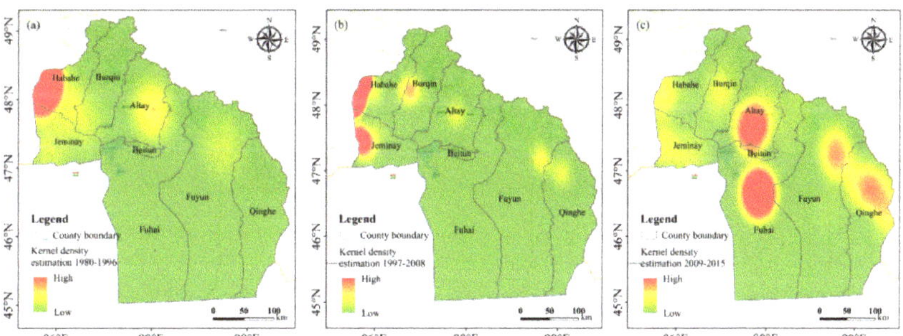

**Figure 5.** Kernel density estimation of the flash floods for the three time periods: (**a**) 1980–1996, (**b**) 1997–2008, and (**c**) 2009–2015.

According to the results of the standard deviational ellipse analysis (Figure 6a,b). The regions with spatial unbalance were mainly distributed in Fuyun and Qinghe from the period to 1980–1996 and 1997–2008. From 1997–2008 to 2009–2015, the occurrence of flash floods became relatively uniform. In general, the orientation and trend of standard deviational ellipses show that more flash floods spread to the southeast of the Altay Prefecture. In addition, according to the evolution track of the gravity center, the displacement of the flash flood gravity center from 1997–2008 to 2009–2015 indicates that the distribution of flash floods has changed significantly. The result of the evolution track of the gravity center is consistent with the result of the standard deviation ellipse.

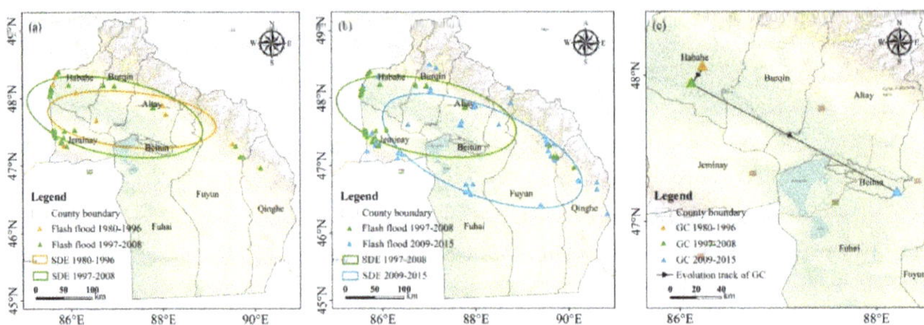

**Figure 6.** Spatial pattern of the standard deviational ellipse at the difficult periods: (**a**) from 1950–1996 to 1997–2008 and (**b**) from 1997–2008 to 2009–2015; (**c**) is the evolution track of the gravity center.

### 4.3. Analysis of the Driving Force of Mountain Flash Flood Kernel Flood

The correlation coefficient and *p*-value between each influencing factor with the kernel density of flash floods in four land cover subregions are listed in Table 3. The interaction between driving factors is shown in Figure 7. We can take Figure 7a,b as examples to explain the specific implementation steps and meanings of the image. The four dimensions in subpicture (a) are the first four principal components of the variable data principal component transformation, and the cumulative contribution rate has reached more than 85%. The number represents the R-squared between the dimensional combination model and the variable data. We know that the R-squared is used to measure the similarity between the regression problem in the linear regression problem. The larger the R-squared, the better the model fitting effect. Therefore, we choose the Dim 3 and 4 whose corresponding largest R-squared is 0.074. Subpicture (b) shows the contribution rate of each variable factor to the model. The model is the linear model that we choose to consist of Dim 3 and 4.

**Table 3.** Correlation coefficient $r_{xy}$ and multiple linear regression *p*-value in the different subregions of the land cover.

| Factors | Farmland | | Forest | | Grassland | | Settlement | |
|---|---|---|---|---|---|---|---|---|
| | $r$ | $p$ | $r$ | $p$ | $r$ | $p$ | $r$ | $p$ |
| H06_20 | 0.030 | 0.563 | −0.580 | <0.001 | 0.084 | 0.007 | 0.351 | 0.007 |
| H24_20 | −0.238 | <0.001 | 0.445 | 0.068 | 0.470 | <0.001 | 0.105 | <0.001 |
| M10_20 | −0.152 | <0.001 | −0.426 | <0.001 | −0.052 | <0.001 | 0.100 | <0.001 |
| M60_20 | 0.047 | <0.001 | −0.497 | <0.001 | 0.133 | 0.002 | 0.371 | <0.001 |
| DEM | 0.107 | <0.001 | −0.377 | 0.196 | −0.176 | 0.255 | 0.263 | 0.179 |
| SLP | 0.007 | 0.719 | −0.248 | 0.037 | −0.149 | 0.945 | 0.144 | 0.373 |
| TR | 0.002 | 0.981 | −0.247 | 0.111 | 0.147 | 0.883 | 0.138 | 0.384 |
| NDVI | 0.008 | 0.543 | −0.188 | <0.001 | 0.064 | <0.001 | −0.086 | <0.001 |
| GDP | 0.122 | <0.001 | 0.227 | <0.001 | 0.130 | <0.001 | 0.154 | <0.001 |
| PD | 0.181 | <0.001 | 0.265 | <0.001 | 0.221 | <0.001 | 0.212 | <0.001 |

The first is farmland. The factors that passed the significance test and have a slight correlation included H24_20 ($r = -0.238$, $p < 0.001$), PD ($r = 0.181$, $p < 0.001$), and M10_20 ($r = -0.152$, $p < 0.001$). The highest adjusted R-squared value of MLR for the first four principal components (accumulative contribution rate was 0.87) was 0.074, which came from Dim-3-4. The variable factors with higher contribution were DEM (18.92%), H24_20 (12.85%), and GDP (12.65%).

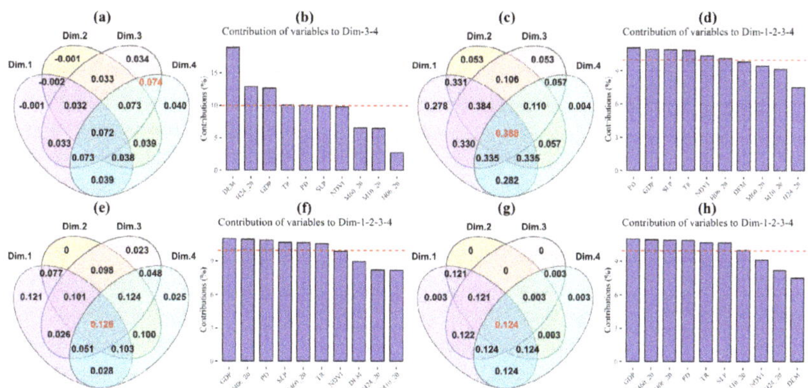

**Figure 7.** Results of interaction and contribution rate of mountain flash flood influencing factors based on the collaborative analysis of PCA and MLR in different land cover subregions. (**a**) The adjusted R-squared value of MLR of corresponding principal components to the kernel density in farmland. (**b**) The contribution of driving factors to Dim.3 and Dim.4 in farmland. (**c**–**h**) Corresponding to forest, grassland, and settlement land cover regions, respectively.

The second was the forest subregion. The correlation coefficient *r* value of some factors was higher than in the forest, including H06_20 ($r = -0.580$, $p < 0.001$), M10_20 ($r = -0.426$, $p < 0.001$), M60_20 ($r = -0497$, $p < 0.001$). The interaction of Dim-1-2-3-4 contributed the highest adjusted R-squared value and the variables that were higher than the mean value include PD (11.04%), GDP (10.91%), SLP (10.89%), TR (10.84%), and NDVI (10.32%).

Behind that was the grassland subregion, only the correlation coefficient *r* value between H24_20 ($r = 0.470$, $p < 0.001$) and PD ($r = 0.221$, $p < 0.001$) was higher than 0.2 and passed the significance test. The highest $R^2$ values were observed for Dim-1-2-3-4, and the main variable were GDP (11.01%), H06_20 (10.94%), PD (10.88%), SLP (10.68%), M60_20 (10.66%), and TR (10.58%). The last region is settlement, the main correlation factors were M60_20 ($r = 0.371$, $p < 0.001$) and PD ($r = 0.212$, $p < 0.001$). The highest adjusted R-squared was determined by the interaction of Dim-1-2-3-4, and GDP (11.01%), M60_20 (10.94%), H06_20 (10.91%), PD (10.89%), TR (10.69%), and SLP (10.69%) were dominant.

We can know that the linear model in the forest areas had the best fit through the results, and terrain and human factors played a more important role in the fitting model. Although there were correlations in each subregion, there were at least four factors that had not passed the significance test. That indicated that the correlation calculation results were poor, and the influence mechanism of the flash flood came from driving factors that cannot be fully demonstrated.

In addition, Table 4 indicates the IncMSE and IncNodePurity of each driving factor with the random forest method in different land cover subregions. The importance of each feature and the impact of all samples are shown in Figure 8. Let us take (a) and (b) as examples to explain the details of Figure 8. It is a model interpretation based on the RF training (25%) and testing (75%) of the variable data model in subpicture (a). By comparing with the prediction when a certain feature takes the baseline value, it is explained that the feature takes a certain value impact. To determine subpicture (a), draw the SHAP value of each feature for each sample, which can be used to better understand the overall pattern and allow the discovery of predicted outliers. IncMSE is equivalent to the mean decrease accuracy, which shows how much the accuracy of our model is reduced if we remove this variable. IncNodePurity is equivalent to the mean decrease in Gini, which is a variable importance metric based on the Gini impurity index. The higher the value of IncMSE or IncNodePurity, the higher the importance of this variable in our model. In the farmland, the DEM had the highest IMSE (IMSE = 54, INP = 56,410). Other higher IMSE values included H06_20 (IMSE = 38, INP = 56,785), M60_20 (IMSE = 36, INP = 40,477),

M10_20 (IMSE = 33, INP = 27,964), and H24_20 (IMSE = 25, INP = 28,333). The highest mean (|SHAP|) was observed for H06_20 (5.69). Other higher means (|SHAP value|) included M60_20 (4.49), DEM (3.76), and M10_20 (2.40). The distribution of the test samples showed a greater change in the SHAP values of H06_20, M60_20, and DEM. Regarding the forest, the highest IMSE values were ranked as follows: M10_20 (IMSE = 35, INP = 20,795), H24_20 (IMSE = 28, INP = 13,659), H06_20 (IMSE = 26, INP = 22,856), M60_20 (IMSE = 25, INP = 25,305), and DEM (IMSE = 19, INP = 11,036). The mean (|SHAP value|) was ranked as follows: M60_20 (3.06), H06_20 (2.00), M10_20 (1.62), and PD (1.31). The test samples of M60_20 and H06_20 were highly suppressive, and the test samples of PD and H24_20 were positively correlated. The most important driving factors were M10_20 (IMSE = 37, INP = 25,405), H06_20 (IMSE = 36, INP = 18,063), H24_20 (IMSE = 3, INP = 25,405), DEM (IMSE = 25, INP = 11,759), and M60_20 (IMSE = 21, INP = 14,071) were the most important driving factors. The main magnitudes of the mean (|SHAP value|) were H24_20 (5.13), M10_20 (2.61), and H06_20 (1.45). The test sample SHAP values of H24_20, M10_20, and H06_20 had a longer span and impact on the model output. The characteristics of the settlement were similar to those of the other three subregions in the IMSE and INP. The higher IMSE values included M10_20 (IMSE = 39, INP = 19,181), H06_20 (IMSE = 37, INP = 37,870), M60_20 (IMSE = 37, INP = 37,460), DEM (IMSE = 37, INP = 36,644), H24_20 (IMSE = 33, INP = 24,294). The SHAP value was determined by the collaboration of M60_20 (4.89), DEM (4.04), and H24_20 (2.23) was dominant.

**Table 4.** IncMSE (IMSE) and IncNodePurity (INP) with the random forest method in four land cover subregions.

| Factors | Farmland | | Forest | | Grassland | | Settlement | |
| --- | --- | --- | --- | --- | --- | --- | --- | --- |
| | IMSE | INP | IMSE | INP | IMSE | INP | IMSE | INP |
| H06_20 | 38 | 56,785 | 26 | 22,856 | 36 | 18,063 | 37 | 37,870 |
| H24_20 | 25 | 28,333 | 28 | 13,659 | 30 | 25,405 | 33 | 24,294 |
| M10_20 | 33 | 27,964 | 35 | 20,795 | 37 | 18,539 | 39 | 19,181 |
| M60_20 | 36 | 40,477 | 25 | 25,305 | 21 | 14,071 | 37 | 37,460 |
| DEM | 54 | 56,410 | 19 | 11,036 | 25 | 11,759 | 37 | 36,644 |
| SLP | 11 | 4590 | 12 | 2824 | 13 | 3445 | 12 | 4223 |
| TR | 9 | 3300 | 10 | 2912 | 10 | 2954 | 10 | 2192 |
| NDVI | 20 | 11,970 | 11 | 2858 | 16 | 6932 | 16 | 5798 |
| GDP | 18 | 7804 | 24 | 6400 | 27 | 6382 | 14 | 5639 |
| PD | 25 | 12,466 | 26 | 8932 | 30 | 7547 | 26 | 14,110 |

Finally, the weight features of the driving factors for disaster points were obtained. We utilized the Extract Values to Points tool of ArcGIS software to extract the driving force factor data to the disaster point. The random forest method was then used to analyze the importance of each driving force factor feature. The IMSE and INP values of the flash flood points are listed in Table 5. The SHAP values and distributions of the feature-test samples are shown in Figure 9. The higher IMSE value included M60_20 (IMSE = 19.23, INP = 13,856.15), DEM (IMSE = 19.03, INP = 4851.71), H06_20 (IMSE = 18.29, INP = 13,796.45), M10_20 (IMSE = 17.40, INP = 5552.94), and H24_20 (IMSE = 17.08, INP = 9781.75). The average impact on the model output magnitude exhibited the following ranking: M10_20 (4.74), H06_20 (3.74), M60_20 (2.83), and PD (2.04), and their SHAP values had a large fluctuation effect.

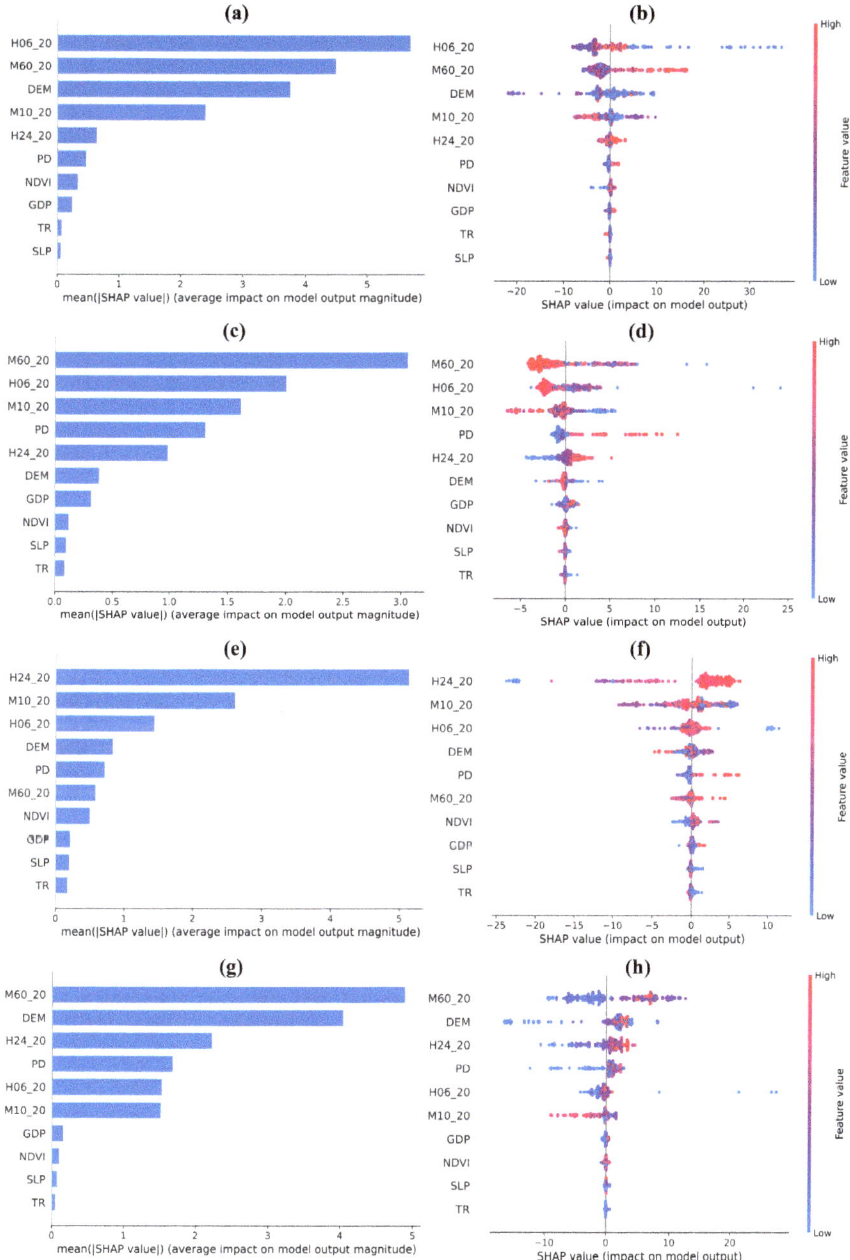

**Figure 8.** The mean SHAP value of each feature and the SHAP value of each test sample in different land cover subregions. (**a**) The sequence that importance of each feature for flash flood kernel density in the farmland. (**b**) The distribution of the SHAP value of each test sample with each feature in the farmland. (**c–h**) Corresponding to forest, grassland, and settlement land cover regions, respectively.

**Table 5.** IncMSE (IMSE) and IncNodePurity (INP) with the random forest method for flash flood point data.

| Factors | IMSE | INP |
| --- | --- | --- |
| H06_20 | 18.29 | 13,796.45 |
| H24_20 | 17.08 | 9781.75 |
| M10_20 | 17.40 | 5552.94 |
| M60_20 | 19.23 | 13,856.15 |
| DEM | 19.03 | 4851.71 |
| SLP | 7.30 | 879.73 |
| TR | 8.65 | 984.11 |
| NDVI | 7.02 | 793.23 |
| GDP | 14.24 | 1991.21 |
| PD | 16.21 | 2565.31 |

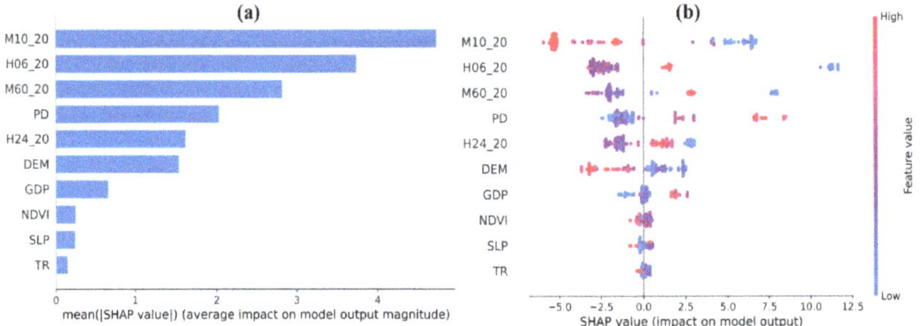

**Figure 9.** The mean ( | SHAP value | ) of the driving force factor and the SHAP value distribution of each feature with flash flood point test data. (**a**) The sequence that importance of each feature for flash flood point data. (**b**) The distribution of the SHAP value of each test sample with each feature in the flash flood point data.

## 5. Discussion

### 5.1. Temporal and Spatial Distribution of the Flash Floods

In this part, we revealed variations in the scale of the period, yearly and monthly for the flash floods in Altay Prefecture. According to the statistical data and the M-K test result, flash floods showed an increasing tendency from 1980 to 2015. The general trend of flash floods was consistent with that of the Sichuan and Fujian provinces of China [30,48]. The possible reasons include yearly intensified human activity and increased extreme precipitation events in the Altay Prefecture [49]. In addition, the possible reason flash floods were rare from 1980–1996 is because these flash floods were not recorded. According to our results and previous studies, precipitation mostly drives the occurrence of flash floods [50]. Precipitation in Altay exhibits obvious diurnal and seasonal trends. Precipitation showed a rapid upward trend from May to July and peaked in July [49]. On a monthly scale, the distribution of flash floods mostly occurred in April, June, and July, and increased from May to July, which is consistent with the distribution of precipitation.

In the spatial pattern analysis part, the kernel density estimation, standard deviational ellipse analysis, and gravity centers analysis were finished and mapped. The results above showed that flash floods had previously been concentrated in Habahe and Jeminay but were currently concentrated in Altay City, Fuhai, Fuyun, and Qinghe. The precipitation, environmental background conditions, and human activities led to the occurrence of flash floods. According to our results, DEM and PD were important factors in flash floods. In Altay, most of the regions near the Altay Mountains are predominately mountainous landscapes [51], which are prone to flash floods. In addition, abundant precipitation zones are mainly formed in the southern slope of the Altay Mountains, which are the main factors affecting the occurrence of flash floods. According to the statistical yearbook of Altay, the population and economy of all the districts of Altay are increasing [52]. The population of Altay City and Fuhai grew faster than those in other regions. Thus, according to the change trajectories of gravity centers, the intensification of human activities may be an important factor leading to the movement of flash floods.

*5.2. Discussion of Driving Factors Results*

Our driving force research shows that precipitation factors and DEM are the main characteristic factors affecting flash flood disasters. The comprehensive results of MLP-PCA and RF characteristic analysis showed that the flood disaster in the Altay region was not significantly affected by a single factor, but by a multiple factor synergistic effect. In the MLP-PCA method (Figure 7), the first three contributions of variables to the corresponding principal component with four land types were farmland (DEM, H24_20, and GDP), forest (PD, GDP, and SLR), grassland (GDP, H06_20, and PD), and settlement (GDP, M60_20, and H06_20). Among them, H24_20, H06_20, and M60_20 can be attributed to precipitation factors. With the increase in extremely heavy precipitation, the stability of the mountain will be greatly impacted, which is highly likely to promote flash flooding. The driving factors were largely the same, but slightly different in different areas of feature types. The contribution of the GDP, SLP, and TR features is high, especially in forest areas. This can be understood as a corresponding increase in GDP as people exploit natural resources, but it undermines the stable structure of some mountain areas, resulting in disasters. In settlement areas, the contribution rate of PD is also above the average line, indicating that the occurrence of disasters caused huge economic losses, especially in densely populated areas. In the SHAP value results (Figure 8), SHAP not only gives the size of the feature's influence but also reflects the positive and negative influence of the feature in each test sample. The larger the mean (|SHAP value|), the more important the feature. Both the precipitation factor and DEM ranked at forefront in terms of importance. M60_20 in the farmland and settlement (Figures 2h and 8b), H24_20 in the grassland (Figure 8f), and M60_20 and H24_20 in the settlement (Figure 8h) showed a positive correlation with the nuclear density value of flood disasters. The larger the value of the SHAP of the feature, the faster the occurrence of flood disasters.

However, the results of PCA and MLR methods were unsatisfactory. Firstly, Table 3 shows the adjusted R-squared value of the MLR of principal components to the flash floods in the four land cover types. The maximum values are 0.074 (Dim.3-4), 0.388 (Dim.1-2-3-4), 0.126 (Dim.1-2-3-4), and 0.124 (Dim.1-2-3-4), respectively. Especially in farmland, all adjusted R-squared values cannot reach 1. In univariate linear regression, the greater the R-squared value, the better the fitting effect. In multiple linear regression, if meaningless variables are added, the adjusted R-squared value will decrease, but the added eigenvalues are significant, and the adjusted R-squared value will increase. This means that the synergistic effects of multiple variables on flash floods were not significant. Secondly, some precipitation factors showed unreasonable SHAP values (Figure 8), such as H06_20 and M10_20 in the farmland (Figure 8b), and M60_20 and H06_20 (Figure 8d) in the forest. These values were expected to be positively correlated, but the results showed a negative correlation, likely due to the difficulty of obtaining geological disasters in the Altay area.

According to the importance of the ten driving force factors based on the SHAP value, we can see that the precipitation factors and topographic factors still had a greater impact on flash flood points. However, the overall precipitation did not receive good results, and there has even been a situation of suppressing flash floods in the forest area. Perhaps different ground features types should have different sensitivity factors to mountain torrents. Subsequent research can consider adding characteristic representative factors for different ground feature types, such as hydrological data in farmland areas and snowfall factors in forest areas. Although we have weighted factors, such as precipitation, topography, and economy, in the selection of factors, there is still room for improvement.

### 5.3. Implications and Limitations

*UNISDR* recently noted that the impact on flood affected economies is increasing in all regions of the world [53]. However, the risk of flooding in developed countries is decreasing due to their increasing incomes and increased capacity for disaster prevention and mitigation. Thus, we should focus more on certain areas in developing countries. Especially, the Altay ecosystem is extremely sensitive to climate change and human activities, due to the fragility and geographical conditions of the Altay ecosystem. Moreover, most current studies of flooding have focused on the humid plains, neglecting arid areas such as Altay, which are also affected by flooding. Currently, there is a need for a more reliable study to identify the spatial and temporal distribution and drivers of flooding in the Altay region, and this study meets that need.

Although some substantial progress was made in this paper, the following limitations remain. (1) Due to the limitations of the number of flash flood points and the availability of data, we did not fully discuss the hydrological processes and simulation methods, etc. For instance, we have not been able to obtain the amount of precipitation data due to the sparse meteorological stations, which leads to the discussion section of spatial-temporal analysis simply stating the effects and analyzing the causes. (2) As mentioned in Section 5.2, the results of PCA and MLR methods may be unsatisfactory due to the difficult availability of geological hazards in the Altay region. (3) Although different methods were used in this study to explore the spatial and temporal variability of flash floods and the driving factors, none of them are novel methods, and future studies can make more interesting explorations in terms of methodological approaches.

## 6. Conclusions

The temporal and spatial analysis and driving force analysis help improve our understanding of flash floods. In this study, we analyzed the distribution of flash floods in the Altay Prefecture from 1980 to 2015. Based on the M-K test, we identified the mutation points in the time series and divided the time series into three time periods. We examined the kernel density, standard deviational ellipse, and spatial gravity center in three periods and analyzed the temporal and spatial variations. The temporal variation in the flash floods showed that the annual quantity of flash floods increased from 1980 to 2015, and the months with the greatest quantity of flash floods were July, June, and April. Habahe and Jeminay had a high incidence of flash floods historically, but Altay City, Fuhai, Fuyun, and Qinghe currently have a high quantity of flash floods. Precipitation and elevation are the main driving factors for mountain torrents based on different land-use types and two driving force analysis methods. "Random forest" is more consistent with the mechanism of mountain torrent disasters than the results obtained by multiple linear regression and principal component analysis. SHAP can better reflect the quality and influence the distribution of the sample data.

**Author Contributions:** Conceptualization, A.A. and Z.D.; methodology, A.A. and Z.D.; investigation, A.A.; formal analysis, A.A. and Z.D.; resources, Z.D.; writing—original draft, A.A.; writing—review and editing, A.A. and Z.D. All authors have read and agreed to the published version of the manuscript.

**Funding:** The Science and Technology Bureau of Altay Region in Yili Kazak Autonomous Prefecture (Y99M4600AL).

**Data Availability Statement:** No new data were created or analyzed in this study. Data sharing was not applicable to this study.

**Acknowledgments:** The authors appreciate the anonymous reviewers for their constructive comments and suggestions that significantly improved the quality of this manuscript.

**Conflicts of Interest:** The authors declare no conflict of interest.

## References


1. Lumbroso, D.; Gaume, E. Reducing the uncertainty in indirect estimates of extreme flash flood discharges. *J. Hydrol.* **2012**, *414*, 16–30. [CrossRef]
2. Borga, M.; Anagnostou, E.N.; Bloschl, G.; Creutin, J.D. Flash floods Observations and analysis of hydro-meteorological controls Preface. *J. Hydrol.* **2010**, *394*, 1–3. [CrossRef]
3. Saharia, M.; Kirstetter, P.E.; Vergara, H.; Gourley, J.J.; Hong, Y. Characterization of floods in the United States. *J. Hydrol.* **2017**, *548*, 524–535. [CrossRef]
4. He, B.S.; Huang, X.L.; Ma, M.H.; Chang, Q.R.; Tu, Y.; Li, Q.; Zhang, K.; Hong, Y. Analysis of flash flood disaster characteristics in China from 2011 to 2015. *Nat. Hazards* **2018**, *90*, 407–420. [CrossRef]
5. Barredo, J.I. Major flood disasters in Europe: 1950–2005. *Nat. Hazards* **2007**, *42*, 125–148. [CrossRef]
6. Pereira, S.; Diakakis, M.; Deligiannakis, G.; Zezere, J.L. Comparing flood mortality in Portugal and Greece (Western and Eastern Mediterranean). *Int. J. Disaster Risk Reduct.* **2017**, *22*, 147–157. [CrossRef]
7. Liu, Y.; Yang, Y. Spatial Distribution of Major Natural Disasters of China in Historical Period. *Acta Geogr. Sin.* **2012**, *67*, 291–300.
8. Lin, Q.; Wang, Y. Spatial and temporal analysis of a fatal landslide inventory in China from 1950 to 2016. *Landslides* **2018**, *15*, 2357–2372. [CrossRef]
9. Xiong, J.N.; Li, J.; Cheng, W.M.; Wang, N.; Guo, L. A GIS-Based Support Vector Machine Model for Flash Flood Vulnerability Assessment and Mapping in China. *ISPRS Int. J. Geo-Inf.* **2019**, *8*, 297. [CrossRef]
10. IPCC. Summary for Policymakers. In *Climate Change 2021: The Physical Science Basis. Contribution of Working Group I to 524 the Sixth Assessment Report of the Intergovernmental Panel on Climate Change*; Masson-Delmotte, V., Zhai, P., Pirani, A., Connors, S.L., Péan, C., Berger, S., Caud, N., Chen, Y., Goldfarb, L., Gomis, M.I., et al., Eds.; Cambridge University Press: Cambridge, UK, 2021. [CrossRef]
11. Xiong, J.N.; Ye, C.C.; Cheng, W.M.; Guo, L.; Zhou, C.H.; Zhang, X.L. The Spatiotemporal Distribution of Flash Floods and Analysis of Partition Driving Forces in Yunnan Province. *Sustainability* **2019**, *11*, 2926. [CrossRef]
12. Duan, Y.; Xiong, J.N.; Cheng, W.M.; Wang, N.; Li, Y.; He, Y.F.; Liu, J.; He, W.; Yang, G. Flood vulnerability assessment using the triangular fuzzy number-based analytic hierarchy process and support vector machine model for the Belt and Road region. *Nat. Hazards* **2021**, 1–26. [CrossRef]
13. Mehr, A.; Akdegirmen, O. Estimation of Urban Imperviousness and its Impacts on Flash floods in Gazipaşa, Turkey. *Knowl.-Based Eng. Sci.* **2021**, *8*, 297.
14. Saber, M.; Kantoush, S.; Sumi, T.; Ogiso, Y.; Hadidi, A. Integrated Study of Flash Floods in Wadi Basins Considering Sedimentation and Climate Change: An International Collaboration Project. In *Wadi Flash Floods*; Springer: Singapore, 2022; pp. 401–422.
15. Budiyono, Y.; Aerts, J.; Brinkman, J.; Marfai, M.A.; Ward, P. Flood risk assessment for delta mega-cities: A case study of Jakarta. *Nat. Hazards* **2015**, *75*, 389–413. [CrossRef]
16. Liu, J.F.; Wang, X.Q.; Zhang, B.; Li, J.; Zhang, J.Q.; Liu, X.J. Storm flood risk zoning in the typical regions of Asia using GIS technology. *Nat. Hazards* **2017**, *87*, 1691–1707. [CrossRef]
17. Martin-Vide, J.P.; Llasat, M.C. The 1962 flash flood in the Rubi stream (Barcelona, Spain). *J. Hydrol.* **2018**, *566*, 441–454. [CrossRef]
18. Varlas, G.; Anagnostou, M.N.; Spyrou, C.; Papadopoulos, A.; Kalogiros, J.; Mentzafou, A.; Michaelides, S.; Baltas, E.; Karymbalis, E.; Katsafados, P. A Multi-Platform Hydrometeorological Analysis of the Flash Flood Event of 15 November 2017 in Attica, Greece. *Remote Sens.* **2019**, *11*, 45. [CrossRef]
19. Papagiannaki, K.; Lagouvardos, K.; Kotroni, V.; Bezes, A. Flash flood occurrence and relation to the rainfall hazard in a highly urbanized area. *Nat. Hazards Earth Syst. Sci.* **2015**, *15*, 1859–1871. [CrossRef]
20. Sun, X.Y.; Zhang, G.T.; Wang, J.; Li, C.Y.; Wu, S.N.; Li, Y. Spatiotemporal variation of flash floods in the Hengduan Mountains region affected by rainfall properties and land use. *Nat. Hazards* **2021**, *21*, 2109–2124. [CrossRef]
21. Noren, V.; Hedelin, B.; Nyberg, L.; Bishop, K. Flood risk assessment—Practices in flood prone Swedish municipalities. *Int. J. Disaster Risk Reduct.* **2016**, *18*, 206–217. [CrossRef]
22. Zelenakova, M.; Ganova, L.; Purcz, P.; Satrapa, L. Methodology of flood risk assessment from flash floods based on hazard and vulnerability of the river basin. *Nat. Hazards* **2015**, *79*, 2055–2071. [CrossRef]
23. Wu, X.S.; Wang, Z.L.; Guo, S.L.; Liao, W.L.; Zeng, Z.Y.; Chen, X.H. Scenario-based projections of future urban inundation within a coupled hydrodynamic model framework: A case study in Dongguan City, China. *J. Hydrol.* **2017**, *547*, 428–442. [CrossRef]
24. Seyoum, S.D.; Vojinovic, Z.; Price, R.K.; Weesakul, S. Coupled 1D and Noninertia 2D Flood Inundation Model for Simulation of Urban Flooding. *J. Hydraul. Eng.* **2012**, *138*, 23–34. [CrossRef]
25. Benito, G.; Thorndycraft, V.R. Use of systematic, palaeoflood and historical data for the improvement of flood risk estimation: An introduction. *Nat. Hazards* **2004**, *31*, 623–643.
26. Fang, J.; Li, M.; Shi, P.J. Assessment and mapping of global fluvial flood risk. *J. Nat. Disasters* **2015**, *24*, 1–8.
27. Fayne, J.V.; Bolten, J.D.; Doyle, C.S.; Fuhrmann, S.; Rice, M.T.; Houser, P.R.; Lakshmi, V. Flood mapping in the lower Mekong River Basin using daily MODIS observations. *Int. J. Remote Sens.* **2017**, *38*, 1737–1757. [CrossRef]
28. Ma, R.; Zhang, D. Assessment of flood risk in Nanning city. *J. Nat. Disasters* **2017**, *26*, 200–2016.
29. Tang, C.; van Asch, T.W.J.; Chang, M.; Chen, G.Q.; Zhao, X.H.; Huang, X.C. Catastrophic debris flows on 13 August 2010 in the Qingping area, southwestern China: The combined effects of a strong earthquake and subsequent rainstorms. *Geomorphology* **2012**, *139*, 559–576. [CrossRef]

30. Xiong, J.N.; Pang, Q.; Fan, C.K.; Cheng, W.M.; Ye, C.C.; Zhao, Y.L.; He, Y.R.; Cao, Y.F. Spatiotemporal Characteristics and Driving Force Analysis of Flash Floods in Fujian Province. *ISPRS Int. J. Geo-Inf.* **2020**, *9*, 133. [CrossRef]

31. Liu, Y.S.; Yuan, X.M.; Guo, L.; Huang, Y.H.; Zhang, X.L. Driving Force Analysis of the Temporal and Spatial Distribution of Flash Floods in Sichuan Province. *Sustainability* **2017**, *9*, 1527. [CrossRef]

32. Zhou, L.; Zhou, C.; Yang, F.; Wang, B.; Sun, D. Spatio-temporal evolution and the influencing factors of PM2.5 in China be-571 tween 2000 and 2011. *Acta Ecol. Sin.* **2017**, *72*, 161–174.

33. Shi, J.; Baozhu, L.I.; Peng, L.I.; Huang, J.; Sun, F.; Liu, B.J. Analisis of Characteristics and Formation Mechanism for the 9·17 Giant Debris Flow in Yuanmou Country, Yunnan Province. *Geol. Rev.* **2018**, *64*, 665–673.

34. Xu, S.; Yuan, Z.; Yang, Z. Based on EKC analysis of landslide and debris flow disasters. *Soil Water Conserv. China* **2015**, *07*, 54–56.

35. Flash Flooding Definition. Available online: https://www.weather.gov/phi/FlashFloodingDefinition (accessed on 1 October 2021).

36. Xiong, J.N.; Wei, F.; Liu, Z. Hazard assessment of debris flow in Sichuan Province. *J. Geo-Inf. Sci.* **2017**, *191*, 604–1161.

37. Gocic, M.; Trajkovic, S. Analysis of changes in meteorological variables using Mann-Kendall and Sen's slope estimator statistical tests in Serbia. *Glob. Planet. Chang.* **2013**, *100*, 172–182. [CrossRef]

38. Kendall, M. The Advanced Theory of Statistics. *Rev. Mex. Sociol.* **1961**, *23*, 310. [CrossRef]

39. Peng, Y.; Song, J.Y.; Cui, T.T.; Cheng, X. Temporal-spatial variability of atmospheric and hydrological natural disasters during recent 500 years in Inner Mongolia, China. *Nat. Hazards* **2017**, *89*, 441–456. [CrossRef]

40. Guo, F.T.; Innes, J.L.; Wang, G.Y.; Ma, X.Q.; Sun, L.; Hu, H.Q.; Su, Z.W. Historic distribution and driving factors of human-caused fires in the Chinese boreal forest between 1972 and 2005. *J. Plant Ecol.* **2015**, *8*, 480–490. [CrossRef]

41. Wang, B.; Shi, W.Z.; Miao, Z.L. Confidence Analysis of Standard Deviational Ellipse and Its Extension into Higher Dimensional Euclidean Space. *PLoS ONE* **2015**, *10*, e118537. [CrossRef]

42. Chai, J.; Wang, Z.Q.; Yang, J.; Zhang, L.G. Analysis for spatial-temporal changes of grain production and farmland resource: Evidence from Hubei Province, central China. *J. Clean. Prod.* **2019**, *207*, 474–482. [CrossRef]

43. Li, M.S.; Ren, X.X.; Zhou, L.; Zhang, F.Y. Spatial mismatch between pollutant emission and environmental quality in China—A case study of NOx. *Atmos. Pollut. Res.* **2016**, *7*, 294–302. [CrossRef]

44. Liu, Y.S.; Yang, Z.S.; Huang, Y.H.; Liu, C.J. Spatiotemporal evolution and driving factors of China's flash flood disasters since 1949. *Sci. China-Earth Sci.* **2018**, *61*, 1804–1817. [CrossRef]

45. Wold, S.; Esbensen, K.; Geladi, P. Principal component analysis. *Chemom. Intell. Lab. Syst.* **1987**, *2*, 37–52. [CrossRef]

46. Irhoumah, M.; Pusca, R.; Lefèvre, E.; Mercier, D.; Romary, R. Diagnosis of induction machines using external magnetic field and correlation coefficient. In Proceedings of the 2017 IEEE 11th International Symposium on Diagnostics for Electrical Machines, Power Electronics and Drives (SDEMPED), Tinos, Greece, 29 August–1 September 2017.

47. Breiman, L. Random forests. *Mach. Learn.* **2001**, *45*, 5–32. [CrossRef]

48. Xiong, J.; Zhao, Y.; Cheng, W.; Guo, L.; Wang, N.; Wei, L. Temporal-spatial Distribution and the Influencing Factors of Mountain-Flood Disasters in Sichuan Province. *Geo-Inf. Sci.* **2018**, *20*, 1443–1456.

49. Bai, S.; Boyuan, L.I.; Huang, X.; Bureau, A.M.; Bureau, Q.M. Climate Characteristics of the Diurnal Precipitation in Altay in the Warm Season. *Desert Oasis Meteorol.* **2015**, *9*, 7.

50. Hou, J.M.; Guo, K.H.; Liu, F.F.; Han, H.; Liang, Q.H.; Tong, Y.; Li, P. Assessing Slope Forest Effect on Flood Process Caused by a Short-Duration Storm in a Small Catchment. *Water* **2018**, *10*, 1256. [CrossRef]

51. He, Y.; Xiong, J.; Abudumanan, A.; Cheng, W.; Ye, C.; He, W.; Yong, Z.; Tian, J. Spatiotemporal Pattern and Driving Force Analysis of Vegetation Variation in Altay Prefecture based on Google Earth Engine. *J. Resour. Ecol.* **2021**, *12*, 729–742.

52. Kabin, H.; Bo, C.; Zhiguo, G. Statistical Yearbook of Altay. 2019. Available online: https://www.chinayearbooks.com/xinjiang-statistical-yearbook-2019.html (accessed on 1 October 2021).

53. Jongman, B.; Ward, P.J.; Aerts, J. Global exposure to river and coastal flooding: Long term trends and changes. *Glob. Environ. Chang.-Hum. Policy Dimens.* **2012**, *22*, 823–835. [CrossRef]

*Article*

# Statistical Evaluation of the Influences of Precipitation and River Level Fluctuations on Groundwater in Yoshino River Basin, Japan

**Linyao Dong, Yiwei Guo, Wenjian Tang *, Wentao Xu and Zhongjie Fan**

Changjiang River Scientific Research Institute, Changjiang Water Resource Commission, Wuhan 430010, China; linyaodonghydro@foxmail.com (L.D.); guoyiwei32ww@gmail.com (Y.G.); xuwt@mail.crsri.cn (W.X.); stephenf88@163.com (Z.F.)
* Correspondence: wjtang@mwr.gov.cn; Tel.: +86-027-82926528

**Abstract:** Precise evaluation of the correlations among precipitation, groundwater and river water enhance our understanding on regional hydrological circulation and water resource management. The innovative and efficient use of wavelet analysis has been able to identify significant interactions in the spatial and temporal domains and to estimate the recharge travel time. In this paper, a wavelet analysis was utilized to analyse 43 years of monthly, and 2 years of daily, precipitation, river level and groundwater level data in the Yoshino River Basin, Japan. There were two main results: (1) There was a significant influence of precipitation and river on groundwater, with a periodicity of 4–128 days, 1 year and 2–7 years. The periodicity of 1 year was correlated with seasonal variability. The significant interaction at 4–128 days mainly occurred in the rainy season. The 2–7-year oscillation of aquifer water levels was determined by precipitation. (2) The recharge-water travel times in the study area estimated from the arrow patterns in the precipitation–groundwater wavelet coherence (WTC) were consistent for each observation well. The response times of the aquifer to precipitation were 1 day and 3–6 days in 2013 and 2014, respectively. The different time lags were likely determined by the timing of maximum daily precipitation.

**Keywords:** precipitation infiltration; groundwater–river interaction; multiscale time analysis; wavelet analysis

**Citation:** Dong, L.; Guo, Y.; Tang, W.; Xu, W.; Fan, Z. Statistical Evaluation of the Influences of Precipitation and River Level Fluctuations on Groundwater in Yoshino River Basin, Japan. *Water* **2022**, *14*, 625. https://doi.org/10.3390/w14040625

Academic Editors: Yuanfang Chen, Dong Wang, Dedi Liu, Binquan Li, Ashish Sharma and Francesco Napolitano

Received: 23 December 2021
Accepted: 14 February 2022
Published: 17 February 2022

**Publisher's Note:** MDPI stays neutral with regard to jurisdictional claims in published maps and institutional affiliations.

## 1. Introduction

As the largest distributed storage of fresh water, groundwater plays an critical role in sustaining ecosystems and facilitating human adaptation to climate change [1]. Water level monitoring in aquifers constitutes the principal means of tracking changes in groundwater storage over time, which provides information regarding the availability of renewable groundwater resources. Temporal groundwater level changes of piezometric levels typically occurs at two distinct time scales: long term (interannual and seasonal variability) and short term (daily or subdaily fluctuations) [2]. The interannual variability of hydrological processes is commonly discussed with regard to large-scale climatic phenomenon, such as solar activity, the El Niño Southern Oscillation (ENSO) and the North Atlantic Oscillation (NAO), while seasonal fluctuations are widely considered to be related to local climatic indices, such as precipitation, temperature, air pressure and humidity. An analysis of aquifer responses to the possible influencing factors could improve our understanding of the regional water circulation and therefore guide water resource management [3].

Climate change has been shown to have direct and indirect influences on subsurface hydrological processes [1,4]. The impacts of large-scale climatic phenomena on subsurface hydrological patterns commonly occur at the decadal and annual scale [3]. Solar activity and climatic anomalies affect the planetary-scale atmospheric circulation in a frequency-dependent manner, which alters regional precipitation characteristics [5]. Precipitation events affect aquifer water levels by changing the recharge patterns [3,6]. Although piezometric levels are commonly influenced by climatic indices, fluctuations in piezometric levels are also driven by various hydrological processes at different spatial and temporal scales. Natural external stresses (air pressure changes, tidal or river water level fluctuations, earth tides, tectonic events) can affect fluctuations in aquifer water level. Dynamic interactions between groundwater and seawater/surface water commonly occur in aquifers where hydraulic connections exist [2,7,8]. The direct, rapid responses of aquifers to external hydraulic stresses serve as an effective tracer to depict hydrological processes. However, the lack of detailed groundwater observations limits our understanding of the interactions between precipitation, external hydrological processes and aquifers, especially over varying temporal scales.

Most previous studies employed spectral analysis to identify the aquifer responses to possible influent factors, such as precipitation variability [9], tidal effects [10,11] and lake level fluctuations [12]. A spectral analysis can be used to evaluate the coherence and phase lag between external hydrological processes and groundwater-level responses in the frequency domains, and to estimate the degree of influence and travel time of external stresses to aquifers. However, a spectral analysis only determines the frequency content of aquifer water levels. Temporal variation in the groundwater-level response to influent factors can be evaluated using a wavelet analysis [13]. The impact of climate and anthropic effects on aquifers has been studied using a wavelet analysis in many parts of the world [3,6,14]. The wavelet analysis can not only evaluate the variability of the aquifer in different spatial and temporal frequencies using the signal amplification function, but also analyse the interaction between possible factors and groundwater using coherence and cross-spectrum functions. The aquifer's teleconnection with climate indices (ENSO index, NAO index et al.), the interaction with local climate variations (precipitation, air temperature, barometric pressure et al.), and the responses to external stresses (tidal effect, pumping et al.) can be assessed based on wavelet techniques. Additionally, the impact of climate and anthropic pressures on groundwater resources can be evaluated to improve our understanding of water resource management. However, the systematic analyses on the interaction between precipitation, rivers and groundwater from short-term to long-term timescales have not been assessed.

The objectives of this study were: (1) to identify the interannual and interdecadal variability of shallow groundwater level fluctuations and the related hydrological processes (precipitation and river level) in the Yoshino River Basin, Japan, at different spatial and temporal scales; and (2) to interpret the interactions between precipitation, river levels and groundwater by analysing coherence and phase lag based on a wavelet technique. Evaluation and quantification of the variability of hydrological processes and their responses to related influencing factors provide guidance on regional water resource management.

## 2. Study Area and Methodology

### 2.1. Study Area and Datum

The study area is located in the lower Yoshino catchment in south Japan (Figure 1). The Yoshino River is 194 km in length and has a watershed of 3750 km$^2$. The study area is a watershed with an area of about 840 km$^2$ located in Tokushima Plain. The plain consists of alluvial delta clay and sand, and about 75% of the study area is covered by forests and meadow. The elevation of the investigated watershed decreases from west to east and ranges from $-10$ to 185 m. Pacific climate patterns dominate in the study area. The mean annual precipitation from 1892 to 2014 at the Tokushima meteorological station was about 1650 mm per year, and the average annual air temperature from 1890 to 2014 was about 15.5 °C. A national meteorological station, hydrological station and four observation wells are located along the Yoshino River. The Ikeda Dam is located about 48 km upstream of the hydrological station, and the Daiju-Zeki Weir Dam is located about 11 km downstream of the station. These two large dams were constructed mainly for domestic and agricultural water usages, and the natural river regimes were consequently altered by dam regulation and artificial water intake. However, this manuscript emphasizes the influence of river level fluctuation on the aquifer in the study area, and the river–aquifer interaction in spatial and temporal domains can be interpreted properly even if the time series of river water levels is influenced artificially.

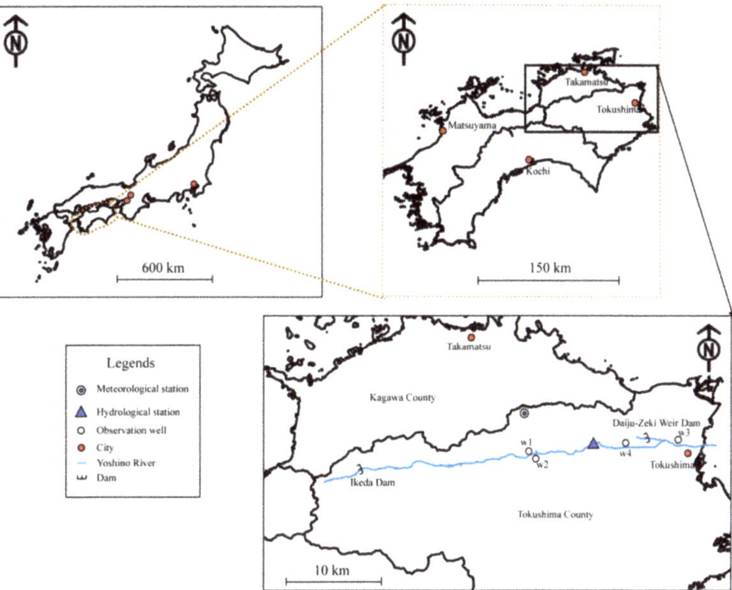



**Figure 1.** Location of the study area.

The data used in the analysis of precipitation, piezometric levels and river levels consist of 43 years (from 1972 to 2014) of daily data. Daily precipitation was recorded by an automatic rain gauge installed at the Tokushima meteorological station. The groundwater level (with reference to the annual mean sea level in Tokyo Bay) was observed using pressure water level gauges installed in the observation wells. The study aquifer is composed of silt, sand and gravel. The screens of the observation wells were positioned at depths from 2 to 10 m, and therefore the study considered the groundwater present in the unconfined aquifer in the study area. The transmissivity of the unconfined aquifer is investigated and calculated as about 1000 m$^2$/day in the study area. No artificial exploitation was detected in the study area, and the aquifer water levels were inferred to be disturbed slightly by human activity. No dams or water diversions were built in the upper reaches. The river level was measured by a bubble-type water gauge installed in the hydrological station, and no artificial channels were found in the station.

A time series of precipitation and aquifer water levels in observation wells 1 to 4 and the river level are presented in Figure 2a–f, respectively. Figure 2 shows that the daily precipitation, piezometric levels and river levels fluctuated periodically. Figure 2c shows the aquifer water levels in well 2 suddenly changed in 2004, which was associated with a shift in the measurement location, and the aquifer water levels after 2004 were adjusted for consistence. There were several water level measurements missing during 1976 and 1978 for well 4. An interpolation process was used to resolve this problem in the subsequent analysis. The time series were standardized before subsequent analyses.

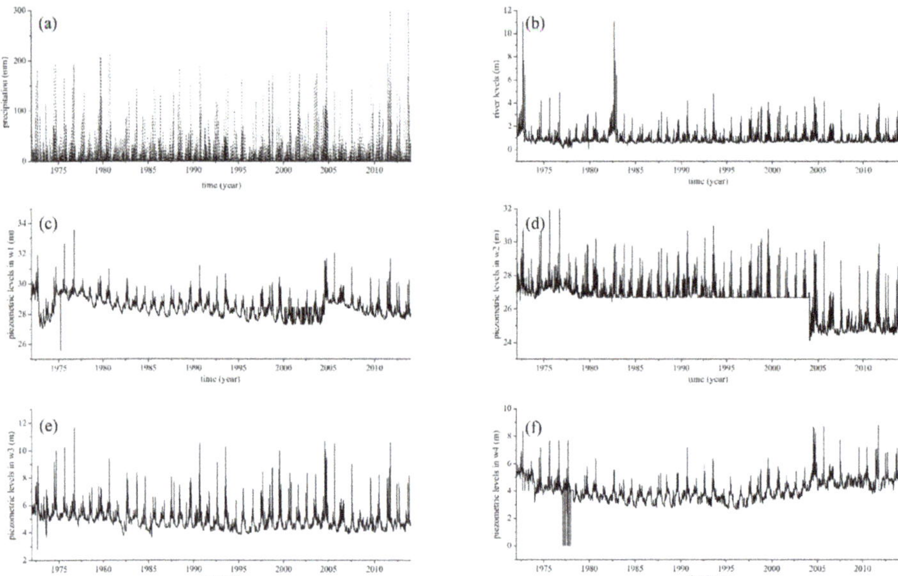

**Figure 2.** The (**a**) precipitation, and piezometric levels (based on average sea level in Tokyo Bay) in observation wells (**b**) 1, (**c**) 2, (**d**) 3 and (**e**) 4, and (**f**) river levels from 1 January 1972 to 30 December 2014 in the study area.

*2.2. Data Analyses Methods*

The multiresolution analysis and continuous wavelet analysis approaches were applied in this study. First, wavelet decomposition was implemented to extract aquifer water levels and their possible impact factors across different resolution levels, and the standard deviations (SD) were calculated to quantify the multiresolution levels. Then, a cross-correlation between aquifer water levels and potential impact variables across different levels was conducted to reveal the dependence of aquifer water level on the possible influencing factors. Finally, a wavelet analysis was performed to quantify the temporal features in different periods of precipitation, and aquifer and river levels. A wavelet coherence (WTC) analysis was applied to detect the aquifer responses to precipitation and river levels in both the time and frequency domains.

A multiresolution analysis can be applied to decompose a time series into a series of successive approximations and details in increasing order of resolution to study signals at different resolutions [14]. Wavelet decomposition returns the wavelet coefficients of the signal at different levels through the implementation of specific wavelets. Wavelet decomposition is performed based on designed signal filters [3]. The details of the algorithm can be found in the paper by Mallat [15].

To quantify the relationship between two signals at different resolution levels, a multiresolution cross-analysis [14] was applied. Cross-correlation evaluates the similarity of two time series as a function of the lag of one series in relation to the other. The cross-correlation function (CCF) analysis was applied to identify the influence of possible factors on aquifer water levels in different frequency domains. The calculation of the CCF is described by Charlier [14].

A wavelet analysis assesses the variation in a signal in both the temporal and frequency domains [13]. In the field of hydrology, this technique is applied in two aspects: to determine the spatial and temporal variation of hydrological factors, including precipitation, river discharge and aquifer water levels [16–19]; and to quantify the interaction between hydrological variations and climatic indices, such as solar activity [11,20], the North Atlantic Oscillation [6,21], and the ENSO [20,22].

The wavelet techniques utilized in this study included the wavelet power spectrum (WPS) and WTC. The WPS represents the magnitude of the variance in a time series at a given frequency and location in time. This technique provides an effective approach to analyse the variability of hydrological processes. The WPS [23] is defined as the square absolute value (or square amplitude) of the wavelet transform coefficients.

To eliminate the distortion of a wavelet analysis in time–frequency domains, the concept of the cone of influence (COI) was introduced, in which the e-folding time function is used to overcome edge effects [23]. The edge effects are negligible for wavelet spectra located in the COI region. The statistical significance of wavelet power can be assessed relative to the null hypothesis that the signal is generated by a stationary process with a given background power spectrum. A significance test was conducted using the Monte Carlo method [23] and a 5% significance level was adopted. The WPS enables us to characterise the degree of complexity of a simplex signal. To assess the relationship between two signals in the time–frequency domains, a cross-wavelet analysis should be introduced [13]. The WTC can determine a significant coherence even though the common power is low [24]. Concerning hydrological issues, this approach has been used to access the influence of ENSO on stream flow [22], construct models to forecast river flow [25] and evaluate timing errors in hydrological predictions [26].

The significance contour for the cross-wavelet analysis can be found using the chi-squared distribution, with details of the algorithm provided by Torrence and Compo [23]. The 5% significance level was considered in this paper. The details of the involved wavelet techniques in this study are described in the relevant references [17,23,24]. All subsequent computations and analyses were conducted within the MATLAB environment.

## 3. Results

### 3.1. Variability of Hydrological Time Series

To interpret the characteristics of hydrological variability in the study area, the mean, maximum, minimum, variance and average annual amplitude of hydrological variables were calculated (Table 1). In a hydrological year in Japan, the maximum river and aquifer water level always occurred in May or June, and the minimum always occurred in December or January; this pattern was associated with the local precipitation. The daily mean, maximum and minimum precipitation amounts from 1972 to 2014 were 4.4, 429.50 and 0.10 mm, respectively. The mean, maximum and minimum water levels in wells 1 and 2 were much higher than those of the other wells and the river. There were no significant differences in the average annual amplitude between the aquifer and river levels.

**Table 1.** Standard deviation of the standardized hydrological signals by multiresolution levels (from level 1 to level 12).

| Level | Aquifer Water Level | | | | Precipitation | River Level | Barometric Pressure | Humidity | Air Temperature | Sunspot Number | SST |
|---|---|---|---|---|---|---|---|---|---|---|---|
| | No. 1 | No. 2 | No. 3 | No. 4 | | | | | | | |
| 1 | 0.087 | 0.095 | 0.139 | 0.124 | 0.616 | 0.258 | 0.272 | 0.434 | 0.094 | 0.071 | 0.026 |
| 2 | 0.123 | 0.144 | 0.199 | 0.140 | 0.510 | 0.322 | 0.410 | 0.481 | 0.126 | 0.097 | 0.030 |
| 3 | 0.155 | 0.169 | 0.238 | 0.141 | 0.398 | 0.331 | 0.373 | 0.383 | 0.130 | 0.172 | 0.042 |
| 4 | 0.192 | 0.183 | 0.284 | 0.169 | 0.299 | 0.333 | 0.271 | 0.283 | 0.107 | 0.310 | 0.063 |
| 5 | 0.215 | 0.180 | 0.279 | 0.178 | 0.212 | 0.300 | 0.227 | 0.228 | 0.084 | 0.212 | 0.070 |
| 6 | 0.221 | 0.162 | 0.268 | 0.188 | 0.153 | 0.267 | 0.148 | 0.200 | 0.081 | 0.157 | 0.103 |
| 7 | 0.228 | 0.141 | 0.280 | 0.205 | 0.112 | 0.213 | 0.158 | 0.165 | 0.092 | 0.155 | 0.180 |
| 8 | 0.446 | 0.221 | 0.554 | 0.401 | 0.150 | 0.357 | 0.662 | 0.460 | 0.953 | 0.124 | 0.261 |
| 9 | 0.227 | 0.094 | 0.193 | 0.166 | 0.072 | 0.263 | 0.067 | 0.077 | 0.096 | 0.104 | 0.557 |
| 10 | 0.309 | 0.114 | 0.158 | 0.177 | 0.036 | 0.260 | 0.052 | 0.061 | 0.092 | 0.152 | 0.621 |
| 11 | 0.277 | 0.119 | 0.114 | 0.151 | 0.029 | 0.246 | 0.062 | 0.078 | 0.087 | 0.547 | 0.284 |
| 12 | 0.267 | 0.184 | 0.143 | 0.219 | 0.031 | 0.157 | 0.072 | 0.041 | 0.105 | 0.497 | 0.191 |

For a better visualization of the annual fluctuation of hydrological variables, the annual amplitudes of aquifer and river levels were calculated and are shown in Figure 3. It was observed that the changing water level amplitudes in the aquifer and river were closely matched. The annual amplitude of the aquifer water level was always higher than the corresponding amplitude of the river level. The annual precipitation amount was also used to investigate the relationships among precipitation, aquifer level and river level. High annual amplitudes of aquifer and river levels were observed in association with high precipitation amounts, while most of the low amplitudes were associated with the low precipitation amounts.

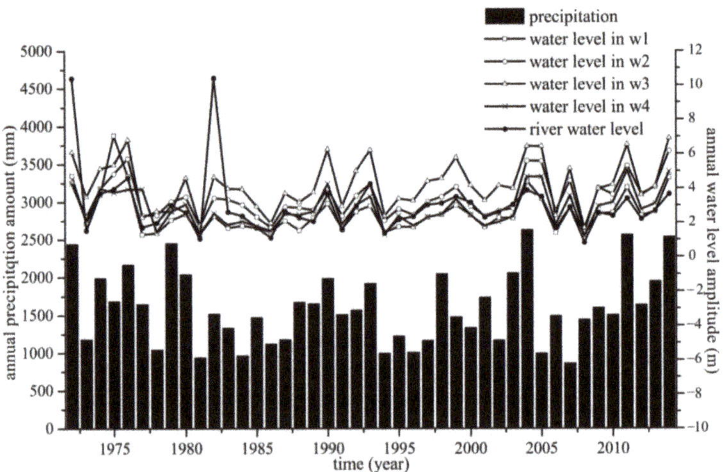


**Figure 3.** Annual precipitation amount, average groundwater level and river level from 1972 to 2014 in the study area.

Wavelet decomposition was used to visualize the spatial distribution of hydrogeological energy in different resolution levels. Multiresolution at 10 levels was performed on the data, and the multiresolution analysis results for precipitation, river levels and aquifer water levels (well 1 is used an example) are shown in Figure 4. The high energy of precipitation was mainly concentrated in the high-frequency domain (corresponding to 1–8 days), indicating that precipitation events lasting several days explained most of the variance in the precipitation signal. Both the river and aquifer water levels had high energy at all levels. From levels 1 to 7, we observed a high energy distribution in the bands of 1972 to 1985 and 1990 to 2005 for river levels. The high energy in aquifer water levels mainly occurred in the bands of 1972 to 1975 and 1990 to 2005 at levels 1 to 7.

To quantify the energy across different multiresolution levels, the SD was calculated for each standardized time series (Table 1). As described previously, the SD of aquifer water levels and river levels had an insignificant peak at level 8, indicating high energy at all levels. Table 1 shows that the SD of precipitation decreased from 0.616 to 0.036 for low-to-high multiresolution levels, indicating that the high energy was concentrated in the high-frequency domain.

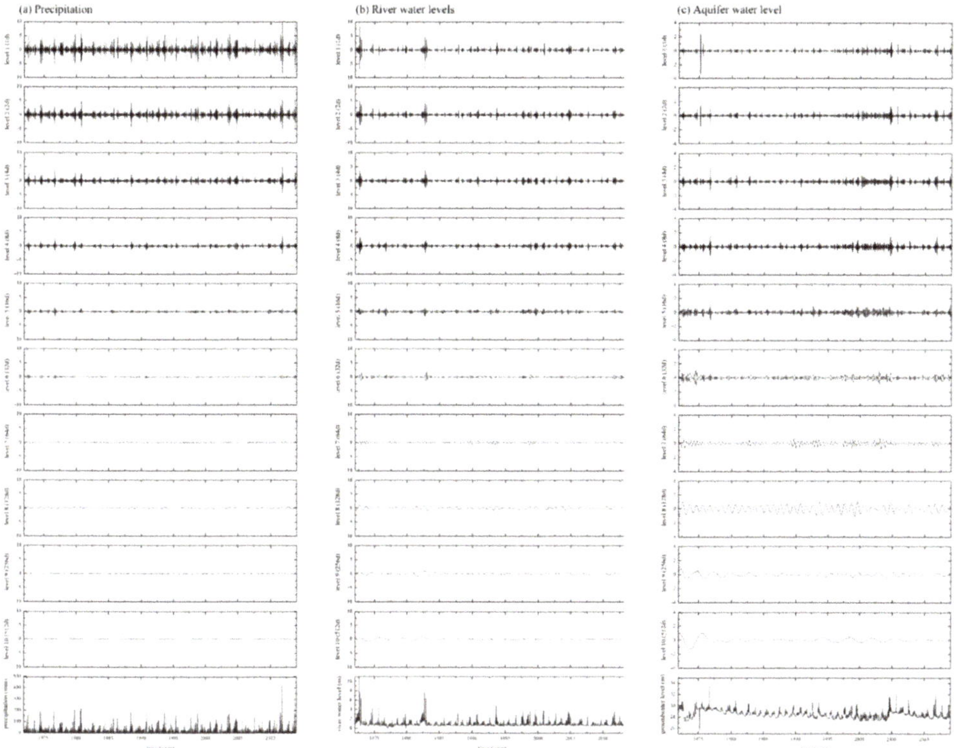

**Figure 4.** Multiresolution analysis results for (**a**) precipitation, (**b**) river level and (**c**) aquifer water level (taking well 1 as an example). The different components of the decomposition correspond (from top to bottom) with short- to long-time-scale processes, with level j corresponding to time scales at $2^{j-1}$.

### 3.2. Correlations of Hydrological Processes

To access the dependence of groundwater level on possible climatic and hydrological variables in the study area, the correlation coefficients between possible impact factors and aquifer water levels were calculated and shown in Table 2. The results show that the groundwater level in the study area shows relatively high negative correlations (absolute value above 0.24) with the barometric pressure and positive correlations (absolute value above 0.24) with precipitation, river level, humidity and air temperature. However, the high correlation with the humidity and air temperature might be explained by the synchronous seasonal variation feature instead of interaction. Relative low correlations were detected between aquifer water levels and sunspot number, SST, indicating the solar activity and ENSO have weak influence on the aquifer in the study area.

**Table 2.** Correlation coefficients between possible impact factors and aquifer water levels.

| | Precipitation | River Level | Barometric Pressure | Humidity | Air Temperature | Sunspot Number | SST |
|---|---|---|---|---|---|---|---|
| w1 | 0.234 | 0.434 | −0.257 | 0.286 | 0.349 | 0.010 | −0.059 |
| w2 | 0.188 | 0.391 | −0.120 | 0.218 | 0.129 | 0.270 | 0.010 |
| w3 | 0.339 | 0.640 | −0.367 | 0.398 | 0.471 | 0.119 | −0.101 |
| w4 | 0.221 | 0.415 | 0.268 | 0.261 | 0.357 | 0.160 | −0.155 |
| average | 0.246 | 0.478 | −0.253 | 0.291 | 0.327 | 0.059 | −0.076 |

To identify the influence of possible factors on aquifer water levels in different frequency domains, we applied the CCF across multiresolution levels. The input signals included aquifer water level, precipitation and river level. Two types of CCF were conducted. The first was a CCF calculation between two signals at the same multiresolution level i [13], while the second was a CCF calculation between the overall input signal and an isolated output signal at a given multiresolution level i [14]. The results are shown in Table 3. Regarding a CCF between precipitation as an input and aquifer water levels as an output, we observed strong positive correlations across all scales except for level 1. The highest correlation coefficient (around 0.85) was obtained for level 8, corresponding to a 128-day resolution period. This indicated that the water levels in the aquifer and precipitation covaried in both the high- and low-frequency domains. The most significant influence of local precipitation on aquifer water level oscillation occurred at the seasonal scale. Regarding a CCF between river levels as an input and water levels in aquifer as an output, high positive correlations were observed across all scales, and the highest correlation coefficients (around 0.85) occurred for levels 7 or 8. It was concluded that the river–aquifer interaction in the study area occurs in both high- and low-frequency domains, and the highest flood event determines the aquifer water level oscillation at the seasonal scale.

**Table 3.** Cross-correlation functions (CCF) applied at different scales of the multiresolution analysis between possible impact factors and aquifer water levels ($x_i$-$y_i$ represents CCF between two signals at the multiresolution level i, and $x_i$-$y_{\text{global}}$ represents CCF between an isolated output signal at a given multiresolution level i and an overall input signal).

| | Resolution Level | Precipitation | | River Level | | Barometric Pressure | | Humidity | | Air Temperature | | Sunspot Number | | SST | |
|---|---|---|---|---|---|---|---|---|---|---|---|---|---|---|---|
| | | $x_i$-$y_i$ | $x_i$-$y_{\text{global}}$ | $x_i$-$y_i$ | $x_i$-$y_{\text{global}}$ | $x_i$-$y_i$ | $x_i$-$y_{\text{global}}$ | $x_i$-$y_i$ | $x_i$-$y_{\text{global}}$ | $x_i$-$y_i$ | $x_i$-$y_{\text{global}}$ | $x_i$-$y_i$ | $x_i$-$y_{\text{global}}$ | $x_i$-$y_i$ | $x_i$-$y_{\text{global}}$ |
| water level in well No. 1 | 1 | −0.12 | −0.01 | 0.36 | 0.03 | 0.08 | 0.01 | 0.02 | 0.00 | −0.01 | 0.00 | 0.01 | 0.00 | 0.01 | 0.00 |
| | 2 | 0.51 | 0.06 | 0.60 | 0.07 | −0.15 | −0.02 | 0.26 | 0.03 | 0.02 | 0.00 | 0.01 | 0.00 | −0.02 | 0.00 |
| | 3 | 0.62 | 0.10 | 0.66 | 0.10 | −0.25 | −0.04 | 0.21 | 0.03 | 0.05 | 0.01 | 0.05 | 0.01 | 0.00 | 0.00 |
| | 4 | 0.57 | 0.11 | 0.67 | 0.13 | −0.15 | −0.03 | 0.19 | 0.04 | −0.07 | −0.01 | 0.02 | 0.00 | 0.04 | 0.01 |
| | 5 | 0.55 | 0.12 | 0.74 | 0.16 | −0.20 | −0.04 | 0.34 | 0.07 | −0.02 | 0.00 | 0.01 | 0.00 | 0.00 | 0.00 |
| | 6 | 0.60 | 0.13 | 0.67 | 0.15 | −0.04 | −0.01 | 0.36 | 0.08 | −0.03 | 0.00 | 0.06 | 0.01 | 0.02 | 0.00 |
| | 7 | 0.47 | 0.11 | 0.82 | 0.19 | −0.17 | −0.04 | 0.21 | 0.05 | 0.21 | 0.05 | −0.13 | −0.03 | −0.17 | −0.03 |
| | 8 | 0.86 | 0.38 | 0.85 | 0.38 | −0.74 | −0.33 | 0.90 | 0.40 | 0.86 | 0.38 | 0.21 | 0.10 | 0.02 | 0.01 |
| | 9 | 0.53 | 0.14 | 0.47 | 0.15 | −0.05 | −0.09 | 0.34 | 0.11 | 0.09 | 0.07 | 0.17 | 0.05 | −0.23 | −0.08 |
| | 10 | 0.45 | 0.16 | 0.40 | 0.16 | 0.12 | 0.09 | 0.54 | 0.21 | −0.04 | −0.07 | 0.10 | 0.04 | −0.08 | −0.01 |
| water level in well No. 2 | 1 | −0.28 | −0.03 | 0.64 | 0.06 | 0.19 | 0.02 | −0.05 | −0.01 | 0.00 | 0.00 | 0.01 | 0.00 | 0.00 | 0.00 |
| | 2 | 0.47 | 0.07 | 0.77 | 0.11 | −0.07 | −0.01 | 0.26 | 0.04 | 0.01 | 0.00 | 0.02 | 0.00 | 0.00 | 0.00 |
| | 3 | 0.69 | 0.12 | 0.82 | 0.14 | −0.26 | −0.04 | 0.27 | 0.05 | 0.08 | 0.01 | 0.02 | 0.00 | 0.00 | 0.00 |
| | 4 | 0.72 | 0.13 | 0.79 | 0.15 | −0.15 | −0.03 | 0.28 | 0.05 | −0.05 | −0.01 | 0.04 | 0.01 | 0.00 | 0.01 |
| | 5 | 0.71 | 0.13 | 0.84 | 0.15 | −0.20 | −0.04 | 0.42 | 0.07 | 0.02 | 0.00 | 0.03 | 0.01 | 0.05 | 0.01 |
| | 6 | 0.66 | 0.11 | 0.74 | 0.12 | −0.01 | 0.00 | 0.42 | 0.07 | −0.07 | −0.01 | 0.00 | 0.00 | 0.06 | 0.01 |
| | 7 | 0.56 | 0.08 | 0.86 | 0.12 | −0.13 | −0.02 | 0.36 | 0.05 | 0.11 | 0.02 | −0.11 | −0.02 | −0.17 | −0.02 |
| | 8 | 0.88 | 0.19 | 0.84 | 0.18 | −0.71 | −0.16 | 0.83 | 0.18 | 0.78 | 0.17 | 0.19 | 0.04 | 0.14 | 0.03 |
| | 9 | 0.67 | 0.06 | 0.59 | 0.07 | 0.07 | 0.01 | 0.70 | 0.08 | 0.03 | −0.03 | 0.01 | 0.01 | −0.36 | −0.05 |
| | 10 | −0.02 | 0.01 | 0.11 | 0.03 | −0.01 | 0.03 | 0.53 | 0.09 | −0.02 | −0.07 | 0.01 | 0.02 | −0.09 | −0.02 |
| water level in well No. 3 | 1 | −0.14 | −0.02 | 0.39 | 0.05 | 0.08 | 0.01 | 0.03 | 0.00 | −0.01 | 0.00 | 0.02 | 0.00 | 0.00 | 0.00 |
| | 2 | 0.50 | 0.10 | 0.65 | 0.13 | −0.13 | −0.03 | 0.23 | 0.05 | 0.00 | 0.00 | 0.03 | 0.01 | 0.00 | 0.00 |
| | 3 | 0.64 | 0.15 | 0.75 | 0.18 | −0.29 | −0.07 | 0.19 | 0.05 | 0.04 | 0.01 | 0.04 | 0.01 | 0.00 | 0.00 |
| | 4 | 0.65 | 0.19 | 0.75 | 0.21 | −0.14 | −0.04 | 0.22 | 0.06 | −0.05 | −0.01 | 0.02 | 0.01 | 0.00 | 0.01 |
| | 5 | 0.67 | 0.19 | 0.81 | 0.23 | −0.21 | −0.06 | 0.36 | 0.10 | 0.00 | 0.00 | 0.05 | 0.01 | 0.04 | 0.01 |
| | 6 | 0.65 | 0.17 | 0.74 | 0.20 | −0.05 | −0.01 | 0.43 | 0.12 | −0.07 | −0.02 | 0.03 | 0.01 | 0.11 | 0.03 |
| | 7 | 0.52 | 0.15 | 0.89 | 0.25 | −0.28 | −0.08 | 0.33 | 0.09 | 0.18 | 0.05 | −0.12 | −0.03 | −0.20 | −0.05 |
| | 8 | 0.88 | 0.49 | 0.85 | 0.17 | 0.81 | 0.45 | 0.94 | 0.52 | 0.92 | 0.51 | 0.27 | 0.15 | 0.08 | 0.04 |
| | 9 | 0.75 | 0.16 | 0.58 | 0.14 | −0.07 | −0.06 | 0.54 | 0.13 | 0.11 | 0.06 | 0.08 | 0.02 | −0.32 | −0.08 |
| | 10 | 0.78 | 0.16 | 0.14 | 0.04 | 0.05 | −0.04 | 0.41 | 0.09 | −0.03 | 0.02 | −0.05 | 0.00 | −0.33 | −0.07 |
| water level in well No. 4 | 1 | −0.05 | −0.01 | 0.23 | 0.03 | 0.03 | 0.00 | 0.04 | 0.00 | 0.01 | 0.00 | 0.00 | 0.00 | 0.00 | 0.00 |
| | 2 | 0.47 | 0.07 | 0.47 | 0.07 | −0.19 | −0.03 | 0.22 | 0.03 | 0.03 | 0.00 | 0.02 | 0.00 | 0.00 | −0.01 |
| | 3 | 0.55 | 0.08 | 0.60 | 0.08 | −0.30 | −0.04 | 0.18 | 0.03 | 0.02 | 0.00 | 0.03 | 0.00 | 0.00 | 0.00 |
| | 4 | 0.60 | 0.10 | 0.67 | 0.11 | −0.18 | −0.03 | 0.25 | 0.04 | −0.06 | −0.01 | 0.02 | 0.00 | 0.00 | 0.01 |
| | 5 | 0.56 | 0.10 | 0.72 | 0.13 | −0.19 | −0.03 | 0.31 | 0.05 | −0.05 | −0.01 | 0.02 | 0.00 | −0.02 | 0.00 |
| | 6 | 0.58 | 0.11 | 0.69 | 0.13 | −0.02 | 0.00 | 0.47 | 0.09 | −0.09 | −0.02 | 0.01 | 0.00 | 0.01 | 0.00 |
| | 7 | 0.50 | 0.10 | 0.83 | 0.17 | −0.18 | −0.03 | 0.20 | 0.04 | 0.28 | 0.06 | −0.14 | −0.03 | −0.21 | −0.04 |
| | 8 | 0.84 | 0.34 | 0.83 | 0.33 | −0.79 | −0.32 | 0.93 | 0.37 | 0.90 | 0.36 | 0.25 | 0.10 | 0.09 | 0.04 |
| | 9 | 0.55 | 0.12 | 0.41 | 0.10 | 0.22 | −0.07 | 0.58 | 0.13 | −0.04 | 0.09 | −0.18 | −0.02 | −0.12 | −0.03 |
| | 10 | 0.62 | 0.14 | 0.24 | 0.07 | −0.06 | −0.19 | 0.42 | 0.11 | 0.13 | 0.17 | −0.46 | −0.08 | −0.15 | −0.06 |

## 3.3. Wavelet Analysis

Multiresolution analysis provides information about frequency domains where the aquifer water levels might be related to precipitation or river levels, but provides no information about the temporal variability involved in the interaction process. Therefore, a wavelet technique was applied to identify the oscillation of the water level in aquifer and its possible responses to precipitation and river level variations in both the spatial and temporal domains. The interdecadal and interannual variability were investigated using 43 years of monthly data and 2 years of daily data.

(1)  Wavelet analysis of interdecadal variability

To eliminate the seasonal components from the hydrological time series, the 12-month moving average data during 1972 and 2014 were analysed using a wavelet analysis. The WPS of the monthly precipitation amount, average river level and aquifer water levels are shown in Figure 5a–f, respectively. The *x*- and *y*-axes represent the time-scale space. The *z*-axis represents the value of wavelet power, with low–high values shown in blue–red colours.

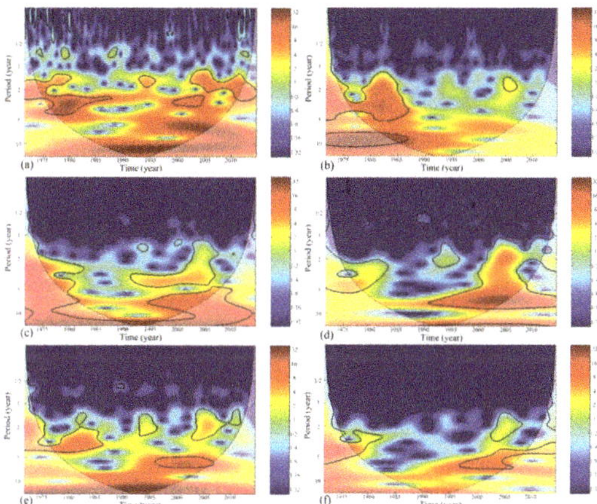

**Figure 5.** Continuous wavelet spectra of monthly (**a**) precipitation, (**b**) river level and (**c**) piezometric levels in observation wells 1, (**d**) 2, (**e**) 3 and (**f**) 4 from 1972 to 2014 in the study area. The thick black contour designates the 5% significance level against red noise, whereas the lighter shade represents the cone of influence (COI), where edge effects might distort the picture.

In the WPS for precipitation (Figure 5a), a high WPS was detected at periodicities of 2–7 and 11 years, indicating strong interdecadal variations in these frequency domains. High-power spectra were also observed in the 2–7- and 11-year bands for river levels (Figure 5b), but they were only concentrated in the period of 1972–1987. The WPS for water levels in the aquifer (Figure 5c–f) showed a similar distribution of wavelet power. Significant power spectra were observed in the 2–7 years band during 1972–1985 and 1995–2012. Significant power spectra with a periodicity of 11 years were only found for aquifer water levels in well w1.

The high CCF between precipitation and groundwater level at multiresolution level 9–10 was calculated in comparison with the river level, suggesting that precipitation was the main cause of interdecadal variability in aquifer water levels. The spatial and temporal features of coherence between precipitation and aquifer water levels were investigated using a WTC. Figure 6a–d shows the WTC between the monthly precipitation amount and water levels in the aquifer. The z-axis represents the value of the coherence, with low–high values in blue–red colours. The WTC highlighted a significant coherence between precipitation and aquifer water levels in the 2–5 years band during 1972–1988 and 1995–2012, and the 7 years band during 1990–2014. A significant coherence between precipitation and aquifer water levels, with a periodicity of 11 years, was only observed in well w3. The in-phase relationship dominated in the domains with significant coherence. Figure 6e–h shows the WTC between river level and aquifer water level in wells 1–4, respectively. A significant coherence was obtained in the 1–2 years band, and in-phase relationships were detected in these domains.

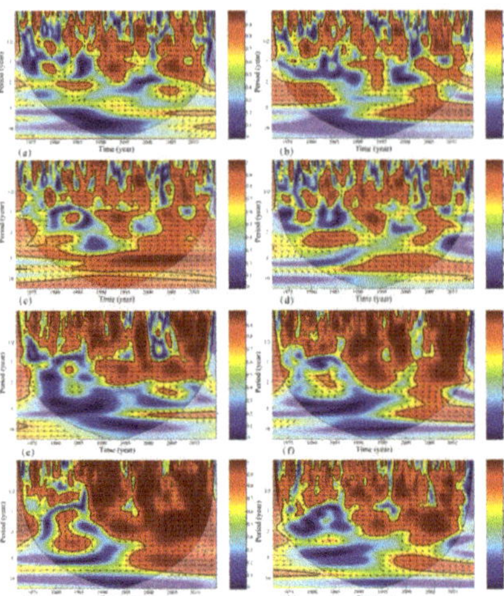

**Figure 6.** Wavelet coherences (WTCs) between the monthly precipitation and aquifer water level in (**a**) wells 1, (**b**) 2, (**c**) 3 and (**d**) 4, and WTCs between the monthly river water and aquifer water level in (**e**) wells 1, (**f**) 2, (**g**) 3 and (**h**) 4 from 1972 to 2014 in the study area. The thick black contour designates the 5% significance level against red noise, whereas the lighter shade represents the COI, where edge effects might distort the picture. The relative phase relationship is shown by the arrows (with the in-phase pointing right and the antiphase pointing left).

(2)    Wavelet analysis of interannual variability

To determine the interannual variability of aquifer water levels, the daily data during 2013 and 2014 were subjected to a wavelet analysis. The WPS of the daily precipitation amount, and average river and aquifer water levels, are shown in Figure 7a–f, respectively. The WPS for the daily precipitation amount (Figure 7a) indicated that a high-power spectrum mainly occurred during July and November in a hydrological year. The WPS for river and water levels in the aquifer (Figure 7b–f) had similar spectral distribution features to those of precipitation, indicating that a strong fluctuation occurred in these domains.

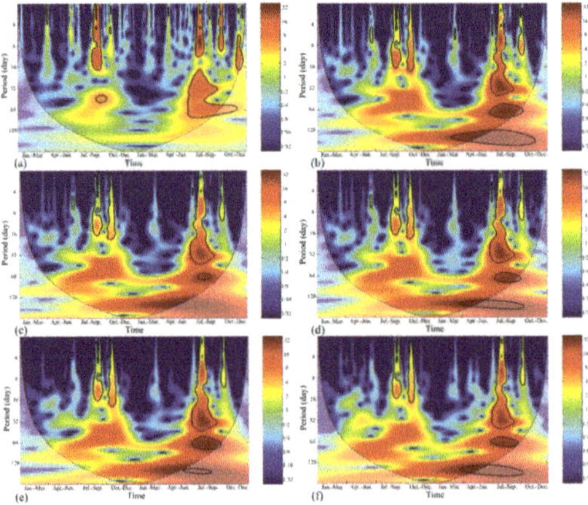

**Figure 7.** Continuous wavelet spectra of daily (**a**) precipitation, (**b**) river level and (**c**) piezometric levels in observation wells 1, (**d**) 2, (**e**) 3 and (**f**) 4 during 2013 and 2014 in the study area. The thick black contour designates the 5% significance level against red noise, whereas the lighter shade represents the COI, where edge effects might distort the picture.

The WTC between the daily precipitation amount and aquifer water levels in the study area are presented in Figure 8a–d. In the low-frequency domains (less than 32 days), a discontinuous coherence was observed. There was a high coherence between precipitation and water levels in the aquifer that was concentrated in both the rainy and dry seasons, with periodicities of 64 and 128 days, respectively. For the periodicity of around 32 days, a significant coherence was obtained only in the rainy season (during July to December). Both in- and anti-phase relationships were observed in the domains with significant coherence. Figure 8e–h shows the WTC between river level and aquifer water level in wells 1–4, respectively. A significant coherence was obtained in almost all domains, and in-phase relationships were detected in these domains.

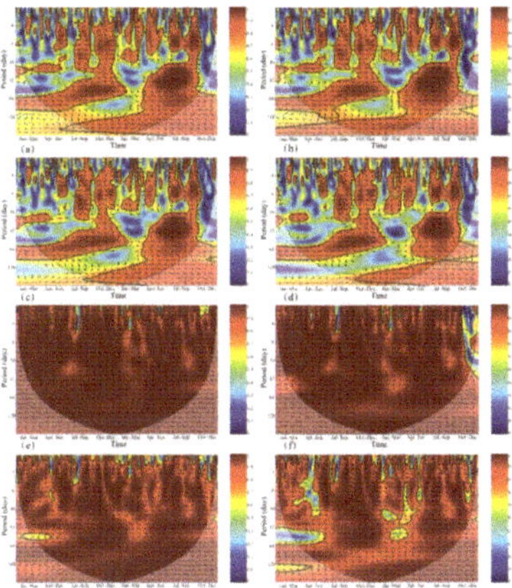

**Figure 8.** The WTCs between the daily precipitation and groundwater level in (**a**) wells 1, (**b**) 2, (**c**) 3 and (**d**) 4, and WTCs between the daily river water and groundwater level in (**e**) wells 1, (**f**) 2, (**g**) 3 and (**h**) 4 from 2013 to 2014 in the study area. The thick black contour designates the 5% significance level against red noise, whereas the lighter shade represents the COI, where edge effects might distort the picture. The relative phase relationship is shown by the arrows (with the in-phase pointing right and the antiphase pointing left).

## 4. Discussion

### 4.1. Influences of Precipitation and River on Groundwater

The influence of precipitation and river level on groundwater was clearly demonstrated by the high CCF between precipitation/river level and groundwater level at different multiresolution levels. Both CCFs had their most significant peaks for an annual frequency, suggesting that seasonal variability dominates in the hydrological processes. The WPS for the 42-year monthly precipitation, river level and aquifer water level data (seasonal components removed) clearly displayed components with a periodicity of 2–7 years, suggesting that the interdecadal interactions might influence this periodicity. The WPS for the 2-year daily data showed significant or high-power spectra in the 4–128-days band during the rainy season, indicating that a strong hydraulic interaction might occur in the rainy season.

The spatial and temporal features of the interaction between precipitation and groundwater at the interdecadal time scale was analysed based on the monthly precipitation–groundwater WTC (Figure 6). Although a significant coherence was detected at a low frequency (less than 2 years), a low cross-interaction power was observed in these domains. Consequently, the interannual coherence displayed in Figure 6 is meaningless. The domains with a significant coherence between precipitation and groundwater (Figure 6) had a high consistency with domains with a significant or high-power spectra. This behaviour suggests that the interdecadal oscillation of shallow aquifer water levels is determined by planetary-scale atmospheric moisture circulation [3,6,14], and the precipitation in the study area reflects these moisture circulation processes. High CCFs between river level and groundwater (Table 3) were observed in levels 2–7, suggesting that the interdecadal variation in precipitation affects the groundwater and river levels in similar domains.

The influences of precipitation and river on groundwater at the interannual time scale were studied based on the daily precipitation/river level–groundwater WTC (Figure 8). A significant WTC was detected in the different spatial–temporal domains between precipitation and groundwater. However, the WPS for the daily precipitation and groundwater data suggested that a strong fluctuation in hydrological processes occurred in the rainy season, and in-phase relationships dominated in this period. This behaviour indicates that the effects of precipitation on the groundwater table mainly occur in the rainy season, and the phase differences can provide a reasonable estimate of the recharge-water travel time. The WTC between river level and groundwater revealed a significant coherence in almost all of the spatial–temporal domains. It could be inferred that the direct hydraulic connection between the river and aquifer occurred across diurnal to annual scales through both recharge and discharge processes. The recharge and discharge processes might be recognized by the direction of the arrows in the river level–groundwater WTC.

*4.2. Recharge Travel Time*

Precipitation recharge to the aquifer occurs with lags that can range from hours to days, and even months [12]. The recharge-water travel time in the study area was estimated from the pattern of the arrows in the precipitation–groundwater WTC at the interannual time scale (Figure 8). The results were consistent for each observation well in the study area. The results of the correlation between precipitation and groundwater in the different spatial and temporal domains indicated that the daily precipitation affected the groundwater at periodicities of 4–16 and 32–64 days during July and September in the study area. In 2013, the area of significant coherence with periodicities of 4–16 days had an in-phasing of about 1/8 of a cycle, suggesting a 1-day delay in the response of the aquifer to precipitation. The phase difference in the periodicities of 32–64 days equated to zero, and this pattern indicated no time lag existed at these time scales. In 2014, the arrow angles with significant coherence in the periodicities of 4–16 days ranged from $-1/3$ to $-2/3\,\pi$, which indicated that the response time of the aquifer to precipitation was about 3–6 days. An in-phasing of about 5/6 of a cycle was detected in the periodicities of 32–64 days, indicating a time lag of about 40 days.

The precipitation led aquifer water levels by up to 6 days in the study area, which differed from other examples of semiconfined aquifers published in the literature [3,6]. This might be a result of the unconfined condition of the study aquifer. The different time lags of groundwater level to precipitation, with periodicities of 4–16 days during 2013 and 2014, might be due to the timing of maximum precipitation. The piezometric levels in the unconfined aquifer responded quickly to the variation in precipitation when the soil pores were not filled. The maximum precipitation occurred in late August in 2013 and early August in 2014, and this behaviour might cause a longer average travel time of precipitation to the aquifer in 2014. The time lags with periodicities of 32–64 days might be associated with other local climatic indices, such as barometric pressure, humidity and air temperature [3].

## 5. Conclusions

The influences of precipitation and river level on groundwater level in different spatial and temporal domains in the Yoshino River Basin, Japan, were assessed using a wavelet analysis. The interannual and interdecadal variability of the hydrological processes were identified using the WPS approach, and the interactions between precipitation, river and groundwater were analysed by multiresolution and WTC analyses. Compared with the spectral analysis, the wavelet technique provided an effective way to detect the correlations among precipitation, river level and aquifer level in the temporal domain. The detailed information of correlations among precipitation, groundwater level and river level provide an efficient implement to improve regional water-resource management efficiency.

The significant fluctuations in precipitation, river level and piezometric levels in the study area were mainly concentrated at periodicities of 4–128 days, 1 year and 2–7 years. The effects of precipitation and river level on piezometric levels were also observed in these periods. The correlation in the 1-year band was recognized as the seasonal variability of the hydrological processes. The significant interaction in the 4–128-days band mainly occurred in the rainy season, and the 2–7-year oscillation of aquifer water levels was determined by the variation in precipitation.

The recharge-water travel time in the study area was estimated from the pattern of arrows in the precipitation–groundwater WTC, and the results were consistent for each observation well. The precipitation led aquifer water levels by up to 6 days in the study area, and the response times of the aquifer to precipitation were 1 day and 3–6 days in 2013 and 2014, respectively. The different time lags of groundwater to precipitation might be caused by the timing of maximum precipitation. The piezometric levels in the unconfined aquifer responded quickly to the variation in precipitation when the soil pores were not filled. The maximum precipitation occurred in late August in 2013 and early August in 2014, which might have resulted in a longer precipitation travel time in 2014.

**Author Contributions:** Conceptualization, L.D. and W.T.; methodology, Y.G.; software, Y.G.; validation, L.D. and W.X.; formal analysis, W.X.; investigation, Z.F.; resources, W.T.; data curation, Y.G.; writing—original draft preparation, L.D.; writing—review and editing, L.D.; visualization, Y.G.; supervision, W.T.; project administration, W.T.; funding acquisition, W.T. All authors have read and agreed to the published version of the manuscript.

**Funding:** This research was funded by [National Key R&D Program of China] grant number [2021YFE0111900], [National Natural Science Foundation of China] grant number [41977171, 41902300]. And the APC was funded by [National Key R&D Program of China] grant number [2021YFE0111900].

**Institutional Review Board Statement:** Not applicable.

**Informed Consent Statement:** Not applicable.

**Data Availability Statement:** Not applicable.

**Acknowledgments:** This study was supported by the National Key R&D Program of China (No. 2021YFE0111900), National Natural Science Foundation of China (No. 41977171, 41902300). The code for the wavelet analysis was provided by Grinsted et al. It is available at http://www.pol.ac.uk/home/research/waveletcoherence/ (accessed on 10 May 2021). Special appreciation is extended for their selfless contributions. The English in this document has been checked by at least two professional editors, both native speakers of English. For a certificate, please see: http://www.textcheck.com/certificate/60iJsZ (accessed on 10 May 2021).

**Conflicts of Interest:** The authors declare no conflict of interest.

# References

1.	Taylor, R.G.; Scanlon, B.; Döll, P.; Rodell, M.; van Beek, R.; Wada, Y.; Longuevergne, L.; Leblanc, M.; Famiglietti, J.S.; Edmunds, M.; et al. Ground water and climate change. *Nature Clm. Change* **2013**, *3*, 322–329. [CrossRef]
2.	Gribovszki, Z.; Szilágyi, J.; Kalicz, P. Diurnal fluctuations in shallow piezometric levels and streamflow rates and their interpretation—A review. *J. Hydrol.* **2010**, *385*, 371–383. [CrossRef]
3.	Dong, L.; Shimada, J.; Kagabu, M.; Fu, C. Teleconnection and climatic oscillation in aquifer water level in Kumamoto plain, Japan. *Hydrol. Process.* **2015**, *29*, 1687–1703. [CrossRef]
4.	Green, T.R.; Taniguchi, M.; Kooi, H.; Gurdak, J.J.; Allen, D.M.; Hiscock, K.M.; Treidel, H.; Aureli, A. Beneath the surface of global change: Impacts of climate change on groundwater. *J. Hydrol.* **2011**, *405*, 532–560. [CrossRef]
5.	Fleming, S.W.; Quilty, E.J. Aquifer Responses to El Niño–Southern Oscillation, Southwest British Columbia. *Groundwater* **2006**, *44*, 595–599. [CrossRef]
6.	Tremblay, L.; Larocque, M.; Anctil, F.; Rivard, C. Teleconnections and interannual variability in Canadian piezometric levels. *J. Hydrol.* **2011**, *410*, 178–188. [CrossRef]
7.	Jiao, J.J.; Tang, Z. An analytical solution of groundwater response to tidal fluctuation in a leaky confined aquifer. *Water Resour. Res.* **1999**, *35*, 747–751. [CrossRef]

8. Dong, L.; Cheng, D.; Liu, J.; Zhang, P.; Ding, W. Analytical Analysis of Groundwater Responses to Estuarine and Oceanic Water Stage Variations Using Superposition Principle. *J. Hydrol. Eng.* **2016**, *21*, 04015046. [CrossRef]
9. Ghanbari, R.N.; Bravo, H.R. Coherence among Climate Signals, Precipitation, and Groundwater. *Groundwater* **2011**, *49*, 476–490. [CrossRef]
10. Shih, D.C.F.; Lee, C.D.; Chiou, K.F.; Tsai, S.M. Spectral analysis of tidal fluctuations in ground water level. *J. Am. Water Resour. Assoc.* **2000**, *36*, 1087–1099. [CrossRef]
11. Dong, L.; Shimada, J.; Kagabu, M.; Yang, H. Barometric and tidal-induced aquifer water level fluctuation near the Ariake Sea. *Environ. Monit. Assess.* **2015**, *187*, 1–16. [CrossRef] [PubMed]
12. Namdar Ghanbari, R.; Bravo, H.R. Evaluation of correlations between precipitation, groundwater fluctuations, and lake level fluctuations using spectral methods (Wisconsin, USA). *Hydrogeol. J.* **2011**, *19*, 801–810. [CrossRef]
13. Labat, D. Recent advances in wavelet analyses: Part 1. A review of concepts. *J. Hydrol.* **2005**, *314*, 275–288. [CrossRef]
14. Charlier, J.B.; Ladouche, B.; Maréchal, J.C. Identifying the impact of climate and anthropic pressures on karst aquifers using wavelet analysis. *J. Hydrol.* **2015**, *523*, 610–623. [CrossRef]
15. Mallat, S.G. A Theory for Multiresolution Signal Decomposition: The Wavelet Representation. *IEEE Trans. Pattern Anal. Mach. Intell.* **1989**, *11*, 674–693. [CrossRef]
16. Kumar, P.; Foufoula-Georgiou, E. Wavelet analysis for geophysical applications. *Rev. Geophys.* **1997**, *35*, 385–412. [CrossRef]
17. Lafrenière, M.; Sharp, M. Wavelet analysis of inter-annual variability in the runoff regimes of glacial and nival stream catchments, Bow Lake, Alberta. *Hydrol. Process.* **2003**, *17*, 1093–1118. [CrossRef]
18. Benke, K.K.; Lowell, K.E.; Hamilton, A.J. Parameter uncertainty, sensitivity analysis and prediction error in a water-balance hydrological model. *Math. Comput. Model.* **2008**, *47*, 1134–1149. [CrossRef]
19. Kang, S.; Lin, H. Wavelet analysis of hydrological and water quality signals in an agricultural watershed. *J. Hydrol.* **2007**, *338*, 1–14. [CrossRef]
20. Fu, C.; James, A.L.; Wachowiak, M.P. Analyzing the combined influence of solar activity and El Niño on streamflow across southern Canada. *Water Resour. Res.* **2012**, *48*. [CrossRef]
21. Massei, N.; Laignel, B.; Deloffre, J.; Mesquita, J.; Motelay, A.; Lafite, R.; Durand, A. Long-term hydrological changes of the Seine River flow (France) and their relation to the North Atlantic Oscillation over the period 1950–2008. *Int. J. Climatol.* **2010**, *30*, 2146–2154. [CrossRef]
22. Zhang, Q.; Xu, C.Y.; Jiang, T.; Wu, Y. Possible influence of ENSO on annual maximum streamflow of the Yangtze River, China. *J. Hydrol.* **2007**, *333*, 265–274. [CrossRef]
23. Torrence, C.; Compo, G.P. A Practical Guide to Wavelet Analysis. *Bull. Am. Meteorol. Soc.* **1998**, *79*, 61–78. [CrossRef]
24. Grinsted, A.; Moore, J.C.; Jevrejeva, S. Application of the cross wavelet transform and wavelet coherence to geophysical time series. *Nonlinear Process. Geophys.* **2004**, *11*, 561–566. [CrossRef]
25. Adamowski, J.F. River flow forecasting using wavelet and cross-wavelet transform models. *Hydrol. Process.* **2008**, *22*, 4877–4891. [CrossRef]
26. Liu, Y.; Brown, J.; Demargne, J.; Seo, D.J. A wavelet-based approach to assessing timing errors in hydrologic predictions. *J. Hydrol.* **2011**, *397*, 210–224. [CrossRef]

*Article*

# Investigating the Linkage between Extreme Rainstorms and Concurrent Synoptic Features: A Case Study in Henan, Central China

**Yu Lang** [1,2,†], **Ze Jiang** [3,†] **and Xia Wu** [4,5,6,*]

1   Institute of Natural Resource Owner's Equity, Chinese Academy of Natural Resources Economics, Beijing 101149, China; yulang@canre.org.cn or ylang2@126.com
2   Harvard Medical School, Harvard University, Cambridge, MA 02138, USA
3   School of Civil and Environmental Engineering, University of New South Wales, Sydney, NSW 2052, Australia; ze.jiang@unsw.edu.au
4   State Key Laboratory of Hydrology-Water Resources and Hydraulic Engineering, Hohai University, Nanjing 210098, China
5   CMA-HHU Joint Laboratory for Hydrometeorological Studies, Hohai University, Nanjing 210098, China
6   College of Hydrology and Water Resources, Hohai University, Nanjing 210098, China
*   Correspondence: xiawu_hydrology@163.com
†   These authors contributed equally to this work.

**Abstract:** Extraordinary floods are linked with heavy rainstorm systems. Among various systems, their synoptic features can be quite different. The understanding of extreme rainstorms by their causative processes may assist in flood frequency analysis and support the evaluation of any changes in flood occurrence and magnitudes. This paper aims to identify the most dominant meteorological factors for extreme rainstorms, using the ERA5 hourly reanalysis dataset in Henan, central China as a case study. Past 72 h extreme precipitation events are investigated, and six potential factors are considered in this study, including precipitable water (PW), the average temperature (Tavg) of and the temperature difference (Tdiff) between the value at 850 hPa and 500 hPa, relative humidity (RH), convective available potential energy (CAPE), and vertical wind velocity (Wind). The drivers of each event and the dominant factor at a given location are identified using the proposed metrics based on the cumulative distribution function (CDF). In Henan, central China, Wind and PW are dominant factors in summer, while CAPE and Wind are highly related factors in winter. For Zhengzhou city particularly, Wind is the key driver for summer extreme rainstorms, while CAPE plays a key role in winter extreme precipitation events. It indicates that the strong transport of water vapor in summer and atmospheric instability in winter should receive more attention from the managers and planners of water resources. On the contrary, temperature-related factors have the least contribution to the occurrence of extreme events in the study area. The analysis of dominant factors can provide insights for further flood estimations and forecasts.

**Keywords:** extreme rainstorms; driver identification; dominant factor; ERA5

**Citation:** Lang, Y.; Jiang, Z.; Wu, X. Investigating the Linkage between Extreme Rainstorms and Concurrent Synoptic Features: A Case Study in Henan, Central China. *Water* 2022, 14, 1065. https://doi.org/10.3390/w14071065

Academic Editors: Maria Mimikou and Yves Tramblay

Received: 28 February 2022
Accepted: 26 March 2022
Published: 28 March 2022

**Publisher's Note:** MDPI stays neutral with regard to jurisdictional claims in published maps and institutional affiliations.

## 1. Introduction

According to the IPCC Sixth Assessment Report, global warming will lead to the increase of extreme weather events; for example, the increase of extremely hot days and heatwaves is very likely to occur on almost all lands, and extreme precipitation magnitude and frequency are very likely to rise in many areas [1]. Extreme precipitation events have a very strong destructive effect. For instance, on 17–21 July 2021, an extremely heavy rainstorm hit Zhengzhou city in Henan, central China. The whole city was flooded, the traffic system was paralyzed, and power was interrupted for more than weeks. The estimated economic loss exceeded CNY 120 billion, with hundreds of lives lost in the end. Extraordinary floods are linked with heavy rainstorm systems, and their synoptic features differ from each other. Therefore, the understanding of heavy rainstorm floods by their causative processes may assist in flood frequency analysis and support the evaluation of any changes in flood occurrence and magnitudes [2].

The method of using numerical weather prediction (NWP) models to simulate the physical process of storm formation and estimate design rainfalls can be found in the National Research Council [3]. This approach links atmospheric processes with rainfall in a quantitative way. Abbs [4] investigated and estimated the probable maximum precipitation (PMP) with an NWP model, and the key message from this study and other studies [5,6] is that the linear assumption between precipitation and available moisture (i.e., precipitable water) of conventional PMP estimates [7] was not always valid. Recently, some researchers used forecasts from NWP models or regional climate models (RCMs) to estimate PMP and pointed out that climate change may affect the estimates of PMPs [8–14]. Other studies applied NWP models to explore the cause of specific extreme precipitation events, particularly in urban areas, such as Mumbai, India, in 2005 [15]; Nashville, Tennessee, the United States, in 2010 [16]; Beijing, China, in July 2012 [17]; Guangzhou, China, in July 2017 [18]; Istanbul, Turkey, in 2017 [19]; North-Rhine Westphalia and adjacent Rhineland-Palatinate, Germany, in 2021 [20]; and Henan, China, in July 2021 [21]. Furthermore, some studies have systematically assessed the relationship between several atmospheric variables (e.g., moisture, wind convergence, and wind vertical velocity) and convective rainfall state [22], peak rainfall intensity [23], or sustained rainfall deficit [24,25]. These studies provided a systematic analysis of the relationship between extreme precipitation and meteorological factors, but most of them only investigated their qualitative relationship for specific events at a given location and did not take enough account of the spatial patterns.

The analysis of spatial patterns necessitates the proper method and high-resolution data. Chen and Hossain [10] investigated the concurrent synoptic features for the extreme rainstorms over the continental United States at a coarse spatio-temporal scale using regional (NARR) and global (ERA-Interim) reanalysis products, which indicated the usefulness of atmospheric reanalysis products and provided a practicable way to explore the spatial variations of the relation between atmosphere variables and extreme rainstorms. Recently, a new generation of ERA (ERA5) data was released, providing hourly estimates of a mass of atmospheric, land, and oceanic climate variables with 30 km grid resolution [26]. Based on this dataset, the relations can be evaluated at a high spatio-temporal resolution, especially over a localized region. Therefore, this paper applies the method of Chen and Hossain [10] and ERA5 data to analyze the relations between the meteorological factors and extreme rainfall events quantitively and investigates the main factors influencing rainstorms in Henan, central China.

While studies on extreme rainfall events have mostly focused on a specific location with historical observations, this study takes advantage of a high spatial-temporal resolution data–ERA5 to analyse the spatial patterns of the key driver for the extreme rainfall and explore the pattern variations in different seasons. The results can provide a more systematic and deeper understanding of the synoptic causes of extreme precipitation, which helps to identify the relevant mechanisms from a regional perspective and improve the forecast accuracy of the extreme rainfall.

The paper is organized as follows: Section 2 introduces the study area, ERA5 reanalysis data, and the diagnosis factors used in this study. Section 3 presents the methods to identify the driver of an extreme event and derive the domain factor at a demo location. Section 4 shows the analysis results, which are the spatial distributions of the dominant factors on extreme rainstorms in the entire period or different seasons from 1981 to 2021. Conclusions are summarized in Section 5.

## 2. Study Area and Dataset

Henan province is located in the central part of China, as shown in Figure 1. This region experienced a severe flood caused by extraordinary heavy rainfall in 1975 [27], and it was hit by another extreme rainstorm recently on 18–22 July 2021. On 20 July 2021, the daily rainfall and hourly rainfall of many national meteorological observatories in Zhengzhou city (represented by red polygon in Figure 1) exceeded the historical largest value since the meteorological record began [21]. In Henan province, more than 65% of the precipitation falls in summer, and most of them are short-duration heavy precipitation. This is also the feature of most watersheds in China. Therefore, this area can be regarded as a classic example to investigate the atmosphere characteristics of extreme storm events in China.

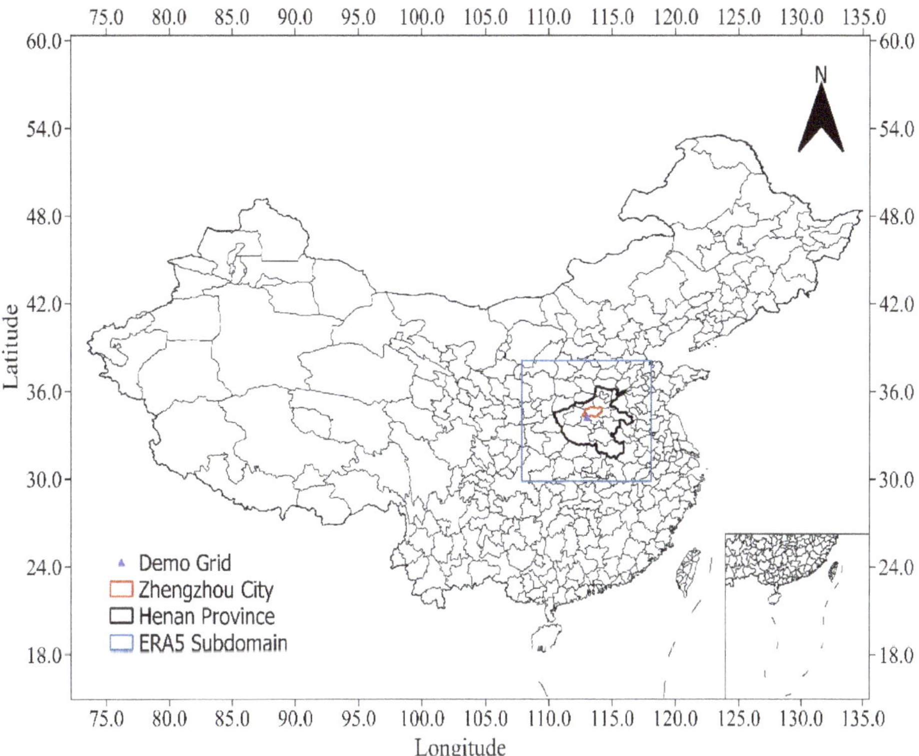


**Figure 1.** The study area of Henan province and Zhengzhou city in central China.

This study applies ERA5 data to implement the investigation. ERA5 data are the latest generation of ECMWF global reanalysis product. From ERA5 data, hourly precipitation estimates, and six atmospheric variables, including precipitable water (PW), the temperature at 850 hPa and 500 hPa (T850 and T500), relative humidity (RH), convective available potential energy (CAPE), and vertical wind velocity (Wind) at 700 hPa are obtained. They cover the period of 1981–2021 and the region represented by a blue box in Figure 1. Average temperature (Tavg $= \frac{T850 + T500}{2}$ ) and temperature difference (Tdiff $=$ T850 $-$ T500) are computed based on the temperature at two different pressure levels. In the end, CAPE, PW, Wind, RH, Tavg, and Tdiff are considered as potential meteorological factors in this study, and these variables are often used in extreme weather events [10,22,28].

## 3. Methodology

In order to find the dominant meteorological factor at a given location, we first needed to identify the driver for each extreme event. The largest 50 rainfall events with a duration of 72 h (i.e., 3 days) from 1981 to 2021 were extracted and used for analysis. Once drivers of all 50 events were identified, the percentage of extreme events that are linked to each driver variable was calculated, and the meteorological variable with the greatest percentage was defined as the dominant factor at the specific location. The following subsections will introduce the metrics used to identify the driver of each extreme event and the dominant factor of a certain region. The details are explained using the extracted extreme events from ERA5 data at a demo grid near Zhengzhou city (the blue triangle in Figure 1).

### 3.1. Extraction of Extreme Events

Analyzing the hourly precipitation data in the entire historic period at a demo grid, the 50 extreme events corresponding to the first 50 largest 72 h accumulated precipitation were selected. Figure 2 demonstrates the time series of precipitation (black line) and the selected extreme events. Each red bar represents one extreme event, while the order of the five largest events is given by a red number. The start time of each event and their accumulated precipitation amount is listed in Table 1.

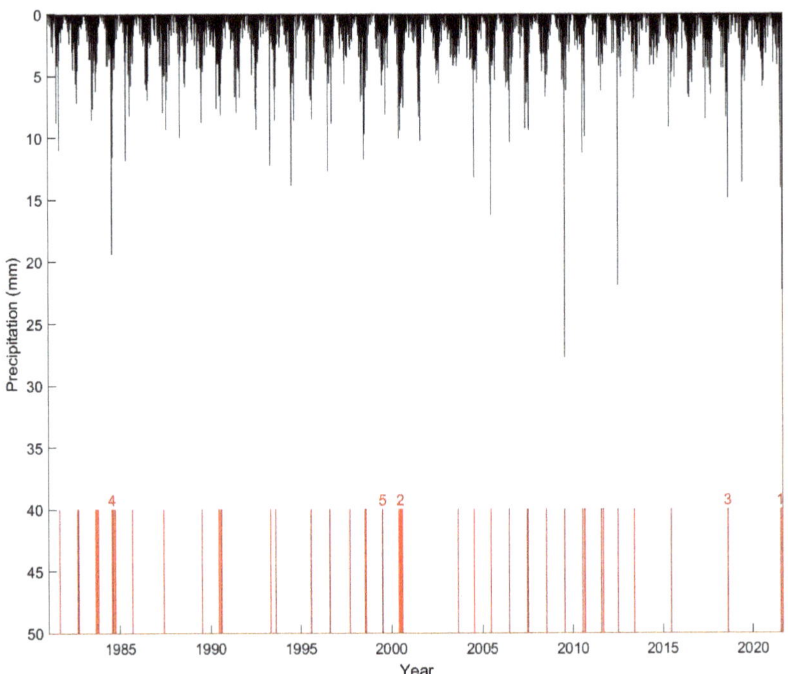

**Figure 2.** Hourly precipitation (black line) and the 50 highest extreme rainstorms at the demo grid near Zhengzhou city from 1981 to 2021. The red number indicates the order of the first five largest events.

It should be noted that the rainfall amount in this grid is underestimated by ERA5. For instance, on 20 July 2021, according to the observation data, the highest hourly rainfall in Zhengzhou city is over 200 mm, but in the ERA5 data, the value in this grid is only 22.4 mm. Some studies focusing on the post-processing of the rainfall data [29–32] may help to deal with this underestimation problem, but this study still implements the analysis on the raw ERA5 data without treatment. If this method is implemented on the more precise observations or post-processing data, the results might be different, but the difference will not be significant. The extreme event is the definition for the relative relationship and the absolute value will not influence the probability of event occurrence. Furthermore, although the total rainfall amount is underestimated, the ERA5 product correctly simulates the timing of the Zhengzhou extreme event beginning on 18 July 2021, which ensures the correct duration of the extreme events.

**Table 1.** List of the largest 50 extreme events with identified drivers at the demo grid.

| Event Order | Start Time | Accumulated 72 h Precipitation (mm) | $M$ [1] | | | | | |
|---|---|---|---|---|---|---|---|---|
| | | | CAPE | PW | Wind | RH | Tavg | Tdiff |
| 1 | 2021-07-18 21:00:00 | 234.92 | 0 | 1 | 1 | 1 | 0 | 0 |
| 2 | 2000-07-03 00:00:00 | 223.94 | 0 | 1 | 1 | 1 | 0 | 0 |
| 3 | 2018-08-16 07:00:00 | 182.18 | 0 | 1 | 1 | 0 | 0 | 0 |
| 4 | 1984-07-16 07:00:00 | 150.53 | 1 | 1 | 1 | 1 | 0 | 0 |
| 5 | 1999-07-03 15:00:00 | 149.99 | 0 | 1 | 1 | 1 | 0 | 0 |
| 6 | 2009-07-20 00:00:00 | 144.32 | 1 | 1 | 1 | 0 | 1 | 0 |
| 7 | 2021-08-19 14:00:00 | 139.27 | 0 | 1 | 1 | 1 | 0 | 0 |
| 8 | 2010-07-16 04:00:00 | 130.87 | 0 | 1 | 1 | 1 | 0 | 0 |
| 9 | 2021-08-28 03:00:00 | 125.42 | 0 | 1 | 1 | 1 | 0 | 0 |
| 10 | 2000-08-03 09:00:00 | 123.12 | 0 | 1 | 1 | 1 | 0 | 0 |
| 11 | 2004-07-14 07:00:00 | 122.64 | 0 | 1 | 1 | 1 | 0 | 0 |
| 12 | 2000-06-24 23:00:00 | 120.97 | 0 | 1 | 1 | 1 | 0 | 0 |
| 13 | 1996-08-02 05:00:00 | 111.64 | 0 | 1 | 1 | 1 | 1 | 0 |
| 14 | 2000-07-12 18:00:00 | 110.28 | 0 | 1 | 1 | 1 | 0 | 0 |
| 15 | 1990-06-17 02:00:00 | 109.47 | 0 | 1 | 1 | 0 | 0 | 0 |
| 16 | 1995-07-22 09:00:00 | 109.13 | 1 | 1 | 1 | 1 | 1 | 0 |
| 17 | 1993-04-29 06:00:00 | 109.00 | 0 | 0 | 1 | 1 | 0 | 0 |
| 18 | 1990-08-13 02:00:00 | 109.00 | 0 | 1 | 1 | 1 | 0 | 0 |
| 19 | 1984-09-21 14:00:00 | 106.25 | 0 | 0 | 1 | 1 | 0 | 0 |
| 20 | 2010-09-04 01:00:00 | 105.82 | 0 | 1 | 1 | 1 | 0 | 0 |
| 21 | 1982-08-11 15:00:00 | 101.03 | 0 | 1 | 1 | 1 | 0 | 0 |
| 22 | 2011-09-11 23:00:00 | 100.94 | 0 | 0 | 1 | 1 | 0 | 0 |
| 23 | 2013-05-24 04:00:00 | 94.25 | 0 | 0 | 1 | 1 | 0 | 0 |
| 24 | 1998-08-03 11:00:00 | 93.30 | 0 | 1 | 1 | 1 | 1 | 0 |
| 25 | 2000-06-01 10:00:00 | 92.92 | 0 | 1 | 0 | 1 | 0 | 0 |
| 26 | 2011-07-31 09:00:00 | 92.74 | 0 | 0 | 1 | 1 | 0 | 0 |
| 27 | 1990-07-19 11:00:00 | 90.09 | 1 | 1 | 1 | 1 | 1 | 0 |
| 28 | 2011-09-04 01:00:00 | 89.85 | 0 | 1 | 0 | 1 | 0 | 0 |
| 29 | 2003-08-27 20:00:00 | 89.11 | 0 | 1 | 1 | 0 | 0 | 0 |
| 30 | 1984-08-06 07:00:00 | 88.28 | 1 | 1 | 1 | 1 | 0 | 0 |
| 31 | 2005-06-24 17:00:00 | 88.15 | 0 | 1 | 1 | 0 | 0 | 0 |
| 32 | 1983-08-09 11:00:00 | 85.46 | 0 | 1 | 1 | 1 | 0 | 0 |
| 33 | 1983-09-04 22:00:00 | 85.25 | 0 | 1 | 1 | 1 | 0 | 0 |
| 34 | 2012-07-04 03:00:00 | 85.16 | 0 | 1 | 1 | 0 | 0 | 0 |
| 35 | 1985-09-13 11:00:00 | 84.97 | 0 | 0 | 1 | 0 | 0 | 0 |
| 36 | 1983-10-03 03:00:00 | 84.93 | 0 | 0 | 1 | 0 | 0 | 0 |
| 37 | 2015-06-22 12:00:00 | 84.71 | 0 | 1 | 1 | 0 | 0 | 0 |
| 38 | 1993-08-12 10:00:00 | 84.34 | 0 | 1 | 1 | 1 | 0 | 0 |
| 39 | 2008-07-20 08:00:00 | 84.26 | 0 | 1 | 0 | 0 | 0 | 0 |
| 40 | 1984-09-06 17:00:00 | 84.18 | 0 | 1 | 1 | 1 | 0 | 0 |
| 41 | 1989-07-04 10:00:00 | 82.94 | 0 | 1 | 0 | 1 | 0 | 0 |
| 42 | 2006-07-01 10:00:00 | 82.75 | 0 | 1 | 1 | 1 | 1 | 0 |
| 43 | 2007-07-18 02:00:00 | 82.00 | 1 | 1 | 1 | 0 | 1 | 0 |
| 44 | 2007-07-03 22:00:00 | 81.92 | 0 | 1 | 1 | 1 | 0 | 0 |
| 45 | 1987-05-31 06:00:00 | 81.82 | 1 | 0 | 1 | 1 | 0 | 1 |
| 46 | 1982-08-28 07:00:00 | 81.56 | 0 | 1 | 1 | 1 | 0 | 0 |
| 47 | 1998-07-14 07:00:00 | 80.70 | 1 | 1 | 1 | 1 | 1 | 0 |
| 48 | 1997-09-12 03:00:00 | 79.52 | 0 | 0 | 1 | 0 | 0 | 0 |
| 49 | 2010-08-22 03:00:00 | 79.25 | 0 | 0 | 1 | 0 | 0 | 0 |
| 50 | 1981-08-09 02:00:00 | 78.79 | 0 | 1 | 0 | 1 | 0 | 0 |

Note: [1] $M = 1$ means the factor is the driver of the extreme rainstorm, while $M = 0$ means it is not.

### 3.2. Driver Identification of an Extreme Event

The concurrence between an extreme synoptic condition and an extreme precipitation event can be quantified by a metric based on the cumulative distribution function (CDF) of the hourly values in the entire period. The proposed metric $M$ for an atmospheric variable $i$ and a storm event $j$ is defined as follows:

$$M_j^i(\theta_1, \theta_2) = \begin{cases} 1, & if\ P\left(X_j^i > X_{\theta_1}^i\right) \geq \theta_2 \\ 0, & otherwise. \end{cases} \tag{1}$$

where $X_j^i$ is the hourly value of atmospheric variable $i$ ($i$ represents CAPE, PW, Wind, RH, Tavg, and Tdiff, respectively, in this study) during the 72 h of the identified storm event $j$. $X_{\theta_1}^i$ is the value of the atmospheric variable $i$ corresponding to the CDF of $\theta_1$ ($\theta_1 = 0.95$ is adopted in the study), which can be regarded as the threshold of the extreme condition. For each specific event, the percentage of the atmospheric variable $i$ reaching the extreme condition ($X_{0.95}^i$) can be calculated. If the percentage exceeds $\theta_2$ ($\theta_2 = 15\%$ is adopted in the study), representing that the extreme condition maintains over a certain duration (72 h $\times$ 15% = 10.8 h), then the atmospheric variable $i$ can be identified as the driver of the extreme event $j$, and its value of metric $M$ is 1.

Taking the first largest event as an example, Figure 3 presents the cumulative probabilities of all the values (blue lines) and the cumulative probabilities of the values in the first largest event (red circles) for each atmospheric variable at the demo grid. It clearly shows that some values of CAPE, PW, Wind, and RH meet the extreme condition (CDF of 95%, represented by a horizontal red dashed line), and the exceeding percentages are 10%, 100%, 59%, and 67%, repressively. According to $\theta_2 = 15\%$, PW, Wind, and RH are regarded as the drivers of the first largest event, and their $M$ values in Table 1 are 1.

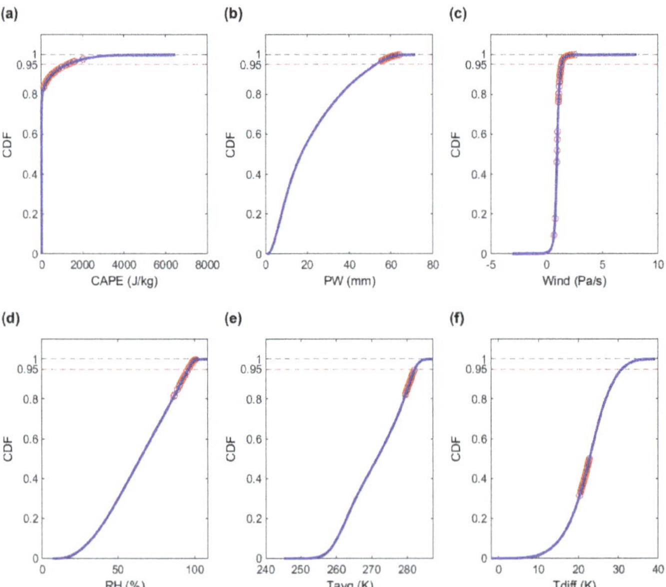

**Figure 3.** Demonstration of driver identification of the first largest event at the demo grid. The result of each atmospheric variable is shown in a subfigure with (**a**) CAPE, (**b**) PW, (**c**) Wind, (**d**) RH, (**e**) Tavg, and (**f**) Tdiff, separately. The blue line presents the cumulative distribution curve (CDF) of all hourly values in the entire period from 1981 to 2021. The red circles present the cumulative probabilities of the hourly values during the 72 h after the first largest event began. The CDF of 95% (horizontal red dashed line) is defined as the threshold of the extreme condition.

### 3.3. Dominant Factor Analysis at a Given Location

Once the driver for each extreme precipitation event is identified by the metric $M$, the dominant factor, $F$, is defined as the driver with the largest percentage of the extreme events that are related to the following:

$$F(\theta_1, \theta_2) = i(\max(R^i)) \tag{2}$$

$$R^i = \frac{\sum_{j=1}^{N} M_j^i(\theta_1, \theta_2)}{N} \tag{3}$$

where $N$ is the total number of extreme events investigated ($N = 50$ in this study). The 50 largest events at the demo grid are shown in Table 1, including their starting time, accumulated precipitation, and associated meteorological drivers. In summary, there are 8, 40, 45, 37, 8, and 1 events related to CAPE, PW, Wind, RH, Tavg, and Tdiff, respectively, and the percentage of the extreme event ($R$) is 16%, 80%, 90%, 74%, 16%, and 2%, respectively. Therefore, Wind with the maximum $R = 90\%$ is regarded as the dominant factor at the demo grid.

The above processes demonstrate how the dominant factor at one specific grid is derived. In the following sections, we will show the spatial variation of dominant factors when extreme precipitation events are taken from the entire period or different seasons.

## 4. Results and Discussions

### 4.1. Spatial Patterns of the Dominant Factor in the Entire Period

As demonstrated in the method section, in order to obtain the dominant factor for each location, we needed to find drivers for each event and calculate the percentage of the extreme events that are related to the corresponding driver. Figure 4 presents the percentage of extreme events that are related to extreme atmospheric variables across the Henan province. It is clearly shown that PW, Wind, and RH are the most common drivers for extreme precipitation events in Henan (black polygon), while CAPE and Tavg are relatively less. There is no obvious link between Tdiff and the occurrence of extreme precipitation events. PW is the indicator of available moisture in the system, mainly affected by Tavg and RH. Figure 4 demonstrates that PW and RH have a close spatial pattern and play a much more significant role in extreme events than Tavg. It illustrates that compared with the temperature indicator Tavg, the moisture indicators RH and PW are more sensitive to extreme rainstorm events. The moisture is the determinant of how much rain will fall, and it is direct to the formation of rainfall. The temperature affects how much moisture can be contained. The moisture here is potential content and not the actual content. That is most likely the reason why Tavg has no obvious relation with extreme precipitation events.

Based on the results of Figure 4, the spatial variation of dominant meteorological factors across the Henan province is shown in Figure 5. It can be seen that most areas, including Zhengzhou city, are dominated by Wind, while few regions are dominated by RH or PW. Vertical wind velocity (referred to as Wind) is the velocity between pressure levels, and the vertical velocity at 700 hPa is most related to precipitation processes [10,23]. Previous studies have shown that the moisture needed for extreme precipitation events cannot be met by PW [10,33]. Vertical wind velocity, representative of the large-scale horizontal convergence, draws moisture from the surrounding area to supply the moisture needed for the extreme event. That is why PW is less dominant than Wind in the formation of extreme rainfall.

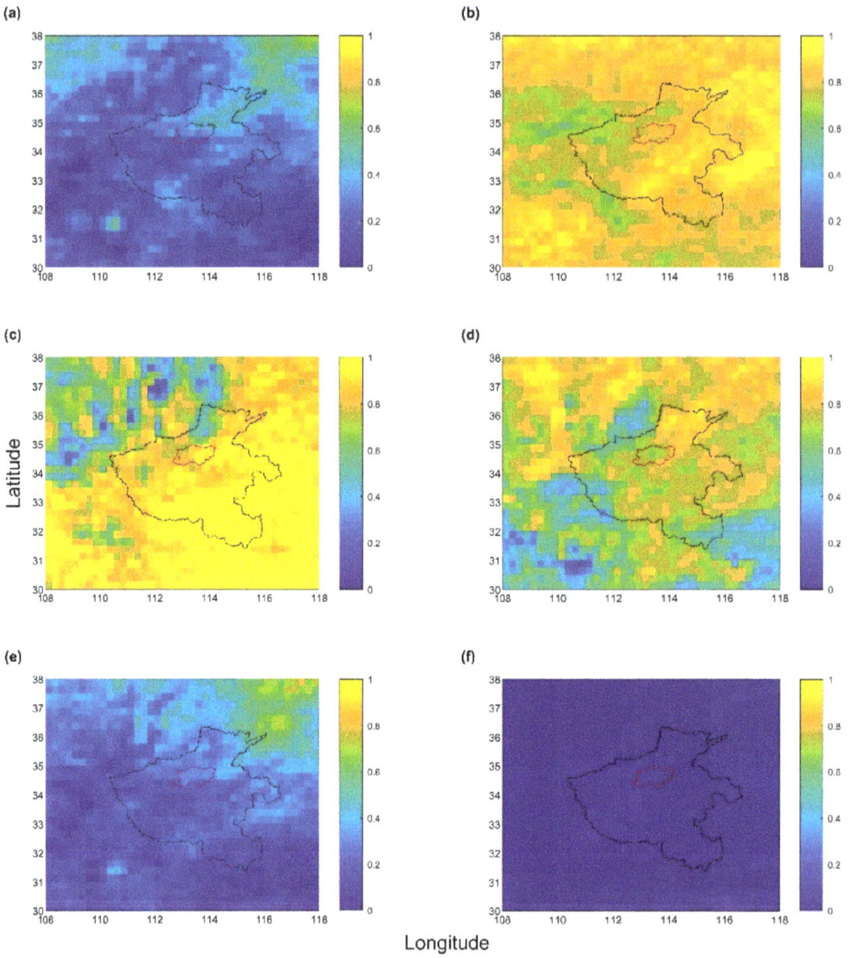

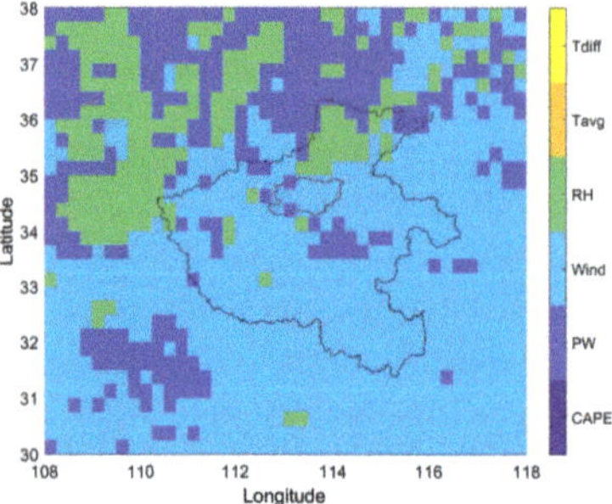

**Figure 4.** Percentage of extreme events that are related to each atmospheric variable across the Henan province in the entire period from 1981 to 2021. (**a**) CAPE; (**b**) PW; (**c**) Wind; (**d**) RH; (**e**) Tavg; and (**f**) Tdiff. The black polygon presents the border of Henan province, while the red polygon is the boundary of Zhengzhou city.

**Figure 5.** Dominant factors of extreme rainstorms across the Henan province. The black polygon presents the border of Henan province, while the red polygon is the boundary of Zhengzhou city.

### 4.2. Spatial Patterns of the Dominant Factor in Different Seasons

Except for the analysis based on extreme events of the whole year, the dominant factors were reanalyzed for different seasons to explore a deeper understanding of the causes of the extreme precipitation. The percentage of extreme events in each season is 4.742% (Spring, March to May), 82.127% (Summer, June to August), 13.128% (Autumn, September to November), and 0.003% (Winter, December to February), respectively. Over 80% of extreme events occur during the summer, while less than 20% of extreme precipitation events mainly happen in the autumn and spring. Due to the effect of the temperate monsoon climate, it is expected that the extreme precipitation event in this region is generally concentrated within the summer season. Therefore, the summer results (Figure 6b) are the most similar to the yearly based analysis (Figure 4). This also indicates that the dominant factors found here are stable, and it agrees with the finding in Chen and Hossain [10]. In addition, regarding the percentage of extreme events that are related to wind and RH, the patterns are stable across seasons. This is because the two drivers show less seasonal variability, and it is relatively easier to reach extreme values in all seasons.

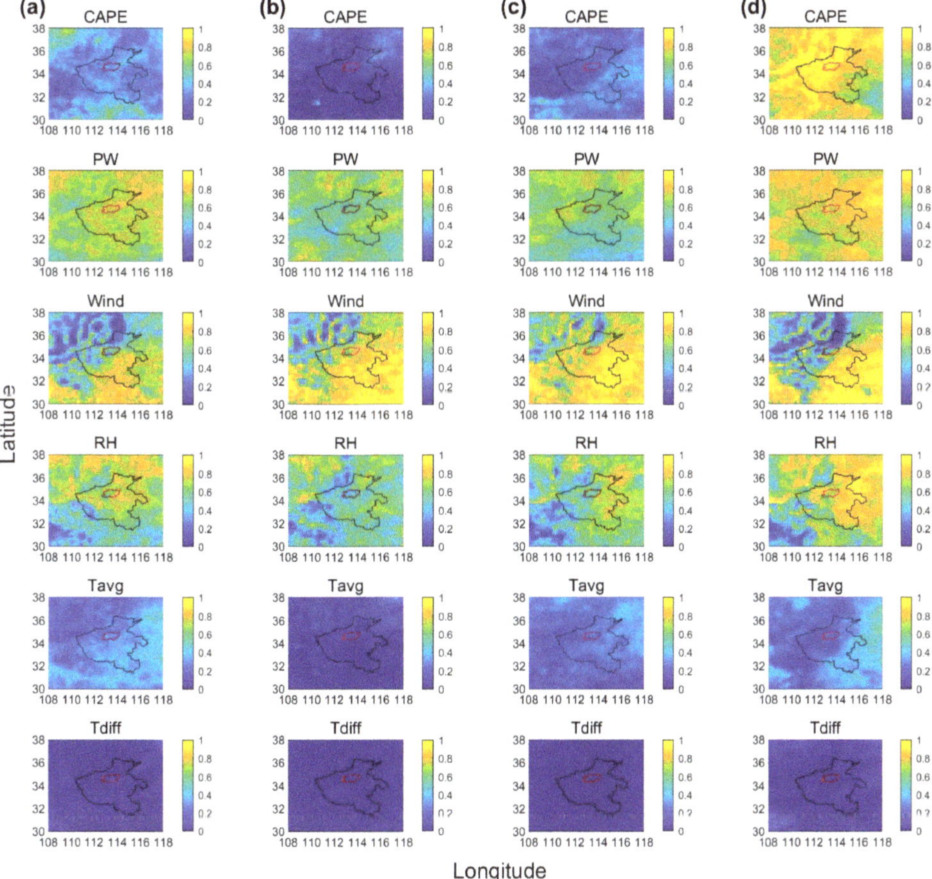

**Figure 6.** Percentage of extreme events that are related to each atmospheric variable across the Henan province for the spring (**a**), summer (**b**), autumn (**c**), and winter (**d**), respectively. The black polygon presents the border of Henan province, while the red polygon is the boundary of Zhengzhou city.

Figure 6 demonstrates that spring, autumn, and summer have similar spatial patterns, but winter is quite different. The key difference focuses on the patterns of CAPE. As documented at the ECMWF website, CAPE is a measurement of atmospheric stability and can be used to assess the potential of convection development. In winter, CAPE is identified as the driver of extreme precipitation events over the study area (Figure 6d), and it is also shown in Figure 7 as the dominant factor in the northern part of Henan province, including Zhengzhou city. However, in summer, CAPE has few effects on extreme events in Figure 6b. This finding is in agreement with the study of Lepore, Veneziano [34]. Most of the precipitation in winter comes from snow, and it has a different causative process from rainstorms in summer. High moisture indicated by PW ensures the initialization of rainstorms, while an unstable weather system caused by a high CAPE and Wind leads to snow in winter [10]. Based on the above analyses at a seasonal scale, the seasonal variability of these physically dominant factors was investigated, and it should be considered in the trend estimation of extreme precipitation due to the various causative processes.

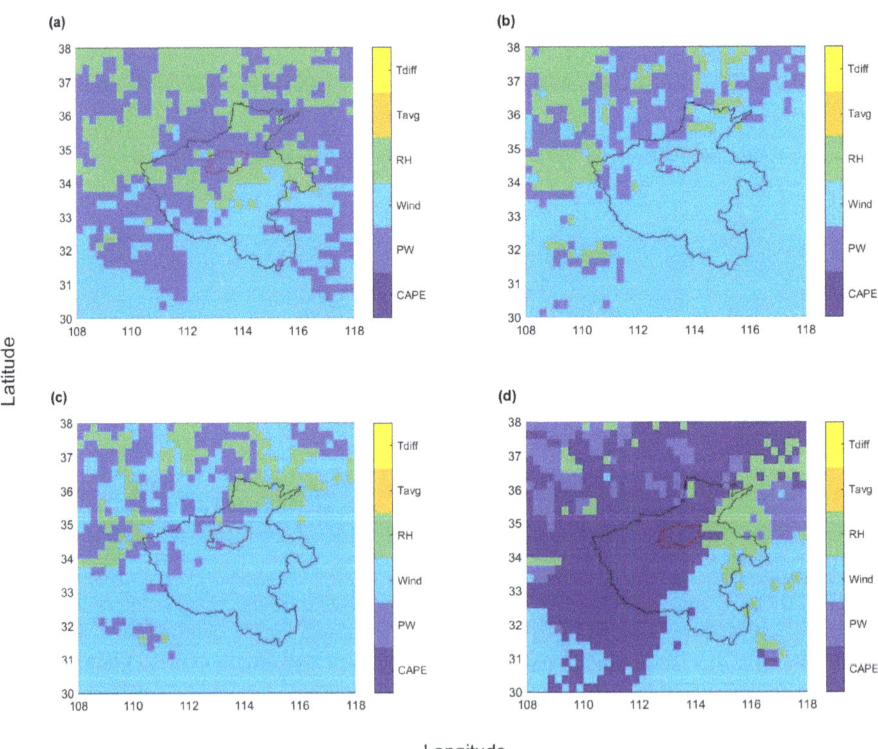

**Figure 7.** Dominant factors of extreme rainstorms across the Henan province in different seasons. (**a**) spring; (**b**) summer; (**c**) autumn; and (**d**) winter. The black polygon presents the border of Henan province, while the red polygon is the boundary of Zhengzhou city.

### 4.3. Considerations of other Factors

In agreement with this study, other studies also found that precipitable water (PW) is often related to precipitation extreme events [35], and wind velocity is more likely to be the dominant factor of summer extreme precipitation events in some regions [10,23]. Besides, other factors, such as topography [36], urbanization-induced urban heat island and aerosol effects [37,38], and global climate change [39], also influenced the occurrence of extreme storms. Compared with variables discussed in this study, those factors are more likely to be static or show little variations in a long time. However, their effects on the spatial pattern are non-negligible and could influence the classification of extreme rainfall events (such as storms, warm/cold front, or tropical cyclones [40]). Therefore, further study is necessary regarding the comprehensive effects of the short-term and long-term variability on extreme precipitation events.

To summarize, it is important to understand how extreme precipitation events are linked to different types of weather circulation patterns, and reanalysis of data can support these subsequent studies due to the high temporal and spatial resolutions.

## 5. Conclusions

This study explores the characteristics of extreme precipitation events over the Henan province, central China using the latest ERA5 dataset. Over a ~41-y period, the largest 50 rainstorm events with the 72 h duration at each grid cell were extracted. The driver of these events was identified using the criteria based on the cumulative distribution function. A single atmospheric variable that controls the largest number of the 50 events was regarded as the dominant factor. The findings from the results are as follows:

- Over the entire study region, extreme precipitation events mostly happen in summer (from June to August).
- For the entire period, PW, Wind, and RH are the most common drivers for extreme precipitation events over the Henan province.
- For the different seasons, across the Henan region, Wind and PW are dominant factors in summer, while Wind and CAPE are highly related factors in winter. For Zhengzhou city particularly, Wind is the key driver for summer, while CAPE plays a key role in winter.
- Temperature-related variables have the lowest contribution to the occurrence of extreme events in the study area.

According to the proposed metric in this study, we can classify each event based on the various atmospheric variables and then identify the dominant factor. The analysis of dominant factors can provide insights for further flood estimations and forecasts. For instance, selecting annual maximum precipitation events with extreme Wind and PW values will likely identify the events that will maximize a storm. Besides, based on this method, further studies can be carried out by considering more factors, such as topography and global warming, to explore more findings on the formation of extreme rainfall and floods.

**Author Contributions:** Conceptualization, Z.J.; methodology, Y.L.; validation, Z.J. and X.W.; formal analysis, Y.L.; writing—original draft preparation, Y.L.; writing—review and editing, Z.J. and X.W.; funding acquisition, X.W. All authors have read and agreed to the published version of the manuscript.

**Funding:** This research was funded by the National Natural Science Foundation of China, grant number 52109015; the Fundamental Research Funds for the Central Universities, grant number B220202030; and Jiangsu Planned Projects for Postdoctoral Research Funds, grant number 2021K046A.

**Institutional Review Board Statement:** Not applicable.

**Informed Consent Statement:** Not applicable.

**Data Availability Statement:** ECMWF Reanalysis v5 (ERA5) dataset provided by the European Centre for Medium-Range Weather Forecasts, and the data can be accessed and downloaded from Copernicus Climate Change Service (C3S) at Website: https://cds.climate.copernicus.eu/ (accessed on 7 December 2021).

**Acknowledgments:** This research has been supported by the National Natural Science Foundation of China (grant no. 52109015), the Fundamental Research Funds for the Central Universities (grant no. B220202030), and Jiangsu Planned Projects for Postdoctoral Research Funds (grant no. 2021K046A).

**Conflicts of Interest:** The authors declare no conflict of interest.

## References


1. Seneviratne, S.I.; Zhang, X.; Adnan, M.; Badi, W.; Dereczynski, C.; Luca, A.D.; Ghosh, S.; Iskandar, I.; Kossin, J.; Lewis, S.; et al. Weather and Climate Extreme Events in a Changing Climate. In *Climate Change 2021: The Physical Science Basis. Contribution of Working Group I to the Sixth Assessment Report of the Intergovernmental Panel on Climate Change*; Masson-Delmotte, V., Zhai, P., Pirani, A., Connors, S.L., Péan, C., Berger, S., Caud, N., Chen, Y., Goldfarb, L., Gomis, M.I., et al., Eds.; Cambridge University Press: Cambridge, UK, 2021; in press.
2. Zhan, D.; Zou, J. *Heavy Rain Storm Floods in China and their Estimation*; Singh, V.P., Ed.; Springer: Dordrecht, The Netherlands, 1987; pp. 35–44. [CrossRef]
3. National Research Council. *Estimating Bounds on Extreme Precipitation Events: A Brief Assessment*; National Academies Press: Washington, DC, USA, 1994.
4. Abbs, D.J. A numerical modeling study to investigate the assumptions used in the calculation of probable maximum precipitation. *Water Resour. Res.* **1999**, *35*, 785–796. [CrossRef]
5. Chen, L.-C.; Bradley, A.A. How does the record July 1996 Illinois rainstorm affect probable maximum precipitation estimates? *J. Hydrol. Eng.* **2007**, *12*, 327–335. [CrossRef]
6. Chen, L.-C.; Bradley, A.A. Adequacy of using surface humidity to estimate atmospheric moisture availability for probable maximum precipitation. *Water Resour. Res.* **2006**, *42*, W09410. [CrossRef]
7. WMO. *Manual on Estimation of Probable Maximum Precipitation (PMP)*; World Meteorological Organization: Geneva, Switzerland, 2009.
8. Chen, X.; Hossain, F. Revisiting extreme storms of the past 100 years for future safety of large water management infrastructures. *Earth's Future* **2016**, *4*, 306–322. [CrossRef]
9. Chen, X.; Hossain, F.; Leung, L.R. Probable Maximum Precipitation in the U.S. Pacific Northwest in a Changing Climate. *Water Resour. Res.* **2017**, *53*, 9600–9622. [CrossRef]
10. Chen, X.; Hossain, F. Understanding Model-Based Probable Maximum Precipitation Estimation as a Function of Location and Season from Atmospheric Reanalysis. *J. Hydrometeorol.* **2018**, *19*, 459–475. [CrossRef]
11. Rouhani, H.; Leconte, R. A novel method to estimate the maximization ratio of the Probable Maximum Precipitation (PMP) using regional climate model output. *Water Resour. Res.* **2016**, *52*, 7347–7365. [CrossRef]
12. Klein, I.M.; Rousseau, A.N.; Frigon, A.; Freudiger, D.; Gagnon, P. Evaluation of probable maximum snow accumulation: Development of a methodology for climate change studies. *J. Hydrol.* **2016**, *537*, 74–85. [CrossRef]
13. Ben Alaya, M.A.; Zwiers, F.; Zhang, X. Evaluation and Comparison of CanRCM4 and CRCM5 to Estimate Probable Maximum Precipitation over North America. *J. Hydrometeorol.* **2019**, *20*, 2069–2089. [CrossRef]
14. Ben Alaya, M.A.; Zwiers, F.W.; Zhang, X. Probable maximum precipitation in a warming climate over North America in CanRCM4 and CRCM5. *Clim. Change* **2020**, *158*, 611–629. [CrossRef]
15. Kumar, A.; Dudhia, J.; Rotunno, R.; Niyogi, D.; Mohanty, U.C. Analysis of the 26 July 2005 heavy rain event over Mumbai, India using the Weather Research and Forecasting (WRF) model. *Q. J. R. Meteorol. Soc.* **2008**, *134*, 1897–1910. [CrossRef]
16. Moore, B.J.; Neiman, P.J.; Ralph, F.M.; Barthold, F.E. Physical processes associated with heavy flooding rainfall in Nashville, Tennessee, and vicinity during 1–2 May 2010: The role of an atmospheric river and mesoscale convective systems. *Mon. Weather Rev.* **2012**, *140*, 358–378. [CrossRef]
17. Liu, L.; Jensen, M.B. Climate resilience strategies of Beijing and Copenhagen and their links to sustainability. *Water Policy* **2017**, *19*, 997–1013. [CrossRef]
18. Yin, J.; Zhang, D.-L.; Luo, Y.; Ma, R. On the Extreme Rainfall Event of 7 May 2017 over the Coastal City of Guangzhou. Part I: Impacts of Urbanization and Orography. *Mon. Weather Rev.* **2020**, *148*, 955–979. [CrossRef]
19. Koç, G.; Natho, S.; Thieken, A.H. Estimating direct economic impacts of severe flood events in Turkey (2015–2020). *Int. J. Disaster Risk Reduct.* **2021**, *58*, 102222. [CrossRef]
20. Fekete, A.; Sandholz, S. Here Comes the Flood, but Not Failure? Lessons to Learn after the Heavy Rain and Pluvial Floods in Germany 2021. *Water* **2021**, *13*, 3016. [CrossRef]
21. Wang, H.; Hu, Y.; Guo, Y.; Wu, Z.; Yan, D. Urban flood forecasting based on the coupling of numerical weather model and stormwater model: A case study of Zhengzhou city. *J. Hydrol. Reg. Stud.* **2022**, *39*, 100985. [CrossRef]
22. Davies, L.; Jakob, C.; May, P.; Kumar, V.V.; Xie, S. Relationships between the large-scale atmosphere and the small-scale convective state for Darwin, Australia. *J. Geophys. Res. Atmos.* **2013**, *118*, 11534–11545. [CrossRef]
23. Loriaux, J.M.; Lenderink, G.; Siebesma, A.P. Peak precipitation intensity in relation to atmospheric conditions and large-scale forcing at midlatitudes. *J. Geophys. Res. Atmos.* **2016**, *121*, 5471–5487. [CrossRef]
24. Jiang, Z.; Rashid, M.M.; Johnson, F.; Sharma, A. A wavelet-based tool to modulate variance in predictors: An application to predicting drought anomalies. *Environ. Model. Softw.* **2021**, *135*, 104907. [CrossRef]
25. Jiang, Z.; Sharma, A.; Johnson, F. Refining Predictor Spectral Representation Using Wavelet Theory for Improved Natural System Modeling. *Water Resour. Res.* **2020**, *56*, e2019WR026962. [CrossRef]
26. Hersbach, H.; Bell, B.; Berrisford, P.; Hirahara, S.; Horányi, A.; Muñoz-Sabater, J.; Nicolas, J.; Peubey, C.; Radu, R.; Schepers, D.; et al. The ERA5 global reanalysis. *Q. J. R. Meteorol. Soc.* **2020**, *146*, 1999–2049. [CrossRef]
27. Zeyi, C.; Zuoshu, W.; Zaitao, P. A numerical study on forecasting the Henan extraordinarily heavy rainfall event in August 1975. *Adv. Atmos. Sci.* **1992**, *9*, 53–62. [CrossRef]

28. Fengmei, Y.; Chongyi, E. Correlation analysis between sand-dust events and meteorological factors in Shapotou, Northern China. *Environ. Earth Sci.* **2010**, *59*, 1359–1365. [CrossRef]

29. Hamill, T.M.; Scheuerer, M. Probabilistic Precipitation Forecast Postprocessing Using Quantile Mapping and Rank-Weighted Best-Member Dressing. *Mon. Weather Rev.* **2018**, *146*, 4079–4098. [CrossRef]

30. Vannitsem, S.; Bremnes, J.B.; Demaeyer, J.; Evans, G.R.; Flowerdew, J.; Hemri, S.; Lerch, S.; Roberts, N.; Theis, S.; Atencia, A.; et al. Statistical Postprocessing for Weather Forecasts: Review, Challenges, and Avenues in a Big Data World. *Bull. Am. Meteorol. Soc.* **2021**, *102*, E681–E699. [CrossRef]

31. Wang, Q.J.; Zhao, T.; Yang, Q.; Robertson, D. A Seasonally Coherent Calibration (SCC) Model for Postprocessing Numerical Weather Predictions. *Mon. Weather Rev.* **2019**, *147*, 3633–3647. [CrossRef]

32. Wu, X.; Marshall, L.; Sharma, A. Quantifying input error in hydrologic modeling using the Bayesian Error Analysis with Reordering (BEAR) approach. *J. Hydrol.* **2021**, *598*, 126202. [CrossRef]

33. Kunkel, K.E.; Karl, T.R.; Easterling, D.R.; Redmond, K.; Young, J.; Yin, X.; Hennon, P. Probable maximum precipitation and climate change. *Geophys. Res. Lett.* **2013**, *40*, 1402–1408. [CrossRef]

34. Lepore, C.; Veneziano, D.; Molini, A. Temperature and CAPE dependence of rainfall extremes in the eastern United States. *Geophys. Res. Lett.* **2015**, *42*, 74–83. [CrossRef]

35. Kim, S.; Sharma, A.; Wasko, C.; Nathan, R. Linking Total Precipitable Water to Precipitation Extremes Globally. *Earth's Future* **2022**, *10*, e2021EF002473. [CrossRef]

36. Su, A.; Lü, X.; Cui, L.; Li, Z.; Xi, L.; Li, H. The Basic Observational Analysis of "7.20" Extreme Rainstorm in Zhengzhou. *Torrential Rain Disasters* **2021**, *40*, 445–454. (In Chinese) [CrossRef]

37. Zhong, S.; Qian, Y.; Zhao, C.; Leung, R.; Wang, H.; Yang, B.; Fan, J.; Yan, H.; Yang, X.Q.; Liu, D. Urbanization-induced urban heat island and aerosol effects on climate extremes in the Yangtze River Delta region of China. *Atmos. Chem. Phys.* **2017**, *17*, 5439–5457. [CrossRef]

38. Zhong, S.; Qian, Y.; Zhao, C.; Leung, R.; Yang, X.-Q. A case study of urbanization impact on summer precipitation in the Greater Beijing Metropolitan Area: Urban heat island versus aerosol effects. *J. Geophys. Res. Atmos.* **2015**, *120*, 10903–10904. [CrossRef]

39. Oscar-Júnior, A.C. Precipitation Trends and Variability in River Basins in Urban Expansion Areas. *Water Resour. Manag.* **2021**, *35*, 661–674. [CrossRef]

40. Cipolla, G.; Francipane, A.; Noto, L.V. Classification of Extreme Rainfall for a Mediterranean Region by Means of Atmospheric Circulation Patterns and Reanalysis Data. *Water Resour. Manag.* **2020**, *34*, 3219–3235. [CrossRef]

*Article*

# Forecasting Summer Rainfall and Streamflow over the Yangtze River Valley Using Western Pacific Subtropical High Feature

Ranran He [1,2], Yuanfang Chen [1,*], Qin Huang [1], Wenpeng Wang [1] and Guofang Li [1]

[1] College of Hydrology and Water Resources, Hohai University, Nanjing 210098, China; heranran2006@163.com (R.H.); hqhhu@outlook.com (Q.H.); wangwenpeng@hhu.edu.cn (W.W.); liguofang@hhu.edu.cn (G.L.)

[2] College of Civil Engineering, Bengbu University, Bengbu 233030, China

* Correspondence: chenyuanfang@hhu.edu.cn

**Abstract:** The western Pacific subtropical high (WPSH) is one of the key systems affecting the summer rainfall over the Yangtze River Valley in China. In this study, the forecasting capacity of the WPSH for summer rainfall and streamflow is evaluated based on the WPSH index (WPSHI) derived from the NCEP/NCAR reanalysis dataset. It has been found that WPSHI can identify extreme flood years with a higher skill than normal wet years. Specifically, exceedance probability forecasting based on WPSHI has higher skills for higher thresholds of rainfall. For streamflow, adding WPSHI as a predictor only enhances the skill for higher thresholds of streamflow relative to models based on antecedent streamflow. Under the same framework, performances of two postprocessing approaches for dynamical forecasts, i.e., the model output statistics (MOS) approach and the reanalysis-based (RAN) approach are compared. Hindcasts from Climate Forecast System version 2 from the National Center for Environmental Prediction (CFSv2) are used to calculate WPSHI, which is used as the predictor for rainfall and streamflow. The result shows that the RAN approach performs better than the MOS approach. This study emphasizes the fact that the forecasting skill of exceedance probability would largely depend on the selected threshold of the predictand, and this fact should be noticed in future studies in the long-term forecasting field.

**Keywords:** western Pacific subtropical high; the Yangtze River Valley; model output statistics (MOS); reanalysis-based (RAN) approach

**Citation:** He, R.; Chen, Y.; Huang, Q.; Wang, W.; Li, G. Forecasting Summer Rainfall and Streamflow over the Yangtze River Valley Using Western Pacific Subtropical High Feature. *Water* **2021**, *13*, 2580. https://doi.org/10.3390/w13182580

Academic Editor: Fernando António Leal Pacheco

Received: 8 August 2021
Accepted: 14 September 2021
Published: 18 September 2021

**Publisher's Note:** MDPI stays neutral with regard to jurisdictional claims in published maps and institutional affiliations.

## 1. Introduction

Managing water resources and controlling risks of flood damages largely depend on the knowledge of the future rainfall and streamflow, leading to a relatively important role of seasonal hydrological forecasting. For the data-driven method, the basic step for making seasonal forecasts is to explore empirical relationships between predictors and rainfall (streamflow). A statistical model that can be directly used for operational prediction must utilize lag relationships, i.e., the relationship between antecedent ocean–atmospheric signals and rainfall (streamflow) in the following season. This method has been frequently used in the seasonal forecasting field [1,2].

At present, postprocessing outputs from dynamical forecasting systems is another frequently used approach for seasonal rainfall forecasting [3–7]. The main reason for postprocessing is that general circulation models (GCMs) often have better skills for forecasting large scale circulations than local precipitation [5,6]. Thus, forecasted circulation variables can be treated as bridges between GCM forecasts and local rainfall [5,7,8]. Streamflow can also be forecasted based on downscaling outputs of GCMs. Specifically, there are two ways for downscaling of streamflow. The first method is to use a two-step procedure, i.e., downscaling GCM outputs to local precipitation and temperature, then using them to force a hydrological model to output streamflow [9]. Another method is to downscale general circulation variables to streamflow directly and skip the hydrological model [10–12].

Considering the postprocessing methodology mentioned above, it is fundamentally important to investigate synchronous relationships between rainfall (streamflow) and circulation variables. The advantage of utilization of synchronous relationships is that their physical mechanism is relatively clearer than lag relationships. In China, it is well known that the western Pacific subtropical high (WPSH) is one of the most important circulation systems affecting summer monsoon rainfall. The spatial distribution of summer rainfall over China largely depends on the location and intensity of the WPSH. When the WPSH extends southwestward, flood often occurs in the Yangtze River Valley, and the summer rain band often locates more southern [13,14]. This mechanism can explain the extreme flood years, such as 1998 in South China. Accordingly, the WPSH is considered as an important factor affecting summer rainfall, and is treated as a key predictor in operational forecasting [15]. Wang, Xiang and Lee [8] have shown that WPSH has higher predictability, and also has a higher potential for seasonal forecasts for summer rainfall.

The Yangtze River Valley is the most important region in China, and the inter-annual variation of the summer monsoon leads to frequent floods in this region. Although the relationship between the WPSH and East Asian summer monsoon has been studied [5,8,14,16], some issues about the seasonal forecasting of summer rainfall and streamflow of this region are still needed to be further explored, which are the main themes of this study. The objectives of this study are summarized as follows.

The first goal of this study is to assess the forecasting skills for both summer rainfall and streamflow over the Yangtze River Valley based on the perfect knowledge about WPSH. For this task, we consider the effect of the definition of the positive event. Specifically, for a given threshold $T$ of the predictand $Y$ (rainfall or streamflow), the positive event can be defined as $Y \geq T$. In this setting, the forecasting procedure will be a binary classification problem. We view the forecasting skill as the function of $T$, and focus on forecasting skills corresponding to different $T$ (in other words, different definitions of the positive event). To the best of the authors' knowledge, limited efforts have been made for understanding the relationship between the forecasting skill and the threshold $T$. If the characteristic of this relationship is well understood, one can define a positive event that can be forecasted with a much higher skill.

The second goal of this study is to compare different postprocessing approaches for dynamical forecasting systems. Two different postprocessing procedures for dynamical forecasts, i.e., the model output statistics (MOS) approach and the reanalysis-based (RAN) approach, are tested and compared for predicting summer rainfall and streamflow over the Yangtze River Valley. A review of the literature suggests that such comparison has not been tried in previous studies in the long-term forecasting field. The Climate Forecast System version 2 from the National Center for Environmental Prediction (CFSv2) is used in this study. The forecasted WPSH index (WPSHI) by CFSv2 is used as the predictor for forecasting summer rainfall and streamflow over the Yangtze River Valley.

The basic technique used in this study is logistic regression, which is used for generating exceedance probability forecasts of rainfall and streamflow. Note that probability forecasting can describe uncertainty of the forecast, which is useful for decision-makers [17]. It should also be noted that summer streamflow is downscaled from WPSHI directly. This approach allows us to downscale seasonal rainfall and streamflow based under the same framework. Based on this framework, probability forecast can be applied for downscaling of streamflow. For streamflow, both antecedent streamflow and WPSHI are used as predictors, for considering both the initial state of the valley and the skill from the climate in the target season (i.e., summer).

The structure of this manuscript is organized as follows. The dataset used in this study and the definition of WPSHI are described in Section 2. In Section 3, we provide an analysis of the predictability of rainfall based on the receiver operator characteristic (ROC) analysis. Sections 4 and 5 present methods and results of a series forecasting experiments. At last, discussions and conclusions are stated in Sections 6 and 7, respectively.

## 2. Dataset Description

### 2.1. Rainfall Data and Streamflow

The NOAA's PRECipitation REConstruction over Land (PREC/L) dataset [18] is used as the observed rainfall data. Summer (June–July–August) streamflow of two stations of the main stream of the Yangtze River, i.e., Hankou station in the middle reaches and Datong station in the lower reaches, is also explored in this study. The record of Hankou and Datong used in this study covers the period of 1960–2018. The location of the Yangtze River Valley and two stations, i.e., Hankou and Datong, are shown in Figure 1.

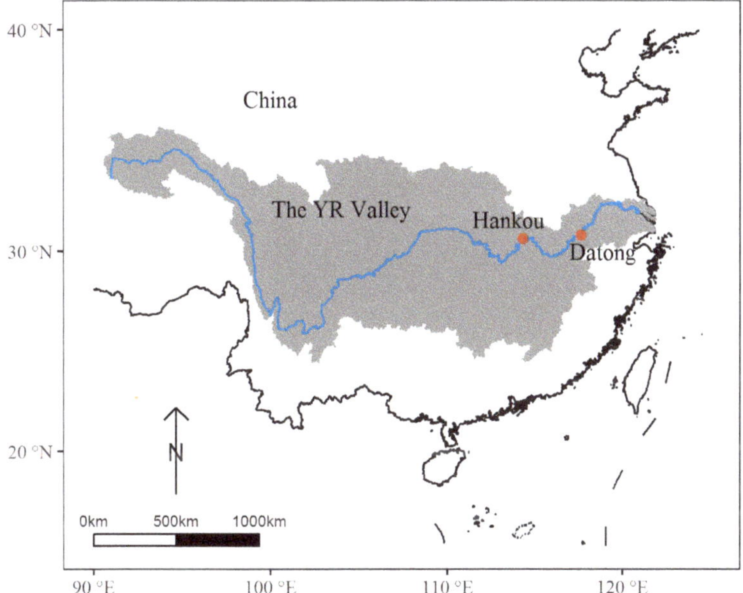


**Figure 1.** Locations of the Yangtze River Valley (YR Valley) in China and two hydrological stations explored in this study.

### 2.2. Reanalysis Dataset and Hindcasts of CFSv2

The NCEP/NCAR reanalysis dataset [19] is used in this study as the reanalysis fields of meteorological variables. The geopotential height of 500-hPa (Z500) forecasted by the CFSv2 system [20] is used as the forecasted fields. The hindcast dataset of CFSv2 from 1982 to 2010 is used in this study, and the operational forecasts from 2011 to 2018 are used to extend the hindcast dataset to cover the period of 1982–2018. The skills of CFSv2 for global and the East Asian summer monsoon have been evaluated by previous studies [21–23]. It has been found that CFSv2 can simulate many features of the East Asian monsoon system [23]. However, CFSv2 often underestimates the intensity of the monsoon system, which is true for both the Southern Asian monsoon and the East Asian monsoon [23].

For the middle time of each month, there are 24 members of forecasts released, which is initiated at successive five days from the previous month (after 7th) to the current month. The ensemble of these 24 models is used in this study. The forecasts with the released dates in February, March, April, and May are selected, and the corresponding leading times are 4 months, 3 months, 2 months, and 1 month, respectively.

### 2.3. Definition of WPSHI

The starting point of our analysis is to define a western Pacific subtropical high index (WPSHI) reflecting the characteristic of WPSH. The Z500 fields from CFSv2 and

NCEP/NCAR reanalysis dataset are used to calculate the time series of WPSHI. The definition of WPSHI used in this study is proposed by Sui et al. [24], who used the area mean value of Z500 in JJA within the region (120° E–140° E, 10° N–30° N) for constructing the WPSHI. This region corresponds to the largest variability of Z500 at the western North Pacific. Figure 2a shows the standardized deviation of the Z500 field of the reanalysis data. The calculation procedure is as follows. Firstly, the 1-order difference operator is applied on the time series data for each grid to remove the low frequency change and only retain the inter-annual component. Then, the standardized deviation is calculated for each grid. At last, the standardized deviation is normalized based on the zonal mean and zonal standardized deviation values. Although based on different procedure and different time range, the region of the largest variability shown in Figure 2 is similar with the result shown in Sui et al. [24].

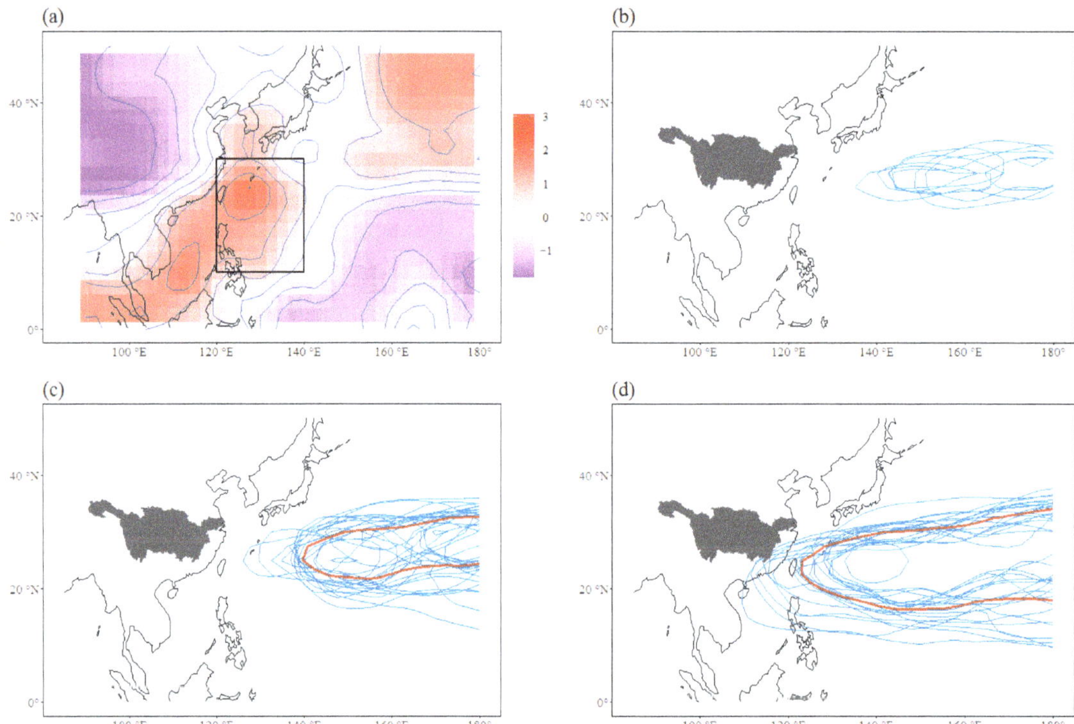

**Figure 2.** (**a**) The standard deviation (SD) of the Z500 field, and the region for defining WPSHI. The SD has been standardized, and the details can be seen in the text; (**b**) the contour lines of the 5880 geopotential metre (gpm) of the 500-hPa geopotential height field, which indicates the location of the WPSH for years with WPSHI < −0.5; (**c**) the 5880 gpm lines for WPSHI between −0.5 and 0.5; (**d**) the 5880 gpm lines for WPSHI > 0.5. For (**b–d**), the thin lines are the 5880 lines for each year in the corresponding grade, and the thick red line is the 5880 gpm line of the corresponding mean Z500 field.

The WPSHI of the reanalysis data (denoted as WPSHI(R)) is standardized based on the mean and standard deviation of the whole period 1960–2018, based on the following equation:

$$X_S = \frac{X - \overline{X}}{Sd} \tag{1}$$

where $X$ is the time series needed to be standardized (here $X$ is mean Z500 within the region (120° E–140° E, 10° N–30° N)), $\overline{X}$ and $Sd$ are the mean and standardized deviation of the $X$, and $X_S$ is the final standardized series. Figure 2b–d show the position of the WPSH for different ranges of WPSHI, and it can be seen that WPSHI has a good indicative capacity for the position of the WPSH.

When we standardize the WPSHI of CFSv2 (denoted as WPSHI(CFS)), the discontinuity of the bias of the CFSv2 forecast is considered. One feature of the outputs of CFSv2 that should be disposed of carefully is the abrupt change in 1999 in the CFSv2 forecast [20,25]. Kumar et al. [25] have shown that this abrupt change comes from the forecast bias for SST in the equatorial Pacific and leads to changes in other variables. Figure 3a shows the WPSHI(CFS) that is standardized based on the mean value of 1982–2018. It can be seen that the forecasting bias is not stationary. Before 1999, there is an apparent larger negative bias, which is true for all leads. Note that stationary bias does not affect the postprocessing procedure, while nonstationary bias does. Thus, the final WPSHI(CFS) is calculated by the following method. First, calculate the average value of the target zone of the Z500 field from the CFSv2 forecast; then for the period of 1982–1998 and 1999–2018, the forecasting climatology of each period are subtracted from sub-series of each period, respectively; at last, the anomaly series is divided by the standard deviation calculated by the whole period. The result of this method is shown in Figure 3b.

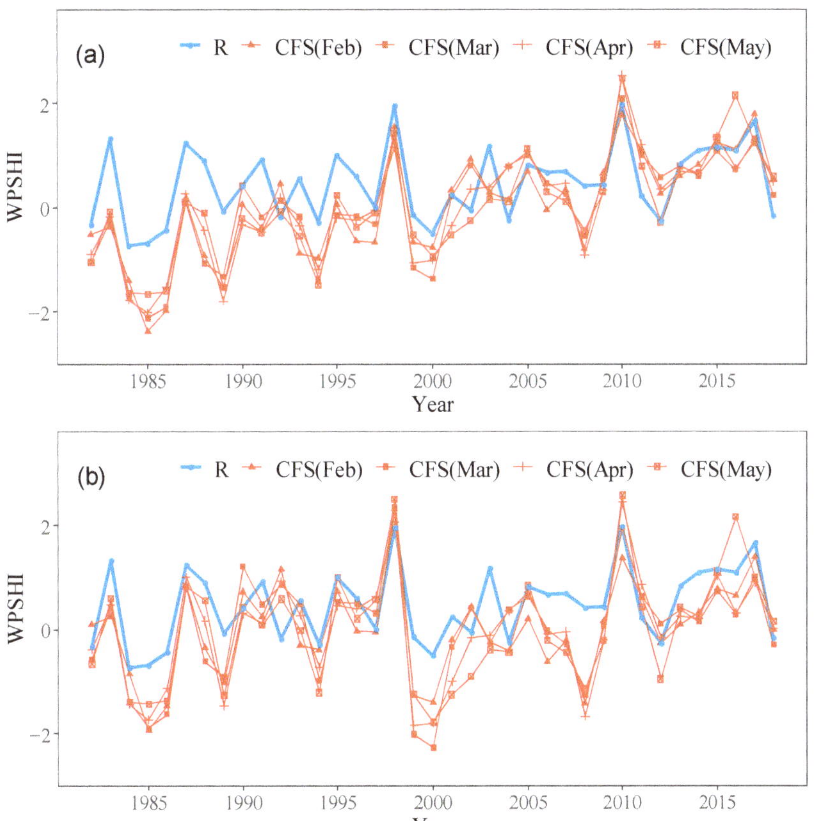



**Figure 3.** Comparison between WPSHI from reanalysis dataset (WPSHI(R)) and from CFSv2 (WPSHI(CFS)) with different released months. WPSHI (CFS) is standardized based on two different methods. (**a**) Standardized based on the mean and the standard deviation of 1982–2018. (**b**) For 1982–1998 and 1999–2018, the mean value of each period is subtracted respectively, then standardized by the standard deviation of the whole period, i.e., 1982–2018.

## 3. ROC Analysis

Several metrics can be used to measure the linkage between the predictor and the predictand, with the most frequently used one likely being the Pearson correlation coefficient. However, using the correlation coefficient neglects some important characteristics of the linkage between predictors and the predictand, which will be discussed here. Specifically, for rainfall of a given location, if a threshold $T$ of rainfall is specified, we can define two classes, i.e., a positive event with the rainfall larger than $T$, and a negative event with the rainfall less that $T$. Then, we can test the capacity of the predictor to distinguish the samples from these two classes. What we want to show is that this ability can be seen as a function of the threshold $T$, i.e., it would change with $T$.

A term which is possible to be confused is the rainfall threshold. Sometimes, it is used in the field of early warning of hydro-geological disasters [26]. In this case, rainfall is the indicator for the target event. While in this paper, the rainfall threshold $T$ is used to define the positive event, i.e., the rainfall larger than $T$, which is the target event identified by WPSHI.

The Receiver Operator Characteristic (ROC) curve is used here for evaluating the forecasting ability. An ROC curve uses the hit rate (also known as the sensitivity) as the y coordinate, versus the false alarm rate as the x coordinate. The area under the ROC curve (AUC) can evaluate the capacity of WPSHI for discriminating between the positive and the negative event. One simple explanation of AUC could be the probability to rank a positive/negative sample pair, which is selected randomly from the sample set [27]. In this approach, building a model for generating a formal probability forecast is not needed. It is claimed that AUC should be treated as the potential skill of the predictor [28]. In this section, we use AUC to evaluate the potential skill of WPSHI for indicating the class of rainfall.

We have calculated the AUC values of WPSHI(R) for indicating class of the standardized anomaly of rainfall with three thresholds, i.e., $-1$, 0, and 1, respectively (Figure 4). It is clearly illustrated that AUC is higher for the threshold 1. This fact enlightens us that we can find a better threshold of the predictand for which the binary classification has a higher predictability. Figure 4 also shows that the grid located at the middle and lower reaches of Yangtze River Valley have higher AUC values, indicating the higher predictability of this region. Considering this fact, in the following analysis, we define the Yangtze River Summer Rainfall Index (YRSRI) as the mean value of JJA rainfall over the box region ($27°$–$32°$ N, $109°$–$120°$ E) shown in Figure 4, which covers most of the middle and lower reaches of the Yangtze River Valley. The YRSRI is also standardized by Equation (1).

The relationship between YRSRI and WPSHI is also analyzed (Figure 5). From Figure 5a, it can be seen that WPSHI(R) and YRSRI are well correlated, and both series show the same abrupt change as the late 1970s. This upward jump of WPSHI means that the WPSH extends southwestward, leading to the wet anomaly over the Yangtze River Valley from the late 1970s [29]. Another important fact is that for extreme flood years such as 1980, 1983, and 1998, the WPSHI(R) has better indicative capacity for the YRSRI. This is also illustrated by Figure 5b, which shows the scatterplot between WPSHI(R) and YRSRI. Clearly, YRSRI only responds to WPSHI at the interval with higher WPSHI values (larger than 0.5). Specifically, for the interval of WPSHI < 0.5 and WPSHI > 0.5, the Pearson correlation coefficient between WPSHI and the YRSRI is 0.09 and 0.85, respectively. Additionally, linear regression lines are fitted for the years of WPSHI < 0.5 and WPSHI > 0.5, respectively, and the slopes are 0.45 and 1.23. AUC is calculated for different thresholds of YRSRI (Figure 5c). Still, it is important to note that the highest AUC is reached when the threshold is near 1. This is consistent with the result shown in Figure 4.

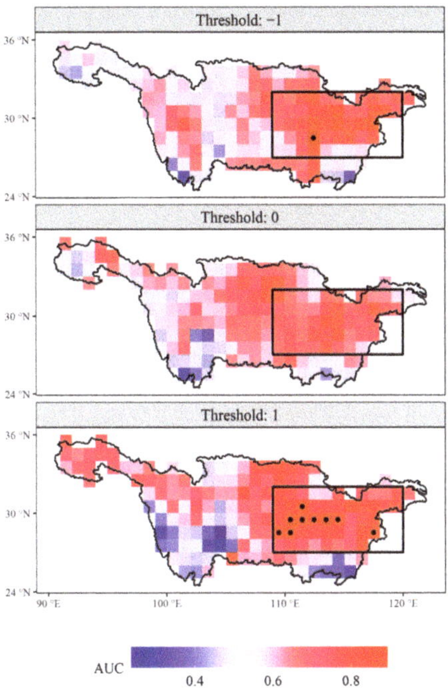

**Figure 4.** AUC for each grid of rainfall over the Yangtze River Valley based on WPSHI(R) as the indicator. Grids with AUC larger than 0.85 are labelled as black points. The box region (27°–32° N, 109°–120° E) shown in the figure is used to calculate the Yangtze River Summer Rainfall Index (YRSRI).

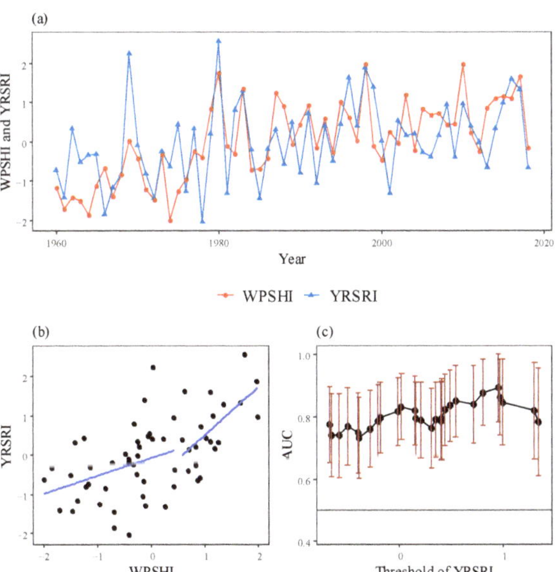

**Figure 5.** (**a**) Time series of the YRSRI and WPSHI (R), and (**b**) the scatter plot of these two variables. (**c**) The AUC values and their 95% confidence intervals corresponding to different thresholds of YRSRI.

All results of this section indicate that the predictability largely depends on the threshold of YRSRI, and the response of summer rainfall to the WPSH is asymmetric and nonlinear. For larger WPSHI, i.e., when the WPSH is westward extending, the rainfall of Yangtze River Valley is more sensitive to the variation of WPSHI. This feature leads to different forecasting skills for different thresholds, and this will be investigated in the following sections.

## 4. Modelling Methodology

Due to the limitation of the relatively small number of samples, we avoid using sophisticated models, and a simple model, i.e., logistic regression, is used as the basic tool for making probability forecasts for the binary classification. Based on logistic regression, three testing procedures based on cross validation are implemented in this study. Technical details of the three testing procedures, logistic regression, and the performance metrics are described as follows.

### 4.1. Three Testing Procedures

In this study, three testing procedures, i.e., predictability assessment (PA), model output statistics (MOS), and the reanalysis-based (RAN) approach are explored. These approaches, except PA, have been discussed in Marzban et al. [30] in the background of weather prediction. For illustrating differences among the above three testing procedures, Figure 6 shows the corresponding schematic diagrams. The details are stated as follows.

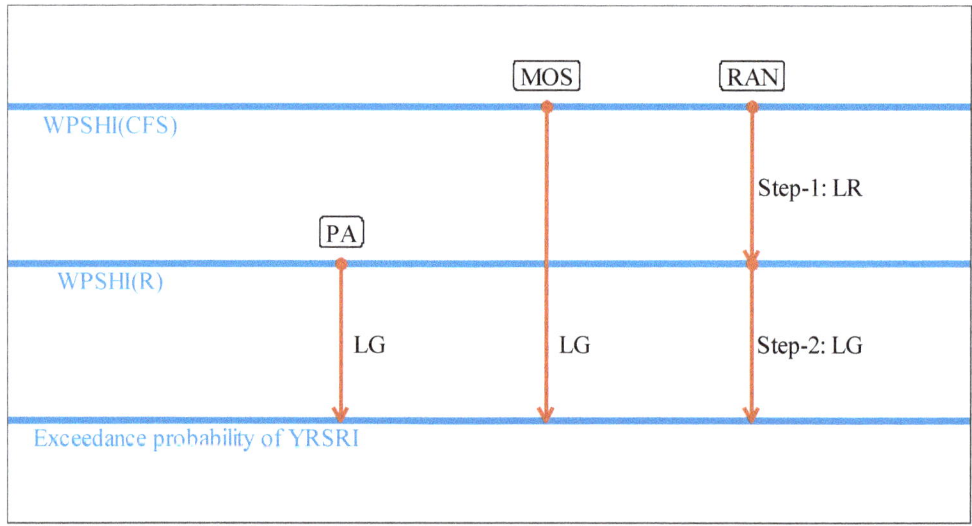


**Figure 6.** Three test procedures explored in this study for YRSRI. In the figure, LG means logistic regression and LR means linear regression. For predicting streamflow, the procedures are similar with what have been shown in this figure, and the only difference is that the antecedent streamflow is used as another predictor for forecasting the exceedance probability of streamflow.

We first describe the procedures for predicting YRSRI. The procedures of PA and MOS are quite straightforward. For PA, the procedure builds the relationship between WPSHI(R) and YRSRI by logistic regression to forecast the exceedance probability of a given threshold. Note that PA is not the real forecast, as the reanalysis data cannot be retrieved for making operational forecasts. The result of PA reflects the predictability of the YRSRI based on the perfect knowledge of the WPSHI in the following summer. Differently from PA, MOS builds the relationship between WPSHI(CFS) and YRSRI. The advantage of MOS is that it is unbiased when making forecasts, which is not true for the perfect prog approach. This fact makes MOS a popular method in the field of seasonal forecasts [31]. However, the disadvantage is that the number of samples for training model is lower depending on the length of hindcasts. In this study, as the hindcasts of CFSv2 are from 1982, only the years of 1982–2018 can be used for building MOS models.

The reanalysis-based (RAN) approach [30] consists of two steps for training the model and making forecasts. The first step is to build an empirical model (linear regression is used here) to map WPSHI(CFS) to WPSHI(R), and the second step is to map WPSHI(R) to the predictand. Note that the error in the first step comes from the model deficiency, while the error in the second step comes from the chaos of the climatic system. The advantage of the RAN approach is that more samples can be used to train the model of the second step, which is independent with the hindcasts of the dynamical model.

The testing procedures for streamflow are similar with that for YRSRI, and the only difference is that the antecedent streamflow is used as another predictor when forecasting JJA streamflow for PA, MOS, and the step 2 of the RAN approach.

All the tests of the above procedures are similar with the leave-one-out cross validation (LOOCV); however, the difference with the LOOCV is that the whole training years and the whole testing years are not the same in some cases. Specifically, the model is tested for each year in the test year set, i.e., 1982–2018, no matter which testing procedure is used. For testing a given year in 1982–2018, the current testing year is excluded from the training set. For PA, the training set includes the years from 1960–2018; for MOS, the training set includes the years from 1982–2018; for RAN, the training set for step 1 includes the years from 1982–2018, while for the step 2, it contains the years from 1960–2018.

*4.2. Logistic Regression*

Logistic regression, which is a frequently used model for making probabilistic classifications, is used in this study to make class forecasts for rainfall and streamflow. An example of an application of the Logistic regression on seasonal rainfall forecast can be seen in Prasad et al. [32].

For a two class problem of a target variable $Y$, suppose that $Y = 1$ means the positive class and $Y = 0$ means the negative class, and $p = P(Y = 1)$, i.e., the probability of the positive class. The logistic model supposes that the logit value, i.e., $\log\left(\frac{p}{1-p}\right)$, is a linear function of the predictor $X$:

$$\log\left(\frac{p}{1-p}\right) = \beta_0 + \beta_1 X \tag{2}$$

The coefficients of models can be estimated by the maximum likelihood estimation method [33]. When the coefficients have been estimated, the probability $p$ can be calculated by:

$$p = \frac{\exp\left(\hat{\beta}_0 + \hat{\beta}_1 X\right)}{1 + \exp\left(\hat{\beta}_0 + \hat{\beta}_1 X\right)} \tag{3}$$

*4.3. Exceedance Probability Forecast*

Exceedance probability forecasts of a given threshold are based on the logistic regression model. Here, we describe the method to generate exceedance probability forecasts of all thresholds. In the following text, the predictand Y is the YRSRI or summer streamflow:

1.  The series of thresholds are selected based on the observation of the predictand $Y$. First, sort the $Y$ values in the samples in 1982–2018 as the descending order $\left\{Y_{[1]}, Y_{[2]}, Y_{[3]}, \ldots, Y_{[n]}\right\}$, where n is the number of all samples. Then, the thresholds used here are $\left\{Y_{[5]}, Y_{[7]}, Y_{[8]}, \ldots, Y_{[n-5]}\right\}$. This setting will make at least 5 samples for the positive or negative class.

2.  Choose one threshold $T$ in step 1 and one test year in the sample set (1982–2018). All samples can be divided into two classes based on the value of the predictand $Y$, i.e.,

years with $Y \geq T$ (the positive class) and years with $Y < T$ (the negative class). Use the training set to fit a logistic regression, and then use the fitted model to forecast $P(Y \geq T)$ for the test year.

3. Repeat step 2 for all threshold $T$ and all test years.

Note that for the threshold $Y_{[i]}$, the exceedance probability of the climatology forecast is $i/n$.

For a larger or smaller threshold $T$, the sample set cut by $T$ is not balanced, i.e., the ratio of the number of positive class and the negative class ($n_+/n_-$) is not 1. In this case, how large $p = P(Y \geq T)$ can allow us to forecast the occurrence of the positive class is a key problem. Note that using the probability output from the model, or simply choose $p = 0.5$ as the decision threshold, will be misleading [34]. The proper selection is that when $\frac{p}{1-p} > \frac{n_+}{n_-}$, we make a positive class forecast. Thus, if the forecasted $p$ is larger than the climatology forecast, the positive class will be forecasted.

It should also be noted that for a given test year, the exceedance probability $P(Y \geq T)$ might not be monotonous decreasing as the threshold $T$ increasing, which must be true in theory. We use the shape constrained P-splines (SCOP-splines) [35] to smooth the exceedance probability curve as monotonous decreasing. The calculation is implemented based on the R package scam.

### 4.4. Skill Metric along the Threshold

Although neglected by other researchers, we want to show that the ability of WPSHI for discriminating the positive/negative classes of rainfall (streamflow) largely depends on the threshold. Thus, the skill for the exceedance probability forecasts is not calculated for each year (as in Piechota et al. [36]) but for each threshold of the YRSRI or streamflow.

For a given threshold $T$, Brier score (BS) is used to calculate skill scores. The definition of BS is:

$$BS = \frac{1}{n} \sum_{i=1}^{n} (f_i - o_i)^2 \tag{4}$$

in which $f_i$ is the $i$th forecast of the probability of the positive class, $o_i$ is the observation of the $i$th sample (1 means positive and 0 means negative), and $n$ is the number of the samples. The value of BS is between 0 and 1. $BS = 0$ means perfect forecast and $BS = 1$ corresponds to the lowest skill forecast. Note that $f_i$ is calculated based on the model trained by the sample set excluding the $i$ sample, as the normal leave-one-out cross validation.

Based on BS, the BS skill score (BSS) can be calculated by:

$$BSS = 1 - \frac{BS}{BS_{CLIM}} \tag{5}$$

where $BS_{CLIM}$ is the BS of the climatic forecast.

For different thresholds, BSS can be calculated respectively. Thus, we can get skill scores for different threshold $T$.

## 5. Results

### 5.1. Results of PA

For evaluating the predictability of the YRSRI based on the WPSHI, skill is tested with WPSHI(R) as the predictor, and this is the test procedure predictability assessment (PA) that has been described in Section 4.1. Figure 7 shows the BSS of the logistic regression models corresponding to various thresholds. One important feature shown in Figure 7 is that, generally, BSS is positive only when the threshold of the YRSRI is larger than 0. Furthermore, BSS reaches the peak value (BSS = 31.5%) when the threshold of YRSRI is 0.97. This result is consistent with the relationship between AUC and the threshold, which has been shown in Figure 5c.

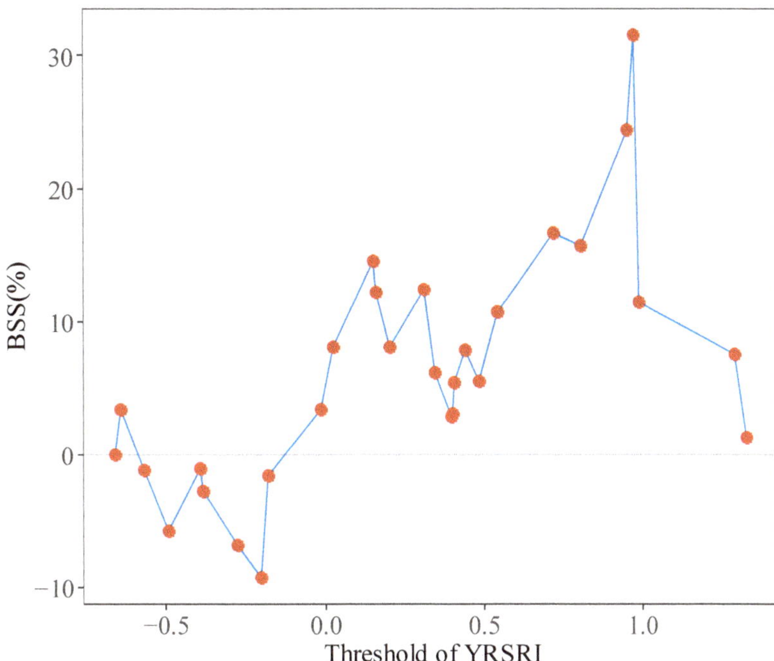

**Figure 7.** Brier skill score (BSS) of the forecast for different thresholds of the Yangtze River Valley summer rainfall index (YRSRI) based on WPSHI(R).

Not surprisingly, the above effect of threshold on rainfall will also impact the predictability of streamflow. As summer streamflow is also affected by the antecedent state of wetness of the basin, two models are built and tested for streamflow of Hankou and Datong station. The first model only uses antecedent streamflow (streamflow of May) as the predictor (denoted as AF hereafter), and the second model utilizes both antecedent streamflow and WPSHI (R) as predictors (denoted as AF + WPSHI hereafter). If the model AF + WPSHI has a higher skill than the model AF, it can thus be concluded that WPSHI provides some skill independent of the memory of the basin.

Figure 8 shows the BSS for summer streamflow of Hankou and Datong. Note that summer streamflow of Datong has relatively higher predictability than Hankou, which is reflected in the BSS of the model AF for all thresholds. The most interesting result is that, for both stations, WPSHI enhances the skill only for higher thresholds of streamflow, and this feature is clearer for Datong than Hankou. The above results are consistent with the result for the YRSRI, i.e., WPSHI shows higher skill for larger thresholds of the predictand. The skill reflected in Figure 8 can also be explained by the coefficients in the logistic regression models (Figure 9). As all predictors have been standardized, the coefficients can reflect the influence of each predictor. The pattern in Figure 9 indicates that the WPSHI plays a much dominant role in classification corresponding to higher thresholds of streamflow.

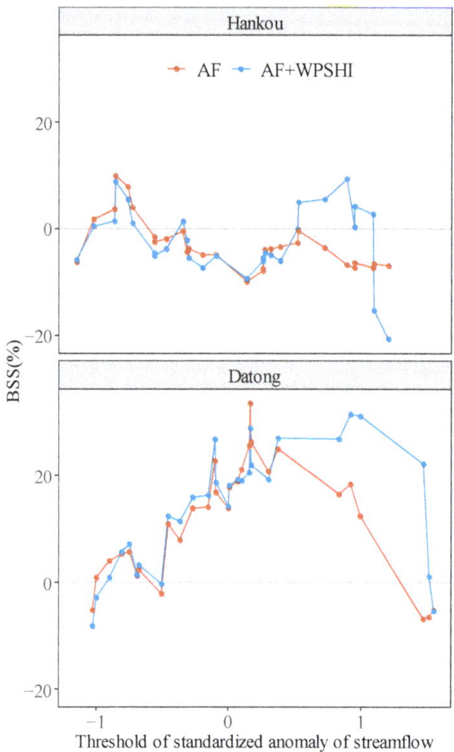

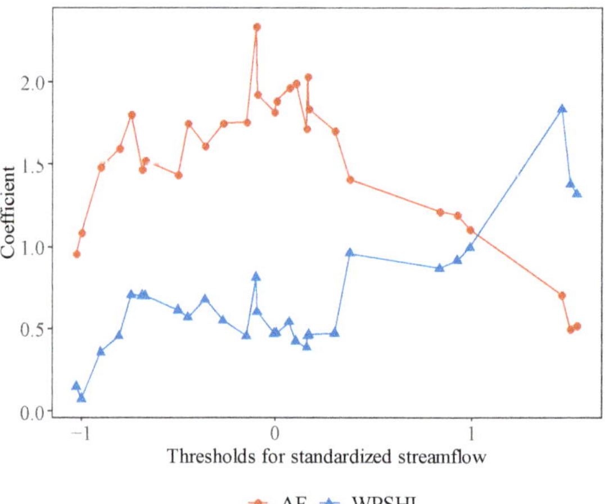

**Figure 8.** Brier skill score (BSS) of the forecast for summer streamflow of Hankou and Datong based on WPSHI(R). Two types of models are shown in the graph, i.e., the model only using antecedent streamflow (AF) and the model using both antecedent streamflow and WPSHI(R) as predictors (AF + WPSHI).

**Figure 9.** Coefficients of the logistic regression model forecasting summer streamflow of Hankou and Datong based on WPSHI(R) and antecedent streamflow (AF). The coefficients are the averaged values of models for all testing years.

### 5.2. Results of MOS and RAN

This section provides the results of two processing methods, i.e., the testing procedure MOS and RAN. As shown in Figure 6, the first step is to forecast WPSHI(R) based on WPSHI(CFS). A linear regression model is used for this task, and a leave-one-out test is used to evaluate the skill of this linear regression model. For the four releasing months (i.e., February, March, April, and May), the Nash–Sutcliffe efficiency coefficient is 0.43, 0.46, 0.48, and 0.62, respectively.

Figure 10 shows the skill scores of different leading times based on various thresholds. Note that in almost all cases, BSS of the RAN approach is larger than the MOS approach, indicating the advantage of the RAN approach. Similar to the characteristic we have shown in Figure 7, BSS is also higher for larger thresholds of YRSRI. This result indicates that WPSHI has higher skills for discriminate extreme events, especially flood summers.

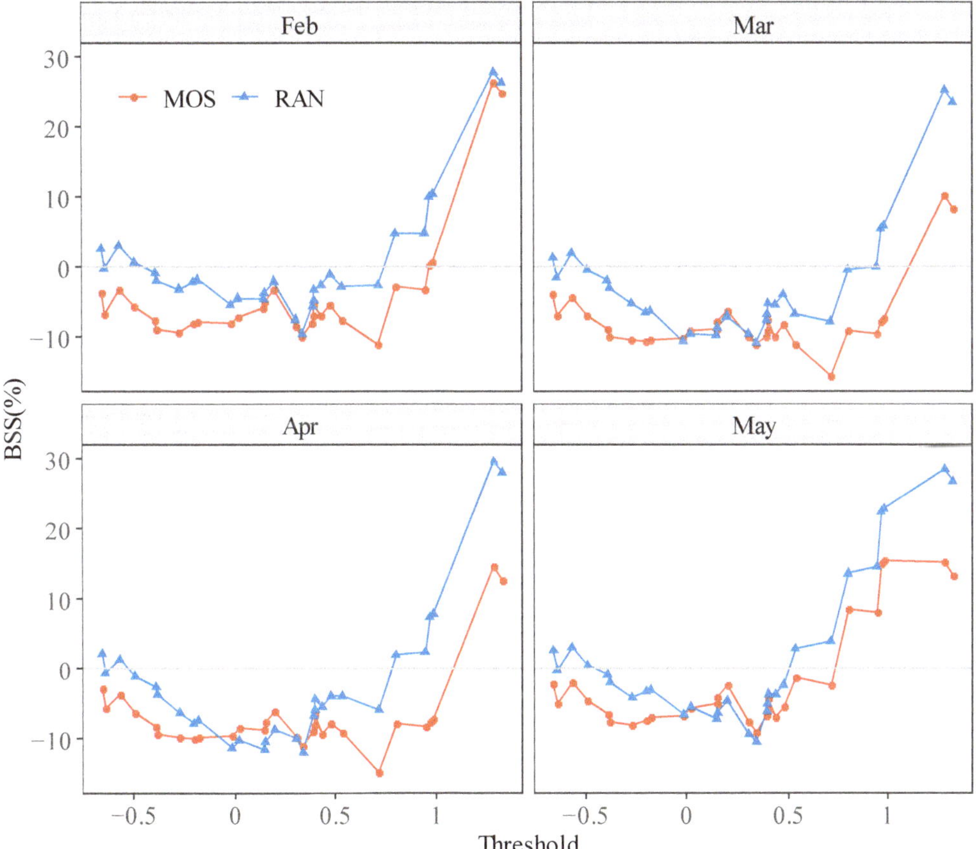

**Figure 10.** BSS by the MOS approach and the RAN approach for the YRSRI corresponding to four releasing months of CFSv2.

The skill of streamflow forecasting based on the RAN approach corresponding to CFSv2 released in May is shown in Figure 11. Not surprisingly, the feature shown in Figure 8 is still obvious in Figure 11, which indicates the enhancement of the skill for larger thresholds of streamflow.

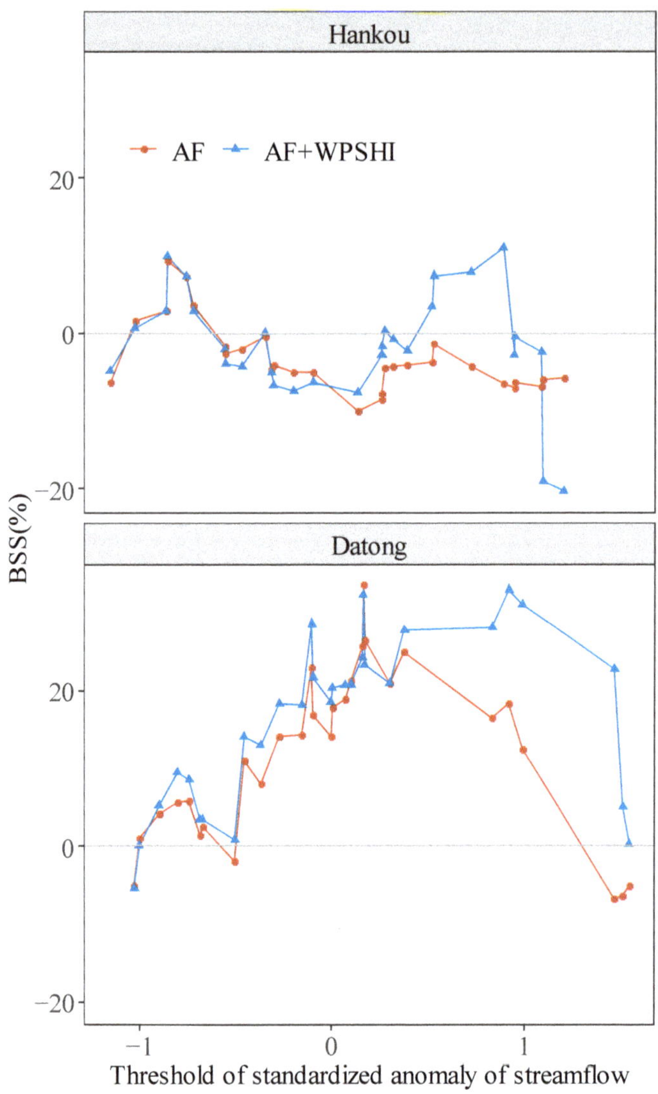

**Figure 11.** BSS by the RAN approach for summer streamflow of Hankou and Datong, corresponding to the forecast of CFSv2 released in May.

## 6. Discussion

Although sophisticated models and various predictors have been used for building seasonal forecasting models for the Yangtze River Valley, the effort to understand the roles of some key predictors based on traditional statistical methods is still quite useful, as this will lead to prediction with better interpretability. In this study, we focus on the predictive capacity of the West Pacific Subtropical High Index (WPSHI) for summer rainfall and streamflow over the Yangtze River Valley. WPSHI can be well forecasted by CFSv2, which makes WPSHI useful as a bridge for generating forecasts of rainfall and streamflow based on postprocessing of outputs of dynamical prediction systems. Thus, exploring the synchronous relationship between WPSHI and rainfall (streamflow) is beneficial to making skillful seasonal predictions.

We have demonstrated that there is a nonlinear response of summer rainfall over the Yangtze River Valley to the WPSHI, and rainfall is more sensitive to WPSHI when the value of WPSHI is higher. Because of this feature, WPSHI only shows higher skills for forecasting the exceedance probability of rainfall corresponding to larger thresholds. Similarly, for streamflow, WPSHI only enhances the skill for higher thresholds relative to the persistence forecast (i.e., the model with antecedent streamflow as the predictor). The above result means that WPSHI is only a good indicator for identifying extreme flood summers of the Yangtze River Valley. These findings allow us to select a new strategy for making long-term hydrological forecasts, i.e., to find a proper definition of the positive event based on selecting a proper threshold with higher predictability. We found that previous studies have not adequately explored exceedance probability forecasting, such as Piechota et al. [36], and this study provides a new perspective to treat probability forecasting.

With the same framework, two post-processing approaches, which have been applied in the field of weather forecasting, are also explored. We have shown that the RAN approach has a better performance than MOS. As discussed in the work of Marzban et al. [30], the main advantage of the RAN approach is that more samples can be used to training the model of step 2 as shown in Figure 6, as step 2 is independent from the dynamical model, thus the number of samples is not limited by the hindcasts available. Although many forecasting models based on downscaling technology have been explored [4,37], little effort has been applied for comparison between different postprocessing approaches. For example, for making forecasts for North China summer rainfall, Guo et al. [4] built a downscaling model based on reanalysis data, and then substituted CFSv2 forecasting values (bias-removed) to make real forecasts. In fact, this is the perfect prog (PP) approach. It is possible that the skill could be enhanced when the circulation variables are not just removed of bias but reforecasted, as what the RAN approach does. More comparisons are still needed in future studies.

This study provides a framework for generating probability forecasting for rainfall and streamflow of the Yangtze River Valley, then represents a contribution for the development of an early warning system (EWS) [38] for the study area. We have shown that the forecast can be skillful for larger thresholds of rainfall even from February. For converting probability forecasts to binary forecasts, tools such as the ROC curve are useful for making the trade-off between the benefit of hit and the cost of a false alarm, which is beyond the topic of this paper.

## 7. Conclusions

In this study, we built forecasting models for summer rainfall and streamflow over the Yangtze River Valley based on the knowledge of the western Pacific subtropical high (WPSH). Several conclusions can be listed here:

1.  The rainfall over the Yangtze River Valley is more sensitive to the variability of WPSHI when WPSHI is high, while when WPSHI is less than 0.5, the rainfall shows low sensitivity. Furthermore, the middle and lower reaches of Yangtze River Valley show higher sensitivity to the variability of WPSHI than other regions. This characteristic leads to higher forecasting skill of exceedance probability forecasts corresponding to larger thresholds of rainfall.
2.  The analysis of predictability of summer streamflow of the Yangtze River Valley shows that WPSHI can only enhance the forecasting skill for binary classification corresponding to larger thresholds of streamflow.
3.  A comparison between two postprocessing approaches shows that the RAN approach shows a higher skill than model output statistics (MOS), as RAN can utilize more samples than MOS.
4.  When building a long-term forecasting model for generating exceedance probability forecasts, one should notice the effect of the threshold, and find a proper threshold with a higher skill.

**Author Contributions:** Conceptualization, R.H., Y.C. and G.L.; formal analysis, R.H.; methodology, R.H., Y.C. and W.W.; software, R.H.; validation, Q.H., W.W. and G.L.; writing—original draft, R.H.; writing—review & editing, Y.C. and Q.H. All authors have read and agreed to the published version of the manuscript.

**Funding:** This research is supported by the National Key Research and Development Program of China (2018YFC1508200, 20175018612), and the National Natural Science Foundation of China (51709075, 41701015).

**Institutional Review Board Statement:** Not applicable.

**Informed Consent Statement:** Not applicable.

**Data Availability Statement:** The data presented in this study are available on request from the corresponding author.

**Conflicts of Interest:** The authors declare no conflict of interest.

## References


1. Shams, M.S.; Anwar, F.A.H.M.; Lamb, K.W.; Bari, M. Relating ocean-atmospheric climate indices with australian river streamflow. *J. Hydrol.* **2018**, *556*, 294–309. [CrossRef]
2. Araghinejad, S.; Burn, D.; Karamouz, M. Long-lead probabilistic forecasting of streamflow using ocean-atmospheric and hydrological predictors. *Water Resour. Res.* **2006**, *42*, W03431. [CrossRef]
3. Kullmann, H.; Baldauf, S.A.; Bakker, T.C.M.; Frommen, J.G. Seasonal forecast for local precipitation over northern taiwan using statistical downscaling. *J. Geophys. Res. Atmos.* **2008**, *113*, D12.
4. Guo, Y.; Li, J.; Li, Y. Seasonal forecasting of north china summer rainfall using a statistical downscaling model. *J. Appl. Meteorol. Climatol.* **2014**, *53*, 1739–1749. [CrossRef]
5. Lee, S.; Lee, J.; Ha, K.; Wang, B.; Schemm, J. Deficiencies and possibilities for long-lead coupled climate prediction of the western north pacific-east asian summer monsoon. *Clim. Dyn.* **2011**, *36*, 1173–1188. [CrossRef]
6. Chen, H.; Sun, J.; Wang, H. A statistical downscaling model for forecasting summer rainfall in china from demeter hindcast datasets. *Weather. Forecast.* **2012**, *27*, 608–628. [CrossRef]
7. Peng, Z.; Wang, Q.J.; Bennett, J.C.; Schepen, A.; Pappenberger, F.; Pokhrel, P.; Wang, Z. Statistical calibration and bridging of ecmwf system4 outputs for forecasting seasonal precipitation over china. *J. Geophys. Res. Atmos.* **2014**, *119*, 7116–7135. [CrossRef]
8. Wang, B.; Xiang, B.; Lee, J.-Y. Subtropical high predictability establishes a promising way for monsoon and tropical storm predictions. *Proc. Natl. Acad. Sci. USA* **2013**, *110*, 2718–2722. [CrossRef]
9. Yuan, X.; Wood, E.F. Downscaling precipitation or bias-correcting streamflow? Some implications for coupled general circulation model (cgcm)-based ensemble seasonal hydrologic forecast. *Water Resour. Res.* **2012**, *48*, 12519. [CrossRef]
10. Tisseuil, C.; Vrac, M.; Lek, S.; Wade, A.J. Statistical downscaling of river flows. *J. Hydrology* **2010**, *385*, 279–291. [CrossRef]
11. Sachindra, D.; Huang, F.; Barton, A.; Perera, B. Least square support vector and multi-linear regression for statistically downscaling general circulation model outputs to catchment streamflows. *Int. J. Climatol.* **2013**, *33*, 1087–1106. [CrossRef]
12. Ghosh, S.; Mujumdar, P.P. Statistical downscaling of gcm simulations to streamflow using relevance vector machine. *Adv. Water Resour.* **2008**, *31*, 132–146. [CrossRef]
13. Huang, R.; Chen, J.; Huang, G. Characteristics and variations of the east asian monsoon system and its impacts on climate disasters in china. *Adv. Atmos. Sci.* **2007**, *24*, 993–1023. [CrossRef]
14. Luo, S.; Jin, Z. Statistical analysis for sea surface temperature over the south china sea, behavior of subtropical high over the west pacific and monthly mean rainfall over the changjiang middle and lower reaches. *Chin. J. Atmos. Sci.* **1986**, *10*, 409–418.
15. Zhang, J.; Chen, X. The operational seasonal forecasting of the summer rainfall in china. *Adv. Atmos. Sci.* **1987**, *4*, 349–362. [CrossRef]
16. Huang, R.; Chen, J.; Wang, L.; Lin, Z. Characteristics, processes, and causes of the spatio-temporal variabilities of the east asian monsoon system. *Adv. Atmos. Sci.* **2012**, *29*, 910–942. [CrossRef]
17. Blschl, G.; Bierkens, M.; Chambel, A.; Cudennec, C.; Zhang, Y. Twenty-three unsolved problems in hydrology (uph)-A community perspective. *Hydrol. Sci. J.* **2019**, *64*, 1141–1158. [CrossRef]
18. Chen, M.; Xie, P.; Janowiak, J.E.; Arkin, P.A. Global land precipitation: A 50-yr monthly analysis based on gauge observations. *J. Hydrometeorol.* **2002**, *3*, 249–266. [CrossRef]
19. Kalnay, E.; Kanamitsu, M.; Kistler, R.; Collins, W.; Deaven, D.; Gandin, L.; Iredell, M.; Saha, S.; White, G.; Woollen, J. The ncep/ncar 40-year reanalysis project. *Bull. Am. Meteorol. Soc.* **1996**, *77*, 437–471. [CrossRef]
20. Saha, S.; Moorthi, S.; Wu, X.; Wang, J.; Nadiga, S.; Tripp, P.; Behringer, D.; Hou, Y.T.; Chuang, H.; Iredell, M. The ncep climate forecast system version 2. *J. Clim.* **2014**, *27*, 2185–2208. [CrossRef]
21. Lang, Y.; Ye, A.; Gong, W.; Miao, C.; Di, Z.; Xu, J.; Liu, Y.; Luo, L.; Duan, Q. Evaluating skill of seasonal precipitation and temperature predictions of ncep cfsv2 forecasts over 17 hydroclimatic regions in china. *J. Hydrometeorol.* **2014**, *15*, 1546–1559. [CrossRef]

22. Silva, G.A.M.; Dutra, L.M.M.; Rocha, R.P.D.; Ambrizzi, T.; Leiva, É. Preliminary analysis on the global features of the ncep cfsv2 seasonal hindcasts. *Adv. Meteorol.* **2014**, *2014*, 117–128. [CrossRef]
23. Yang, S.; Zhang, Z.; Kousky, V.E.; Higgins, R.W.; Yoo, S.H.; Liang, J.; Fan, Y. Simulations and seasonal prediction of the asian summer monsoon in the ncep climate forecast system. *J. Clim.* **2008**, *21*, 3755–3775. [CrossRef]
24. Sui, C.H.; Chung, P.H.; Li, T. Interannual and interdecadal variability of the summertime western north pacific subtropical high. *Geophys. Res. Lett.* **2007**, *34*, 93–104. [CrossRef]
25. Kumar, A.; Chen, M.; Zhang, L.; Wang, W.; Xue, Y.; Wen, C.; Marx, L.; Huang, B. An analysis of the nonstationarity in the bias of sea surface temperature forecasts for the ncep climate forecast system (cfs) version 2. *Mon. Weather. Rev.* **2011**, *140*, 3003–3016. [CrossRef]
26. Luca, D.L.D.; Versace, P. Diversity of rainfall thresholds for early warning of hydro-geological disasters. *Adv. Geosci.* **2017**, *44*, 53–60. [CrossRef]
27. Fawcett, T. An introduction to roc analysis. *Pattern Recognit. Lett.* **2006**, *27*, 861–874. [CrossRef]
28. Wilks, D.S. *Statistical Methods in the Atmospheric Sciences*; Elsevier: Amsterdam, The Netherlands, 2011.
29. Gong, D.Y.; Ho, C.H. Shift in the summer rainfall over the yangtze river valley in the late 1970s. *Geophys. Res. Lett.* **2002**, *29*. [CrossRef]
30. Marzban, C.; Sandgathe, S.; Kalnay, E. Mos, perfect prog, and reanalysis. *Mon. Weather. Rev.* **2006**, *134*, 657–663. [CrossRef]
31. Salimun, E.; Tangang, F.T.; Juneng, L.; Zwiers, F.W.; Merryfield, W.J. Skill evaluation of the cancm4 and its mos for seasonal rainfall forecast in malaysia during the early and late winter monsoon periods. *Int. J. Climatol.* **2016**, *36*, 439–454. [CrossRef]
32. Prasad, K.; Dash, S.K.; Mohanty, U.C. A logistic regression approach for monthly rainfall forecasts in meteorological subdivisions of india based on demeter retrospective forecasts. *Int. J. Climatol.* **2010**, *30*, 1577–1588. [CrossRef]
33. Hastie, T.; Tibshirani, R.; Friedman, J.; Hastie, T.; Friedman, J.; Tibshirani, R. *The Elements of Statistical Learning*; Springer: New York, NY, USA, 2009; Volume 2.
34. Provost, F. Machine learning from imbalanced data sets 101. In Proceedings of the AAAI'2000 Workshop on Imbalanced Data sets, Austin, TX, USA, 31 July 2000; pp. 1–3.
35. Pya, N.; Wood, S.N. Shape constrained additive models. *Stat. Comput.* **2015**, *25*, 543–559. [CrossRef]
36. Piechota, T.C.; Chiew, F.H.; Dracup, J.A.; McMahon, T.A. Development of exceedance probability streamflow forecast. *J. Hydrol. Eng.* **2001**, *6*, 20–28. [CrossRef]
37. Liu, Y.; Fan, K.; Zhang, Y. A statistical downscaling model for summer rainfall over china stations based on the climate forecast system. *Chin. J. Atmos. Sci.* **2013**, *37*, 1287–1296.
38. Kelman, I.; Glantz, M.H. Early warning systems defined. In *Reducing Disaster: Early Warning Systems for Climate Change*; Singh, A., Zommers, Z., Eds.; Springer: Dordrecht, The Netherlands, 2014; pp. 89–108.

*water*

*Article*

# A Stacking Ensemble Learning Model for Monthly Rainfall Prediction in the Taihu Basin, China

Jiayue Gu [1], Shuguang Liu [1,2,*], Zhengzheng Zhou [1,*], Sergey R. Chalov [3] and Qi Zhuang [1]

[1]  Department of Hydraulic Engineering, Tongji University, Shanghai 200092, China; 1910008@tongji.edu.cn (J.G.); 1932631@tongji.edu.cn (Q.Z.)

[2]  Key Laboratory of Yangtze River Water Environment, Ministry of Education, Tongji University, Shanghai 200092, China

[3]  Hydrology Department, Faculty of Geography, Lomonosov Moscow State University, 119991 Moscow, Russia; srchalov@geogr.msu.ru

*  Correspondence: liusgliu@tongji.edu.cn (S.L.); 19058@tongji.edu.cn (Z.Z.)

**Abstract:** The prediction of monthly rainfall is greatly beneficial for water resources management and flood control projects. Machine learning (ML) techniques, as an increasingly popular approach, have been applied in diverse climatic regions, showing their respective superiority. On top of that, the ensemble learning model that synthesizes the advantages of different ML models deserves more attention. In this study, an ensemble learning model based on stacking approach was proposed. Four prevalent ML models, namely k-nearest neighbors (KNN), extreme gradient boosting (XGB), support vector regression (SVR), and artificial neural networks (ANN) are taken as base models. To combine the outputs from the base models, the weighting algorithm is used as second-layer learner to generate predictions. Large-scale climate indices, large-scale atmospheric variables, and local meteorological variables were used as predictors. $R^2$, RMSE and MAE, were used as evaluation metrics. The results show that the performance of base models varied among the nine stations in the Taihu Basin, while the stacking approach generally performed better than the four base models. The stacking model showed better performance in spring and winter than in summer and autumn. During wet months, the accuracy of model prediction varied more significantly. On the whole, based on performance evaluation measures, it is concluded that the proposed stacking ensemble multi-ML model can provide a flexible and reasonable prediction framework applicable to other regions.

**Keywords:** rainfall; prediction; machine learning; stacking model; Taihu basin

**Citation:** Gu, J.; Liu, S.; Zhou, Z.; Chalov, S.R.; Zhuang, Q. A Stacking Ensemble Learning Model for Monthly Rainfall Prediction in the Taihu Basin, China. *Water* **2022**, *14*, 492. https://doi.org/10.3390/w14030492

Academic Editors: Yuanfang Chen, Dong Wang, Dedi Liu, Binquan Li and Ashish Sharma

Received: 30 December 2021
Accepted: 5 February 2022
Published: 7 February 2022

**Publisher's Note:** MDPI stays neutral with regard to jurisdictional claims in published maps and institutional affiliations.

## 1. Introduction

Rainfall is an essential component in the hydrological cycle. Rainfall prediction is a fundamental issue in hydrological application. Reliable rainfall prediction is principal for water resource management, agriculture and flood control projects [1–3]. In the current context of climate change [4] and intense human activity, rainfall pattern becomes more complicated; thus, rainfall prediction remains a significant and demanding problem [5,6].

Generally, for modeling precipitation, numerical models based on the physical mechanisms and the statistical models were commonly employed [7,8]. The numerical models are based on the physical equations, including the complex process of atmosphere, ocean and land [9,10]. A large amount of data, such as temperature, pressure and moisture are acquired to drive the numerical models, which expends a lot of calculation costs. The statistical model is an approach of acquiring the features of historical rainfall time series and then predicting the evolution based on these features. The autoregressive model (AR) [11], the autoregressive moving average (ARMA) model [12,13] and the autoregressive moving integrated average (ARIMA) model [14,15] have been widely used for hydrological series predicting.

Machine learning (ML) techniques, as an increasingly popular approach, provide an attractive alternative to traditional methods for rainfall prediction [16], driven by flexible predictor datasets [17,18]. It can take advantage of all kinds of information, including, but not limited to, atmospheric, geographical and oceanic factors to predict the target [19,20]. Multiple machine learning methods have been employed for predicting rainfall. Yu et al. [21] compared the effectiveness of support vector regression (SVR) and random forest (RF) in radar-derived rainfall forecasting in three reservoir catchments in Taiwan and found that SVR was more accurate in the estimation of rainfall. Cramer et al. [22] compared to the application of ML techniques in 20 cities around Europe and 22 cities in the United States, and found that ANN, SVR, and genetic programming (GP) showed better agreement than Markov chain, radial basis neural networks (RBNN), M5 rules, M5 model trees and K-nearest neighbors (KNN). Pour et al. [23] predicted seasonal rainfall extremes in Malaysia, and found that Bayesian artificial neural networks (BANN) performed the best, followed by SVR and RF. Sachindra et al. [24] compared the effectiveness of relevance vector machine (RVM) with ANN, SVR, and GP for downscaling reanalysis data to monthly rainfall in Australia. This research shows that RVM is recommended over GP, ANN or SVR in developing downscaling models. Diez-Sierra and del Jesus [19] predicted long tern term daily rainfall, showing that neural networks (NN) presented significantly better results when predicting the intensity of rainfall, followed by SVM, KNN and RF, with slightly worse values of R and RMSE than NN in Spain. Zeynoddin et al. [3] demonstrated that a hybrid model by integrating a linear model and non-linear ELM model was powerful for monthly rainfall prediction in a tropical region. Zhou et al. [25] compared RF, gradient boosting regression (GBM), SVR, ANN and dual-stage attention-based recurrent neural network (DA-RNN) in predicting monthly rainfall in Yangtze River Delta, China, showing that RF performed better in terms of MAE, and that RF and ANN proved to be favorable in terms of $R^2$, RMSE.

Previous studies have generally investigated an individual ML method with single structure, demonstrating their respective superiority. Considering that rainfall is affected by different factors, as well as that it shows different statistical characteristics, the individual ML model with a specific structure possesses limited ability to present the complex relationship between rainfall and diverse predictors in varying climatic regions. In recent years, ensemble learning methods, which can combine multiple ML models, have shown their advantages [26]. The stacking ensemble model is a popular one among them [27–29]. 'Stacking' is a specific type of ensemble learning which can take advantage of different base model structures to generate theoretically more promising prediction [30]. Zounemat et al. [31] summarized research on the application of ensemble learning approaches in a hydrological field, and claimed that using ensemble strategies is superior over individual machine learning models. Li et al. [32] integrated SVR, RF, elastic net regression (ENR) and extreme gradient boosting (XGB), through the stacking ensemble approach for mid-term streamflow forecasting. It was found that the application of the stacking strategy improved the ability of individual models. Wang et al. [33] compared stacking model with individual models for beach water quality prediction, finding the stacking model is the most robust one for 3 beaches in 5-year prediction. Nevertheless, the potential of stacking ensemble model in rainfall prediction has less explored.

The main objective of this study is to develop a stacking ensemble model for monthly rainfall prediction with multiple predictors and to examine the performance of the model. Specially, four machine learning models (KNN, XGB, SVR, ANN) were utilized as base learners due to their high popularity and good performance on previous studies. By means of assigning weights, the four base learners were combined to the stacking ensemble model. The performance of the stacking ensemble model is assessed by evaluation metrics $R^2$, RMSE, MAE. The predicted results are examined on an annual aggregated scale, seasonal scale, dry/intermediate/wet month months and months of extreme rainfall.

The rest of this paper is organized as follows: Section 2 introduces the study area and data. Section 3 presents a brief introduction of four machine learning models, the stacking ensemble framework, hyper-parameter optimization, evaluation metrics and categorization of dry/intermediate/wet months. Section 4 presents the results and discussions, including the comparison of model performances, the examination of the performance at different time scales, and the discussion of prediction results. Section 5 presents a summary and conclusions.

## 2. Study Area and Data

The Taihu basin (ranging from latitude 30°28′ N to 32°15′ N and longitude 119°11′ E to 121°53′ E) is located in the Yangtze River Delta, on the southeast coast of China, as shown in Figure 1. The total area of the watershed is approximately 36,895 km$^2$, comprising of parts of Jiangsu Province, Zhejiang Province and Anhui Province and Shanghai City. Around 80% of the Taihu basin is plain, and the remaining 20% is occupied by low hills in the western part of the Taihu basin [34], with rivers and lakes accounting for 17% of the total area of the basin [35]. The Taihu basin is located in a subtropical monsoon zone, with the average annual precipitation is 1218.1 mm [36]. Cyclonic storms and convectional rainfall frequently occurring in flood season (May to September), are the main triggers for flood events that, consequently, affect infrastructure and human lives.

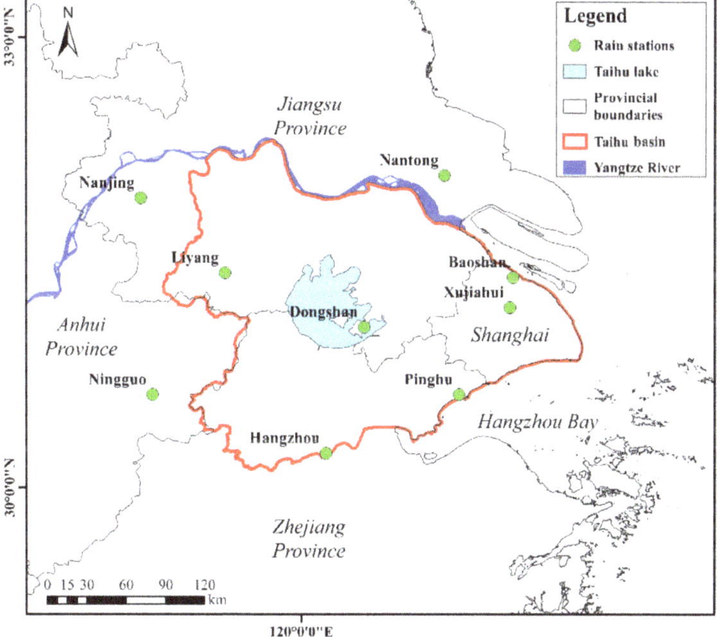


**Figure 1.** Map of the study region and location of rain stations.

For the monthly rainfall prediction, nine stations located in the Taihu Basin and its surroundings were selected, as shown in Figure 1. Since the long-term rainfall series data in the Taihu basin for access are limited, three stations (Nanjing, Nantong and Ningguo) within about 30 km from the Taihu basin were used in this study. The monthly rainfall at these adjacent stations are also subject to the similar climatic condition [37,38]. The monthly rainfall datasets for the period 1961–2019 were obtained from the China Meteorological Data Service Centre, China Meteorological Administration (CMA) (http://data.cma.cn/data/cdcdetail/dataCode/SURF_CLI_CHN_MUL_DAY_V3.0.html (accessed on 27 February 2021)). Table 1 provides the geographic details and climatic properties of the nine stations.

**Table 1.** The geographic details and climatic characteristics of the nine stations in the study.

| No. | Station | Abbr. | Longitude (°E) | Latitude (°N) | Altitude (m) | Monthly Precipitation | | |
|---|---|---|---|---|---|---|---|---|
| | | | | | | Mean (mm) | Maximum (mm) | Coefficient of Variation ($C_v$) |
| 1 | Xujiahui | XJH | 121.43 | 31.20 | 4.6 | 101.3 | 725.5 | 0.809 |
| 2 | Baoshan | BS | 121.45 | 31.40 | 5.5 | 94.5 | 570.9 | 0.834 |
| 3 | Dongshan | DS | 120.43 | 31.07 | 17.5 | 95.8 | 696.6 | 0.764 |
| 4 | Liyang | LY | 119.48 | 31.43 | 7.7 | 97.3 | 521.3 | 0.820 |
| 5 | Pinghu | PH | 121.08 | 30.62 | 5.4 | 103.4 | 569.3 | 0.788 |
| 6 | Hangzhou | HZ | 120.17 | 30.23 | 41.7 | 119.2 | 611.0 | 0.712 |
| 7 | Nanjing | NJ | 118.90 | 31.93 | 35.2 | 90.5 | 661.5 | 0.952 |
| 8 | Nantong | NT | 120.98 | 32.08 | 4.8 | 91.6 | 604.4 | 0.909 |
| 9 | Ningguo | NG | 118.98 | 30.62 | 87.3 | 120.8 | 783.2 | 0.730 |

A total of 14 variables, including large-scale climate indices, large-scale atmospheric variables, and local meteorological variables, were used as predictors (Table 2).

The large-scale climate indices in the prediction were the Nino 3.4 index (Nino 3.4), the southern oscillation index (SOI), the Western Pacific subtropic high intensity (WPSH) and the Southern Hemisphere annular mode Index (SAMI). Nino 3.4 is identified as the average sea surface temperatures (SST) anomaly in the region of 5° N–5° S and 170° W–120° W. The southern oscillation index (SOI) is typically calculated using the Troup's method using the values of pressure differences from Tahiti and Darwin. Nino 3.4 and SOI are el nino southern oscillation (ENSO) indictors, which is one of the most important global atmospheric phenomena, influencing rainfall and temperature across the globe. The Western Pacific subtropic high intensity (WPSH) is measured by the geopotential height at 500 hPa in the region of 110° E–180° E and 10° N to the north [39]. The Southern Hemisphere annular mode index (SAMI) is defined as the difference in the normalized monthly zonal-mean sea level pressure between 40° S and 65° S [40]. Previous studies [39–42] demonstrated that WPSH and SAMI significantly impact the summer rainfall in the lower Yangtze River basin. The climate indices with the lag month (up to 6 months lagged) of the highest correlation coefficient were utilized as predictors, as shown in Figure S1.

The large-scale atmospheric variables used in this study were sea level pressure (SLP) and meridional wind at 850 mb (V-wind), representing large-scale circulation anomalies [43]. The sea level pressure (SLP) in the Indian Ocean is relevant to rainfall in the study region [42]. The meridional wind at 850 mb (V-wind) is commonly used as the large-scale atmospheric predictor for rainfall in varying regions [5,43–45]. Correlation coefficient between the large-scale atmospheric variables and rainfall was used to select the spatial grid and the lag month of the large-scale atmospheric variables. As shown in Figure 2, the spatial grids of SLP were selected by the interactive correlation analysis provided by the Physical Sciences Division in the Earth System Research Laboratory (ESRL 2008) (https://psl.noaa.gov/data/correlation/ (accessed on 7 December 2021)), and the correlation coefficient between the selected SLP with 4 months lagged and rainfall was −0.464. All the selected large-scale atmospheric variables were highly correlated with rainfall in the study region of over 0.001 statistical significance level.

The local meteorological predictors for each station were monthly maximum temperature ($T_{max}$), monthly minimum temperature ($T_{min}$), monthly mean temperature ($T_{mean}$), monthly mean pressure ($P_{mean}$), monthly mean water pressure ($e_{mean}$), monthly mean relative humidity ($d_{mean}$) and monthly sunshine duration ($D_{sun}$). These predictors were selected for representing local scale characteristics.

**Table 2.** Summary of candidate predictors for the stacking model.

| No. | | Multiscale Predictors | Data Source |
|---|---|---|---|
| 1 | Large-scale climate indices | Nino 3.4 index (Nino 3.4) | Hadley Centre Global Sea Ice and Sea Surface Temperature (Had-ISST). (https://psl.noaa.gov/gcos_wgsp/Timeseries/Data/nino34.long.data (accessed on 17 March 2021)) |
| 2 | | Southern Oscillation Index (SOI) | Climatic Research Unit, University of East Anglia. (https://crudata.uea.ac.uk/cru/data/soi/ (accessed on 8 March 2021)) |
| 3 | | Southern Hemisphere annular mode index (SAMI) | (http://ljp.gcess.cn/dct/page/65609 (accessed on 15 June 2021)) |
| 4 | | Western Pacific subtropic high intensity (WPSH) | National Climate Center (https://cmdp.ncc-cma.net/Monitoring/ (accessed on 3 June 2021)) |
| 5 | Large-scale atmospheric variables | sea level pressure (15° S to 25° S, 55° E to 70° E) (SLP) | Reanalysis data of NCEP/NOAA [46] (http://www.esrl.noaa.gov/psd/cgi-bin/data/timeseries/timeseries1.pl (accessed on 17 June 2021)) |
| 6 | | meridional wind (20° N to 47.5° N, 105° E to 125° E) ($V$-wind$_{(1)}$) | |
| 7 | | meridional wind (32.5° N, 120° E) ($V$-wind$_{(2)}$) | |
| 8 | Local meteorological variables | Monthly mean air temperature (°C) ($T_{mean}$) | China Meteorological Data Service Centre, China Meteorological Administration (CMA) (http://data.cma.cn/data/cdcdetail/dataCode/SURF_CLI_CHN_MUL_DAY_V3.0.html (accessed on 27 February 2021)) |
| 9 | | Monthly maximum air temperature (°C) ($T_{max}$) | |
| 10 | | Monthly minimum air temperature (°C) ($T_{min}$) | |
| 11 | | Monthly mean air pressure ($P_{mean}$) | |
| 12 | | Monthly mean vapor pressure ($e_{mean}$) | |
| 13 | | Relative humidity ($d_{mean}$) | |
| 14 | | Sunshine duration ($D_{sun}$) | |

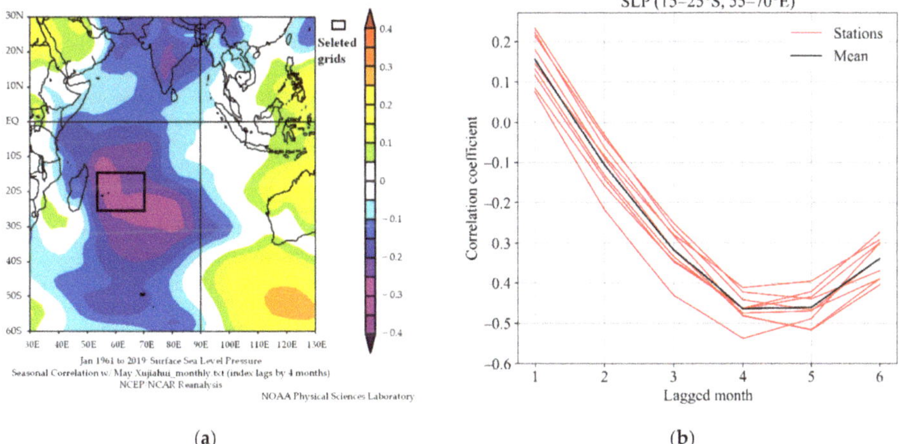


(a)      (b)

**Figure 2.** The correlation coefficient between the sea level pressure (SLP) and rainfall in the study region: (**a**) The correlation map for the spatial grids selection; (**b**) The correlation of the time series between SLP and rainfall for the lagged months selection.

## 3. Methodology

The models are trained and evaluated using above predictors. Since regional rainfall is related to multiscale climatic and meteorological features, the 14 predictors utilized represent the factors with multiple scales associated with rainfall in the study region. In addition, rainfall data from 9 rain stations are employed, keeping nearly 90% of each station for fitting the models (training), and the remaining roughly 10% for evaluating their prediction skill (testing) [47,48]. A fifty-nine years-long time series for each station are split in two sets (shown in Figure 3): the training set for the period of 1961–2012, containing 52 years of data and the testing set for the period of 2013–2019, containing the remaining 7-year data. Predictive performance is evaluated over the testing set, which is not learned in any methods.

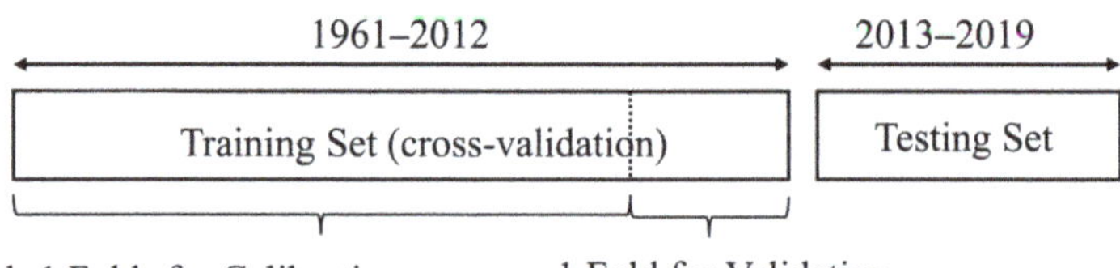



**Figure 3.** Methodological scheme of training and testing set division to fit and evaluate the models.

### 3.1. Machine Learning Methods

#### 3.1.1. K-Nearest Neighbors (KNN)

K-nearest neighbors (KNN) was proposed by Cover T.M. and Hart P.E. [49]. K-nearest neighbors (KNN) is a non-linear method whose predictions are computed through the weighted mode (classification) or the weighted mean (regression) of the k nearest points to the one being predicted. The Euclidean distance metric and Manhattan distance metric are commonly used metrics for finding the closest k neighbors in the training set. Then, the predicted target is obtained by averaging these neighbors, or the weighted average according to the distance. More details on the KNN algorithm can be found in [50].

#### 3.1.2. Extreme Gradient Boosting (XGB)

Extreme gradient boosting (XGB, also known as XGBoost) proposed by Chen and Guestrin [51], is a new application of gradient boosting machines. As the gradient boosting machines, XGB is developed through an additive training strategy. The predictions are made from weak learners that continuously develop over the mistakes from the former learners. The difference is that the gradient boosting algorithm is a negative gradient that learns a weak learner to approximate the loss function. XGB first finds the second-order Taylor approximation of the loss function at that point, and then minimizes the approximation loss function to train the weak learner. XGB can process sparse data automatically, and it is generally more than ten times faster than the conventional gradient boosting technique. For more information, readers are referred to [52].

#### 3.1.3. Support Vector Regression (SVR)

Support vector regression (SVR) is a kind of support vector machine (SVM) [53] for performing the regression task. The general concept of SVR is that it nonlinearly maps the feature data into the high-dimensional feature space. The objective of SVR is to find a hyperplane that maximizes the margins by separating samples belonging to different groups. The data points that support the margin at a close distance from the hyperplane are known as support vectors. In SVR, mapping the feature set into the high-dimensional feature space is achieved by the kernel function. The detailed description on various kernels can be found in [47]. Previous hydrological studies of SVR application demonstrated that the radial basis function kernel was found to be effective [5,20,54,55].

#### 3.1.4. Artificial Neural Network (ANN)

Artificial neural network (ANN) is inspired by the neurological structure of the human brain [56]. A common ANN architecture used in this study is the multiple layer perceptrons (MLP). The mathematical description of the method can be found in [57]. As a brief description, MLP is a feedforward network that consists of an input layer, hidden layer(s) and an output layer. The input layer receives external data and the output one produces the final result. The hidden layers are neurons nodes between the input and out layer, providing nonlinearity. More complex problems can be solved by increasing the hidden neurons or layers used. A neuron is a computational unit that receives input from other neurons that are interconnected with weight. The 'activation function' that each neuron uses receives the linear combination of inputs to produce the results in non-linear transformation.

In the present study, the traditional backpropagation algorithm [58] was adopted as the learning algorithm.

### 3.2. Stacking Ensemble Learning

The stacking ensemble learning is proposed by Wolpert [26], taking advantage of mutual complementarity among the base models to enhance generalization ability. The process occurs by, firstly, obtaining the results predicted by a set of diverse base models, and then optimally combining the outputs from the base models using a meta-learner to generate the final prediction. To prevent overfitting, the outputs from the base models are not directly learned by the meta-model. The leave-one-out cross validation method is used in this ensemble learning strategy. The validation folds are stacked as the new dataset for the meta-model to learn, which is the reason this strategy is called "stacking". How to integrate the base models is important. Multiple linear regression ML models such as RF, can be used as a meta-model. In our study, weights were assigned to the base models to constitute the stacking model prediction. The mathematical expression can be presented as:

$$y_{P,i} = \sum_{m=1}^{M} \omega_m f_{m,i} \tag{1}$$

where $\omega_m$ (m = 1, 2, ... , M) is the weight assigned for each base models, $f_{m,i}$ represents the prediction of the model m for the ith observation.

To obtain the optimal final prediction, the set of stacking weights were estimated by minimizing the mean square linear regression. Thus, the objective function under two constraints are as follows:

$$\Omega = \operatorname{argmin} \sum_{i=1}^{N} \left[ y_{O,i} - \sum_{m=1}^{M} \omega_m f_{m,i} \right]^2 \tag{2}$$

$$\omega_m \geq 0 \quad m = 1, 2, \ldots, M \tag{3}$$

$$\sum_{m=1}^{M} \omega_m = 1 \quad m = 1, 2, \ldots, M \tag{4}$$

where $\Omega = \{\omega_1, \omega_2, \ldots, \omega_M\}$ is the set of weights assigned to the base models. Two constraints are: (i) weights should be larger than or equal to zero, and (ii) the sum of the weights equals to one. This leads to a quadratic minimization problem [59], and the python package 'qpsolvers' was used to solve it. Through calculating the weights of the base models, the stacking model was integrated to generate the final prediction. The construction of the proposed stacking model and the overall flowchart of the adopted methodology in this study is presented in Figure 4.

### 3.3. Hyper-Parameter Optimization

Hyper-parameter tuning is commonly used to construct an appropriate model for a specific prediction. The model performance varies with different selection of hyper-parameter values. Table 3 summarizes the main hyper-parameters of the four machine learning models applied in this study. Taking SVR and ANN as examples, Figure 5 shows the process of hyper-parameter tuning of the two models. The hyper-parameters were tuned and evaluated over the training set by k-fold cross-validation [60]. K-fold cross-validation leveraging information in a small dataset helps to avoid overfitting and to produce a model that performs well on new data [61]. Figure 5a,b illustrates how the performance of SVR varies with the hyper-parameter Cost © and Gamma($\gamma$). For SVR, the cost C and $\gamma$ with the radial basis function kernel are significant hyper-parameters. It illustrates a proper value range of the cost C; $\gamma$ were nearly $10^{-2}$ to $10^{-1}$ and 10 to 100, respectively. For ANN, the size of the hidden layer is an essential hyper-parameter, indicating the complexity of the learning model. In Figure 5c, ANN with a hidden layer of (8) and (8,8) were compared. Earlier convergence (nearly 190 epochs) and higher performance ($R^2$ of

0.53) over the validation set were shown on the ANN with the layer of (8) compared to one with a hidden layer of (8,8). This indicates that the relatively smaller size of hidden layer ANN has enough learning capacity, and that too large of a size of hidden layer will cause overfitting. For the machine learning model, multiple important hyper-parameters impact the model performance comprehensively; the grid search approach was utilized to optimize the combination of hyper-parameters within the specified range in this study. Then, the models with hyper-parameters tuned were applied in the testing set. There was no notable higher performance in the training set than the testing one, indicating that the models built are reasonable and capable of generalization.

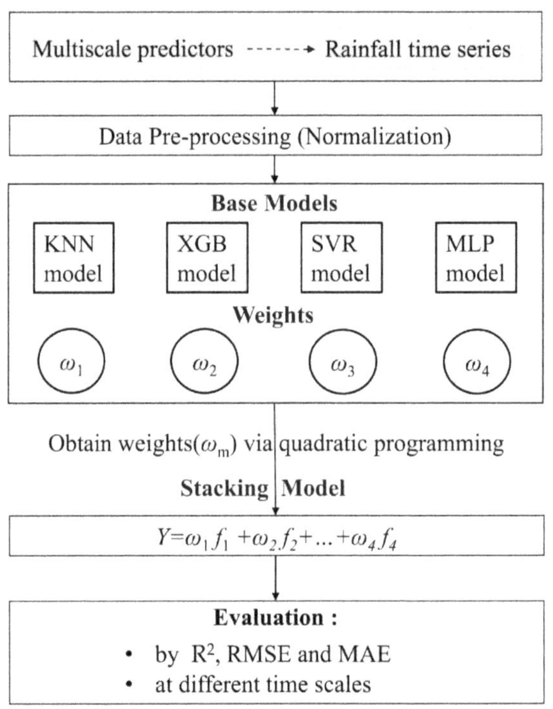


**Figure 4.** Flowchart of the stacking-based methodology in the study.

**Table 3.** Summary of the hyper-parameters of the four machine learning models.

| Machine Learning Model | Hyper-Parameters |
| --- | --- |
| K-nearest neighbors (KNN) | Number of neighbors<br>Weights |
| Extreme gradient boosting (XGB) | Number of estimators<br>Learning rate<br>Max depth |
| Support vector regression (SVR) | Cost $C$<br>Parameter of Gaussian Kernel—Gamma($\gamma$) |
| Artificial neural network (ANN) | Size of hidden layer<br>Activation function<br>Learning rate<br>Batch size |

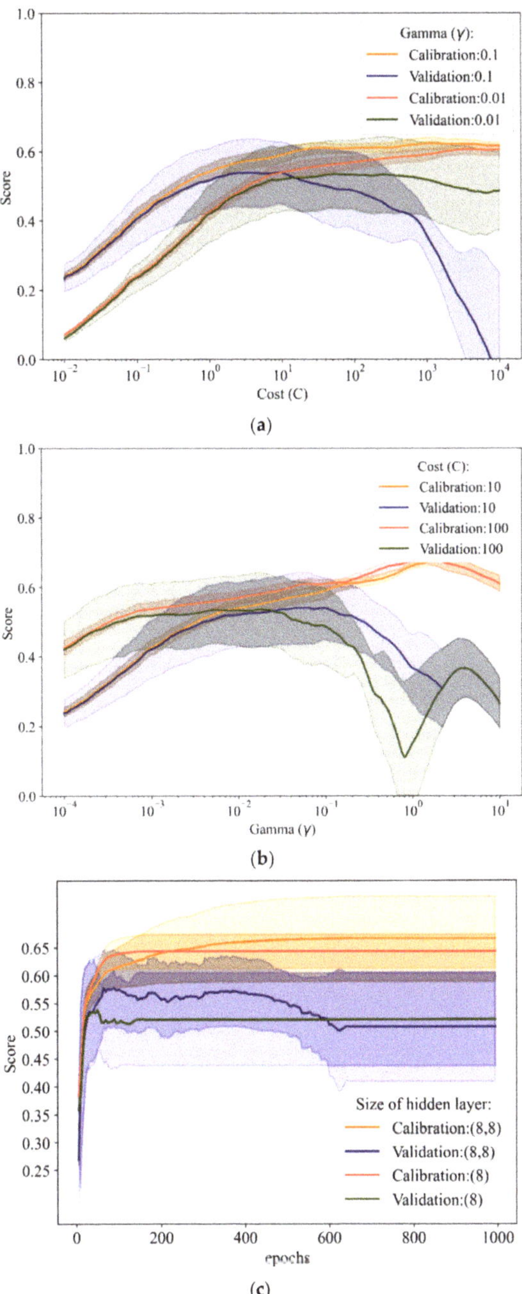

**Figure 5.** The $R^2$ score of hyper-parameters tuning at Ningguo station. (**a**) Cost (C) of SVR; (**b**) Gamma ($\gamma$) of SVR; (**c**) size of hidden layer of ANN. The shaded areas include 5-flod cross-validation results.

### 3.4. Performance Evaluation

The performances of the above machine learning models were evaluated by the commonly used statistic metrics: (1) Coefficient of determination ($R^2$), (2) root mean square error (RMSE), (3) mean absolute error (MAE). $R^2$ measures the proportion of variance explained by the model. The best possible score is 1.0; a larger value represents a better fit. RMSE evaluates the residual between observed and predicted values and is particularly sensitive to the large errors, since the errors are squared before they are averaged. the MAE is less sensitive to extreme values than the RMSE [62]. The mathematical formulas are as follows:

Coefficient of determination ($R^2$)

$$R^2 = 1 - \frac{\sum\limits_{i=1}^{N}\left(y_{P,i} - y_{O,i}\right)^2}{\sum\limits_{i=1}^{N}\left(y_{O,i} - \bar{y}_{O,i}\right)^2} \tag{5}$$

Root mean square error (RMSE)

$$\mathrm{RMSE} = \sqrt{\frac{1}{N}\sum\limits_{i=1}^{N}\left(y_{P,i} - y_{O,i}\right)^2} \tag{6}$$

Mean absolute error (MAE)

$$\mathrm{MAE} = \frac{1}{N}\sum\limits_{i=1}^{N}\left|y_{P,i} - y_{O,i}\right| \tag{7}$$

where $y_{P,i}$ and $y_{O,i}$ are the predicted and observed monthly precipitation in test period $t$ (test slice), respectively, i is the month of the dataset and N (= 84) is the length (number of samples in the test set) in period t (2013–2019), $\bar{y}_{O,i}$ is the mean values of the series $y_{O,i}$.

### 3.5. Categorization of Dry, Intermediate and Wet Months in Terms of Standardized Precipitation Index (SPI)

For measuring the model performance on normal, below and above normal monthly rainfall prediction, the standardized precipitation index (SPI) proposed by McKee et al. [63] was used to designate the monthly precipitation into the dry/intermediate/wet classifications. SPI was calculated using the available program from the National Drought Mitigation Centre (https://drought.unl.edu/droughtmonitoring/SPI/SPIProgram.aspx (accessed on 29 July 2021)). The SPI calculated in this study is based on representing the historical monthly precipitation record with a gamma distribution. Positive SPI values represent wet conditions; the higher the SPI, the more unusually wet a month is. Negative SPI values represent dry conditions; the lower the SPI, the more unusually dry a month is. The detailed methodology and the computation process of SPI can be found in Angelidis et al. [64].

SPI was obtained based on the observed monthly rainfall series. The calculated SPI fall into three categories, namely, 'dry' (SPI < −1), 'intermediate' (−1 ≤ SPI ≤ 1), and 'wet' (SPI > 1). The performance of the models above was assessed respectively in terms of the three categories.

## 4. Results and Discussion

### 4.1. Intercomparison of Model Performances

Four base models and the stacking model are constructed at nine stations in the Taihu basin for prediction of monthly rainfall. Prediction is independent for each station. The observed and predicted monthly precipitation series of all the models at the nine stations are shown in Figure S2.

Figure 6 demonstrates the prediction skills of all the models at the nine rainfall stations. Among the four base models, the model performances vary in terms of $R^2$, RMSE, and MAE. The $R^2$ ranges from 0.29 to 0.70. The RMSE and MAE range from 48 mm to 79 mm and from 35 mm to 51 mm, respectively. It presents analogous ranges of the evaluation metrics with the previous predictions at the lower reach of the Yangtze River [25], illustrating the models in this study perform in the reasonable range. Among the base models, ANN at Xujiahui had the best prediction accuracy with the highest $R^2$ and the smallest RMSE and MAE, while the accuracy of KNN was the worst in terms of the three metrics at almost all the stations. There was no base model that performed best at all the stations.

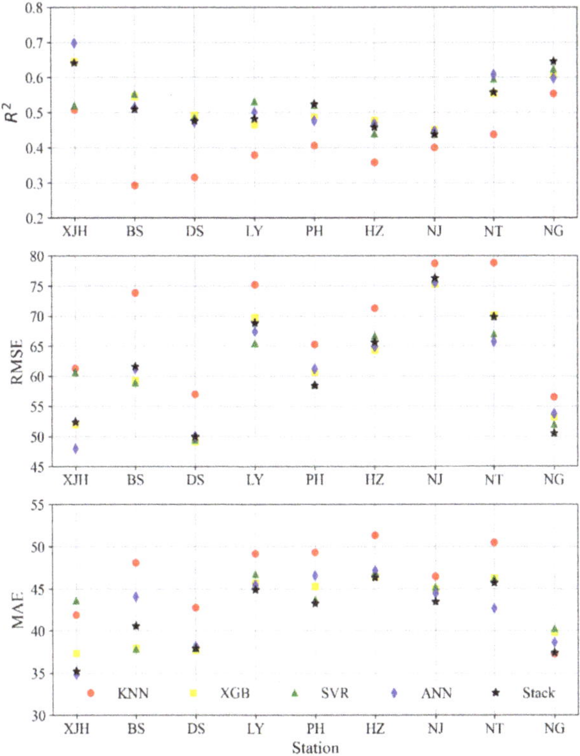

**Figure 6.** Comparison of model overall performance for the 9 stations using $R^2$; RMSE and MAE.

We then compared the performance of the base models and the stacking model. The best models selected in terms of $R^2$ and RMSE were same at the nine stations (shown in Table S1), and the stacking model performed best at two stations. In terms of MAE, the stacking model performed best at four stations. This implies that, through combining ML models of diverse structures, the stacking model has the potential to over-perform all its base models. At the other stations, the stacking model showed analogous accuracy with the best base models. It should be noted that, though the stacking model was not selected as the best one at all the nine stations, the variation of each metric was lower, implying that the stacking model can produce more robust predictions at regional scale. Additionally, as shown in Table 4, the stacking strategy reduced MAE more effectively than RMSE, since MAE evaluates the average magnitude, while RMSE is more sensitive to the large errors, which are squared before they are averaged. This indicates that, except for the magnification of the large errors generally occurring at extreme rainfall samples [65], the stacking model appeared to be more favorable in the measurement of average performance in the entire rainfall series prediction. The best model for each station selected in terms of $R^2$, RMSE and MAE is shown in Table S1.

**Table 4.** Evaluation metrics averaged over the nine stations at different time scales.

| | Evaluation Metrics | KNN | XGB | SVR | ANN | Stack |
|---|---|---|---|---|---|---|
| All months | $R^2$ | 0.407 | 0.526 | 0.523 | 0.532 | 0.526 |
| | RMSE (mm) | 68.72 | 61.57 | 61.65 | 60.92 | 61.51 |
| | MAE (mm) | 46.34 | 42.41 | 43.16 | 42.47 | 41.65 |
| Annual aggregation scale | RMSE (%) | 26.12 | 22.33 | 22.12 | 24.63 | 23.34 |
| | MAE (%) | 21.39 | 18.31 | 18.61 | 21.02 | 19.40 |
| Spring | RMSE (mm) | 43.82 | 45.74 | 45.79 | 47.16 | 44.22 |
| | MAE (mm) | 33.86 | 37.18 | 36.88 | 37.89 | 35.58 |
| Summer | RMSE (mm) | 95.50 | 87.56 | 87.06 | 85.19 | 87.44 |
| | MAE (mm) | 73.85 | 66.46 | 66.21 | 66.72 | 66.82 |
| Autumn | RMSE (mm) | 77.17 | 64.35 | 64.02 | 62.89 | 65.27 |
| | MAE (mm) | 50.55 | 44.21 | 44.03 | 42.61 | 43.15 |
| Winter | RMSE (mm) | 34.45 | 27.46 | 29.52 | 28.06 | 26.22 |
| | MAE (mm) | 27.09 | 21.77 | 25.54 | 22.67 | 21.05 |
| Dry months | RMSE (mm) | 61.05 | 47.56 | 49.45 | 43.03 | 46.78 |
| | MAE (mm) | 49.90 | 37.47 | 40.39 | 32.33 | 35.87 |
| Intermediate months | RMSE (mm) | 36.98 | 40.53 | 42.27 | 43.94 | 38.77 |
| | MAE (mm) | 28.08 | 32.26 | 33.06 | 33.98 | 30.21 |
| Wet months | RMSE (mm) | 121.23 | 101.22 | 99.03 | 96.82 | 103.43 |
| | MAE (mm) | 97.86 | 73.15 | 72.94 | 71.33 | 76.69 |
| Months of extreme rainfall | RMSE (mm) | 197.70 | 172.36 | 164.65 | 157.80 | 173.26 |
| | MAE (mm) | 188.36 | 162.22 | 153.26 | 143.32 | 163.38 |

### 4.2. Prediction Skills at Different Time Scales

It is also of importance to predict annual, seasonal and other scales in the water resources management. Thus, we examined the model performance at annual aggregated scale, seasonal scale, dry/intermediate/wet months and months of extreme rainfall.

At the annual aggregation scale, Table 4 shows the evaluation metrics (RMSE and MAE) of the five models at nine rainfall stations over the study region. The RMSE of the stacking model at the annual aggregation scale was 157.5–399.7 mm (accounting for 15–35% of the annual precipitation averaged over the 1961–2019 period), and MAE was 157.6–336.7 mm (accounting for 11–30%). Among the base models, SVR performed satisfactorily at the annual aggregation scale, with an RMSE of 135.7–333.8 mm (accounting for 10–31%) and MAE of 110.9–299.6 mm (accounting for 9–25%). Generally, in terms of the performance at the annual aggregation scale, the stacking model and ML models, such as SVR and XGB, showed good ability in readily applying to long-term rainfall prediction for regional water resources management.

Over four seasons, rainfall shows significantly seasonal variability in the study region. The average monthly rainfall (1961–2019) at the stations was 81.5–137.1 mm in spring (from March to May), 151.9–194.8 mm in summer (from June to August), 65.0–96.8 mm in autumn (from September to November), and 39.9–73.0 mm in winter (from December to February). Thus, evaluation metrics (RMSE and MAE), the percentage of which accounts for average monthly rainfall over four seasons, were evaluated at the seasonal scale, shown in Figure 7. In terms of RMSE and MAE, the prediction in winter was the most accurate, followed by spring and autumn. The evaluation metrics were highest in summer considering its largest amount of rainfall over four seasons. While, in terms of the percentage of RMSE and MAE, the prediction in spring was the most accurate, similar in summer and winter, but worst in autumn. Generally, the stacking model performed better in spring and winter than in summer and autumn. It is noted that previous studies [40,42,66] have highlighted the importance of accurately predicting summer rainfall. Future work is needed to explore suitable models and the main factors for summer rainfall prediction in this region.

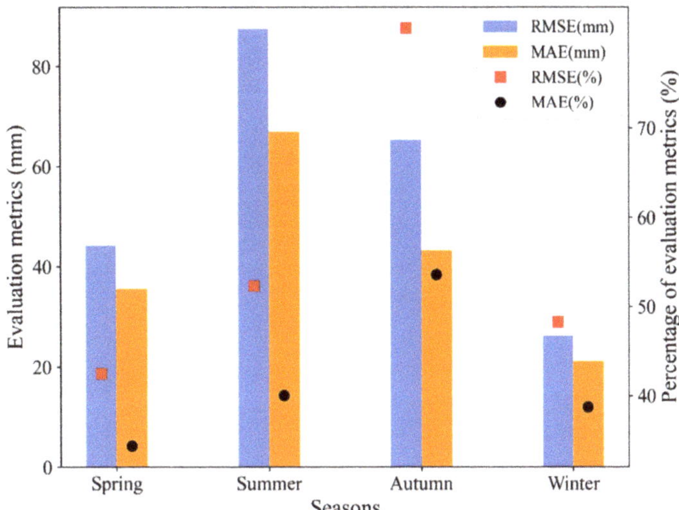


**Figure 7.** The value and percentage of evaluation metrics (RMSE and MAE) of the stacking model at the nine stations in four seasons.

The prediction from the above models was further compared in terms of dry/intermediate/ wet months. As mentioned earlier, SPI was used as the index for classifying the categorization. The SPI was calculated based on observed monthly rainfall series, and divided all months into three categories, namely, 'dry' (SPI < −1), 'intermediate' (−1 ≤ SPI ≤ 1), and 'wet' (SPI > 1). The scatter plots of the stacking model are presented in Figure 8. The results of base models are shown in Figure S3. It revealed that all the models underestimated rainfall for the wet months, and slightly overestimated rainfall for the dry months.

The predictions for intermediate and dry months were within a minor error range. The prediction error on wet conditions was high, and rainfall prediction for wet JJA (June-July-August) months was the most underestimated, indicating that the wet feature is the most difficult for the machine learning models to capture. The evaluation metrics (RMSE and MAE) for dry/intermediate/wet months by the stacking model and the base models shown in Table 4 also offered the same indication. Similar results were also found in other climate regions [5,24]. Further work is needed to pay attention on wet JJA rainfall prediction, which is crucial to the regional flood prevention.

Extreme precipitation deserves special attention in the Taihu basin, considering that intensive precipitation during the 'Plum Rain Season' (the rainy season from late June to early July in the Yangtze Plain) and typhoon season may cause flooding [35]. We compared the prediction skill of the above models on precipitation above 300 mm, which is considered as extreme rainfall in the study region. Since the samples of extreme rainfall are a tiny part in the series (nearly 3%), the extreme rainfall is generally underestimated by the above models. Such a feature seems difficult for models to capture. The evaluation metrics on extreme rainfall are shown in Table 4. They indicated that ANN showed the greatest predictive ability, followed by SVR. The stacking model performed comparable to XGB. KNN showed poor predictive power on extreme rainfall, since the stacking model with the weight-distributed strategy is influenced by all the base models. One of the base models with poor performance may reduce the prediction ability of the stacking one. Other ML models can be utilized as an alternative in the flexible stacking framework for enhancing the predictive skill.

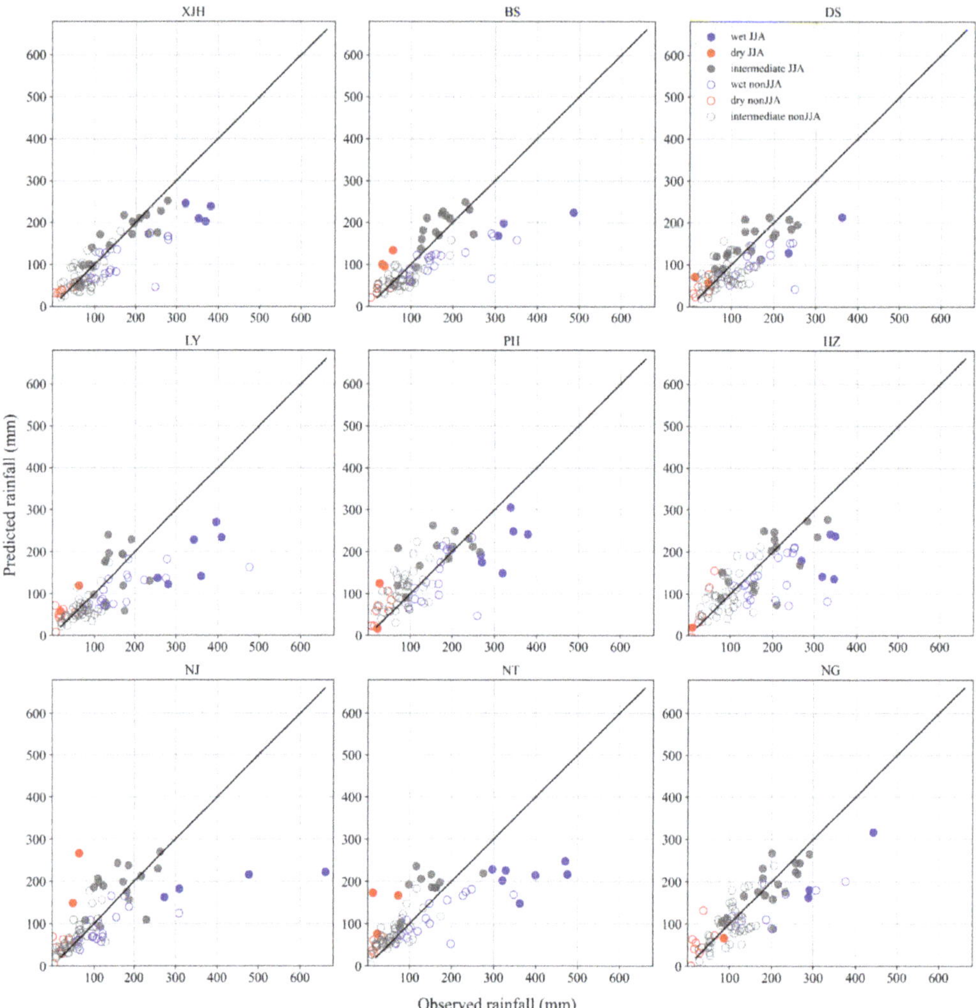

**Figure 8.** Scatter plot showing the association between observed and predicted rainfall of the stacking model for the testing period (2013–2019).

## *4.3. Discussions*

Irrespective of the models used in prediction, there are quite differing prediction effects shown among the nine stations. Higher performance was shown at Xujiahui and Ningguo station, with $R^2$ of 0.642 and 0.645, respectively, while lower performance was shown at Nanjing, with $R^2$ only reaching 0.438 by the stacking model, as depicted in Figure 9. The certain possible reasons that may impact the performance are addressed as follows.

One of the crucial reasons is likely associated with different characteristics in the rainfall series among these stations. We used $C_V$ and probability density of time series as examples to demonstrate the various features. Figure 9 shows the performance of the stacking model contrast to the coefficient of variation ($C_V$) at the nine stations. The higher $C_V$ indicated a more disperse rainfall distribution, which may increase the difficulty of the series prediction. In Table S2, lower $C_V$ are shown in all predicted series than in observational ones, which indicates that the dispersion feature in the time series is difficult to capture. Figure 10 shows the probability density distribution of observations and predictions by the stacking model at the nine stations. Lower probability density in distribution tails and excessive distribution around 100 mm also show that the predicted rainfall is prone to concentrate on moderate values, making the dispersion of the series difficult to reproduce. Further research is needed to examine the main features of time series that impact the prediction performance.

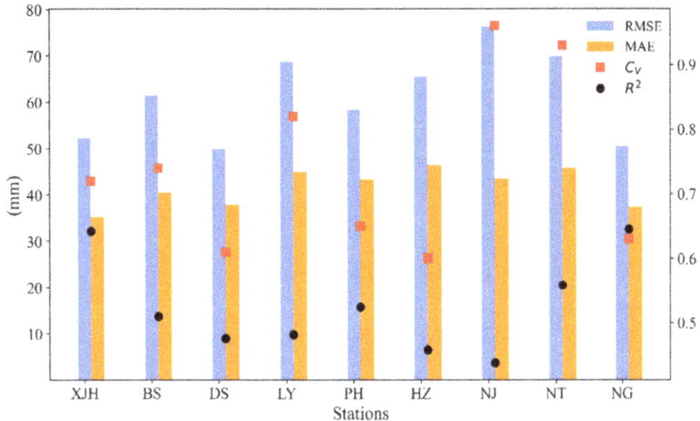

**Figure 9.** Evaluation metrics ($R^2$, RMSE and MAE) on the stacking model and the coefficient of variation ($C_V$) of the rainfall series at the nine stations for the testing period (2013–2019).

Another characteristic that may affect the prediction is the discrepancy between the rainfall distributions for the training set (1961–2012) and the testing set (2013–2019). Figure 10 shows that the probability distribution in the range of 0–100 mm significantly reduces, while the monthly rainfall larger than 200 mm occurs more frequently during the testing period (2013–2019) at most of stations. In comparison, there are similar probability distributions in the training and testing period at Xujiahui and Ningguo station, conducive to high prediction accuracy at these stations. It implies that the characteristics of training and testing sets have a notably high impact on the prediction accuracy. Further works can consider the statistical characteristics in the ML prediction model construction to enhance the predictive ability.

The division of training and testing sets is an inevitable issue in time series prediction. Generally, for building a statistical predictive model, the training set and the testing set are required to contain the same distribution [67], which is conducive to achieving good prediction results. However, due to the complexities in the change of rainfall characteristics [4] which is caused by natural and anthropogenic factors, the physical factors that impact rainfall characteristics are needed in the models as prediction factors in the long term rainfall prediction to reveal this change. In addition, other climatic and meteorological variables utilized as predictors also show non-stationarity and complexity in dynamic climate systems [68]. Identifying major drivers of regional rainfall for mapping relationship construction is also important to enhance the predictive ability.

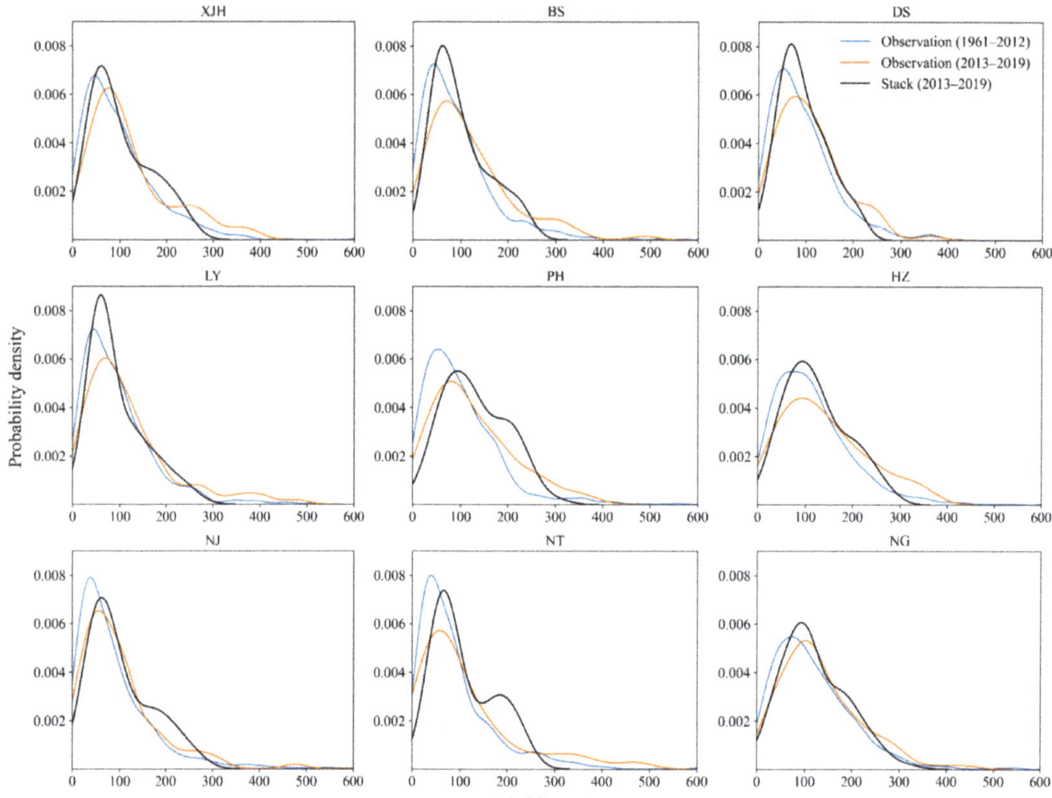

**Figure 10.** Probability density distribution of observations and predictions by the stacking model at the nine stations.

## 5. Conclusions

In this study, a stacking ensemble learning model and its base models were compared for the prediction of monthly rainfall at nine stations in the Taihu basin, China, using large-scale climate indices, large-scale atmospheric variables, and local meteorological variables as predictors. Principal conclusions of the study are as follows:

(1)     Through combining models of diverse structures, the stacking model showed the potential to over-perform all the base models. In terms of different evaluation metrics, the results varied among the models. In terms of $R^2$ and RMSE, the stacking model performed best at two stations (Pinghu and Ningguo). In terms of MAE, the stacking model performed best at four stations (Liyang, Pinghu, Hangzhou and Nanjing). At the other rainfall stations, the stacking approach also showed satisfactory performance, close to the best one of the individual base models, and especially showed favorable results in term of MAE. Thus, the proposed stacking model can produce reasonable predictions for the entire rainfall series.

(2)     At the annual aggregation scale, the stacking model and ML models (SVR and XGB) performed satisfactorily, showing good ability in applying long-term rainfall prediction for regional water resources management. Over four seasons, the stacking model generally showed better performance in spring and winter than in summer and autumn. In terms of dry/intermediate/wet months, the models showed a greater minor

error range in dry and intermediate months than wet months, with underestimation of the wet months and slight overestimation of the dry months.

(3)    In terms of extreme rainfall, ANN outperformed the stacking model. The ML models generally undervalue extreme rainfall. ANN, relatively, generated the closest prediction, showing the potential to capture the extreme wet condition. Further work is needed to explore ML methods to enhance the ability of predicting extreme rainfall, especially in regions vulnerable to flooding.

In this study, a stacking ensemble model of combining different machine learning model structures was proposed in rainfall prediction. In this flexible stacking framework, the attempts to improve base-learners and meta-learners were promising to enhance the prediction ability in further research. In addition to the model structures, the difference between training and testing data distributions also affected the prediction performance. Further study should focus on the variability in rainfall series, the identification of important drivers to enhance the prediction ability and the examination of more ML models, such as recurrent neural network (RNN) [58], under the ensemble framework. The data-driven model with the stacking ensemble framework is readily generalized to other climatic regions, using climatic, meteorological and diverse information.

**Supplementary Materials:** The following supporting information can be downloaded at: https://www.mdpi.com/article/10.3390/w14030492/s1, Figure S1: The correlation between lagged climate indices and rainfall at the 9 stations in the Taihu basin. Climate indices are: (a) Nino 3.4; (b) SOI; (c) WPSH; (d) SAMI, Figure S2: Prediction results of monthly rainfall at the stations in the Taihu basin, Figure S3: Scatter plot showing the association between observed and predicted rainfall of the base models for the testing period (2013–2019), Table S1: The best model selected in terms of $R^2$, RMSE and MAE at the nine stations, Table S2: The coefficient of variation deviation ($C_V$) of the observed and the predicted monthly rainfall series at the nine stations.

**Author Contributions:** J.G.: conceptualization, methodology, formal analysis, visualization, writing—original draft preparation; S.L.: conceptualization, writing—review and editing, supervision; Z.Z.: conceptualization, methodology, writing—review and editing, supervision; S.R.C.: conceptualization, writing—review and editing; Q.Z.: conceptualization, formal analysis, writing—review and editing. All authors have read and agreed to the published version of the manuscript.

**Funding:** This research was funded by the National Key Research and Development Program of China (Grant no.2018YFD1100401), National Natural Science Foundation of China (51909191, 52111530045 and 51961145106) and Russian Fund for Basic Research—National Natural Science Foundation of China (21-55-53039).

**Institutional Review Board Statement:** Not applicable.

**Informed Consent Statement:** Not applicable.

**Data Availability Statement:** The public archived datasets in the study can be accessed by the links in Section 2, or requesting on the corresponding author.

**Acknowledgments:** We are grateful to the NOAA/OAR/ESRL PSL, Boulder, Colorado, USA for providing us with NCEP Reanalysis Derived data.

**Conflicts of Interest:** The authors declare no conflict of interest. The funders had no role in the design of the study; in the collection, analyses, or interpretation of data, or in the writing of the manuscript.

## References

1.    Ali, M.; Deo, R.C.; Downs, N.J.; Maraseni, T. Multi-stage hybridized online sequential extreme learning machine integrated with Markov Chain Monte Carlo copula-Bat algorithm for rainfall forecasting. *Atmos. Res.* **2018**, *213*, 450–464. [CrossRef]

2.    Bagirov, A.; Mahmood, A.; Barton, A. Prediction of monthly rainfall in Victoria, Australia: Clusterwise linear regression approach. *Atmos. Res.* **2017**, *188*, 20–29. [CrossRef]

3.    Zeynoddin, M.; Bonakdari, H.; Azari, A.; Ebtehaj, I.; Gharabaghi, B.; Madvar, H.R. Novel hybrid linear stochastic with non-linear extreme learning machine methods for forecasting monthly rainfall a tropical climate. *J. Environ. Manag.* **2018**, *222*, 190–206. [CrossRef] [PubMed]

4.  Climate Change 2014: Synthesis Report. Contribution of Working Groups I, II and III to the Fifth Assessment Report of the Intergovernmental Panel on Climate Change. Geneva. 2014. Available online: https://www.ipcc.ch/report/ar5/syr (accessed on 20 December 2021).

5.  Das, P.; Chanda, K. Bayesian Network based modeling of regional rainfall from multiple local meteorological drivers. *J. Hydrol.* **2020**, *591*, 125563. [CrossRef]

6.  Abbot, J.; Marohasy, J. Input selection and optimisation for monthly rainfall forecasting in Queensland, Australia, using artificial neural networks. *Atmos. Res.* **2014**, *138*, 166–178. [CrossRef]

7.  Shahrban, M.; Walker, J.; Wang, Q.; Seed, A.; Steinle, P. An evaluation of numerical weather prediction based rainfall forecasts. *Hydrol. Sci. J.* **2016**, *61*, 2704–2717. [CrossRef]

8.  Ali, M.; Deo, R.C.; Downs, N.J.; Maraseni, T. Chapter 3—Monthly Rainfall Forecasting with Markov Chain Monte Carlo Simulations Integrated with Statistical Bivariate Copulas. In *Handbook of Probabilistic Models*; Samui, P., Tien Bui, D., Chakraborty, S., Deo, R.C., Eds.; Butterworth-Heinemann: Boston, MA, USA, 2020; pp. 89–105. ISBN 978-0-12-816514-0.

9.  Giebel, G.; Kariniotakis, G. Wind power forecasting—A review of the state of the art. In *Renewable Energy Forecasting: From Models to Applications*; Woodhead Publishing: Cambridge, UK, 2017; ISBN 978-0081005040.

10. Yu, W.; Nakakita, E.; Jung, K. Flood Forecast and Early Warning with High-Resolution Ensemble Rainfall from Numerical Weather Prediction Model. *Procedia Eng.* **2016**, *154*, 498–503. [CrossRef]

11. Carlson, R.F.; MacCormick, A.J.A.; Watts, D.G. Application of Linear Random Models to Four Annual Streamflow Series. *Water Resour. Res.* **1970**, *6*, 1070–1078. [CrossRef]

12. Burlando, P.; Rosso, R.; Cadavid, L.G.; Salas, J.D. Forecasting of short-term rainfall using ARMA models. *J. Hydrol.* **1993**, *144*, 193–211. [CrossRef]

13. Valipour, M.; Banihabib, M.E.; Behbahani, S.M.R. Comparison of the ARMA, ARIMA, and the autoregressive artificial neural network models in forecasting the monthly inflow of Dez dam reservoir. *J. Hydrol.* **2012**, *476*, 433–441. [CrossRef]

14. Rahman, M.A.; Yunsheng, L.; Sultana, N. Analysis and prediction of rainfall trends over Bangladesh using Mann–Kendall, Spearman's rho tests and ARIMA model. *Arch. Meteorol. Geophys. Bioclimatol. Ser. B* **2016**, *129*, 409–424. [CrossRef]

15. Lana, X.; Rodríguez-Solà, R.; Martínez, M.D.; Casas-Castillo, M.C.; Serra, C.; Kirchner, R. Autoregressive process of monthly rainfall amounts in Catalonia (NE Spain) and improvements on predictability of length and intensity of drought episodes. *Int. J. Clim.* **2020**, *41*. [CrossRef]

16. Basha, C.Z.; Bhavana, N.; Bhavya, P. Rainfall Prediction Using Machine Learning Amp; Deep Learning Techniques. In Proceedings of the 2020 International Conference on Electronics and Sustainable Communication Systems (ICESC), Coimbatore, India, 2–4 July 2020; pp. 92–97.

17. Ortiz-García, E.; Salcedo-Sanz, S.; Casanova-Mateo, C. Accurate precipitation prediction with support vector classifiers: A study including novel predictive variables and observational data. *Atmos. Res.* **2014**, *139*, 128–136. [CrossRef]

18. Grace, R.K.; Suganya, B. Machine Learning Based Rainfall Prediction. In Proceedings of the 2020 6th International Conference on Advanced Computing and Communication Systems (ICACCS), Coimbatore, India, 6–7 March 2020; pp. 227–229.

19. Diez-Sierra, J.; del Jesus, M. Long-term rainfall prediction using atmospheric synoptic patterns in semi-arid climates with statistical and machine learning methods. *J. Hydrol.* **2020**, *586*, 124789. [CrossRef]

20. Tian, D.; He, X.; Srivastava, P.; Kalin, L. A hybrid framework for forecasting monthly reservoir inflow based on machine learning techniques with dynamic climate forecasts, satellite-based data, and climate phenomenon information. *Stoch. Hydrol. Hydraul.* **2021**, 1–23. [CrossRef]

21. Yu, P.-S.; Yang, T.-C.; Chen, S.-Y.; Kuo, C.-M.; Tseng, H.-W. Comparison of random forests and support vector machine for real-time radar-derived rainfall forecasting. *J. Hydrol.* **2017**, *552*, 92–104. [CrossRef]

22. Cramer, S.; Kampouridis, M.; Freitas, A.; Alexandridis, A.K. An extensive evaluation of seven machine learning methods for rainfall prediction in weather derivatives. *Expert Syst. Appl.* **2017**, *85*, 169–181. [CrossRef]

23. Pour, S.H.; Wahab, A.K.A.; Shahid, S. Physical-empirical models for prediction of seasonal rainfall extremes of Peninsular Malaysia. *Atmos. Res.* **2019**, *233*, 104720. [CrossRef]

24. Sachindra, D.; Ahmed, K.; Rashid, M.; Shahid, S.; Perera, B. Statistical downscaling of precipitation using machine learning techniques. *Atmos. Res.* **2018**, *212*, 240–258. [CrossRef]

25. Zhou, Z.; Ren, J.; He, X.; Liu, S. A comparative study of extensive machine learning models for predicting long-term monthly rainfall with an ensemble of climatic and meteorological predictors. *Hydrol. Process.* **2021**, *35*, e14424. [CrossRef]

26. Wolpert, D.H. Stacked generalization. *Neural Netw.* **1992**, *5*, 241–259. [CrossRef]

27. Rice, J.S.; Emanuel, R.E. How are streamflow responses to the El Nino Southern Oscillation affected by watershed characteristics? *Water Resour. Res.* **2017**, *53*, 4393–4406. [CrossRef]

28. Zhai, B.; Chen, J. Development of a stacked ensemble model for forecasting and analyzing daily average PM2.5 concentrations in Beijing, China. *Sci. Total Environ.* **2018**, *635*, 644–658. [CrossRef] [PubMed]

29. Sun, W.; Trevor, B. A stacking ensemble learning framework for annual river ice breakup dates. *J. Hydrol.* **2018**, *561*, 636–650. [CrossRef]

30. Breiman, L. Stacked Regressions. *Mach. Learn.* **1996**, *24*, 49–64. [CrossRef]

31. Zounemat-Kermani, M.; Batelaan, O.; Fadaee, M.; Hinkelmann, R. Ensemble machine learning paradigms in hydrology: A review. *J. Hydrol.* **2021**, *598*, 126266. [CrossRef]

32. Li, Y.; Liang, Z.; Hu, Y.; Li, B.; Xu, B.; Wang, D. A multi-model integration method for monthly streamflow prediction: Modified stacking ensemble strategy. *J. Hydroinform.* **2019**, *22*, 310–326. [CrossRef]
33. Wang, L.; Zhu, Z.; Sassoubre, L.; Yu, G.; Liao, C.; Hu, Q.; Wang, Y. Improving the robustness of beach water quality modeling using an ensemble machine learning approach. *Sci. Total Environ.* **2020**, *765*, 142760. [CrossRef]
34. Peng, D.; Qiu, L.; Fang, J.; Zhang, Z. Quantification of Climate Changes and Human Activities That Impact Runoff in the Taihu Lake Basin, China. *Math. Probl. Eng.* **2016**, *2016*, 1–7. [CrossRef]
35. Wu, J.; Wu, Z.-Y.; Lin, H.-J.; Ji, H.-P.; Liu, M. Hydrological response to climate change and human activities: A case study of Taihu Basin, China. *Water Sci. Eng.* **2020**, *13*, 83–94. [CrossRef]
36. Liang, W.; Yongli, C.; Hongquan, C.; Daler, D.; Jingmin, Z.; Juan, Y. Flood disaster in Taihu Basin, China: Causal chain and policy option analyses. *Environ. Earth Sci.* **2010**, *63*, 1119–1124. [CrossRef]
37. Ge, Q.; Bian, J.; Zheng, J.; Liao, Y.; Hao, Z.; Yin, Y. The climate regionalization in China for 1981-2010. *Chin. Sci. Bull.* **2013**, *58*, 3088–3099. [CrossRef]
38. Tao, L.; He, X.; Qin, J. Multiscale teleconnection analysis of monthly total and extreme precipitations in the Yangtze River Basin using ensemble empirical mode decomposition. *Int. J. Clim.* **2020**, *41*, 348–373. [CrossRef]
39. Liu, Y.; Li, W.; Ai, W.; Li, Q. Reconstruction and Application of the Monthly Western Pacific Subtropical High Indices. *J. Appl. Meteorol. Sci.* **2012**, *23*, 414–423.
40. Nan, S.; Li, J. The relationship between the summer precipitation in the Yangtze River valley and the boreal spring Southern Hemisphere annular mode. *Geophys. Res. Lett.* **2003**, *30*. [CrossRef]
41. Tang, Y.; Huang, A.; Wu, P.; Huang, D.; Xue, D.; Wu, Y. Drivers of Summer Extreme Precipitation Events Over East China. *Geophys. Res. Lett.* **2021**, *48*. [CrossRef]
42. Fan, K.; Wang, H.; Choi, Y.-J. A physically-based statistical forecast model for the middle-lower reaches of the Yangtze River Valley summer rainfall. *Chin. Sci. Bull.* **2008**, *53*, 602–609. [CrossRef]
43. Guo, Y.; Li, J.; Li, Y. Seasonal Forecasting of North China Summer Rainfall Using a Statistical Downscaling Model. *J. Appl. Meteorol. Clim.* **2014**, *53*, 1739–1749. [CrossRef]
44. Wang, C.; Jia, Z.; Yin, Z.; Liu, F.; Lu, G.; Zheng, J. Improving the Accuracy of Subseasonal Forecasting of China Precipitation with a Machine Learning Approach. *Front. Earth Sci.* **2021**, *9*. [CrossRef]
45. Babel, M.S.; Sirisena, T.A.J.G.; Singhrattna, N. Incorporating large-scale atmospheric variables in long-term seasonal rainfall forecasting using artificial neural networks: An application to the Ping Basin in Thailand. *Water Policy* **2016**, *48*, 867–882. [CrossRef]
46. Kalnay, E.; Kanamitsu, M.; Kistler, R.; Collins, W.; Deaven, D.; Gandin, L.; Iredell, M.; Saha, S.; White, G.; Woollen, J.; et al. The NCEP/NCAR 40-Year Reanalysis Project. *Bull. Am. Meteorol. Soc.* **1996**, *77*, 437–472. [CrossRef]
47. Hofmann, T.; Schölkopf, B.; Smola, A.J. Kernel methods in machine learning. *Ann. Stat.* **2008**, *36*, 1171–1220. [CrossRef]
48. Marsland, S. *Machine Learning: An Algorithmic Perspective*, 2nd ed.; Chapman and Hall/CRC: New York, NY, USA, 2014; ISBN 978-0-429-10250-9.
49. Cover, T.; Hart, P. Nearest neighbor pattern classification. *IEEE Trans. Inf. Theory* **1967**, *13*, 21–27. [CrossRef]
50. Ahmadi, A.; Moridi, A.; Lafdani, E.K.; Kianpisheh, G. Assessment of climate change impacts on rainfall using large scale climate variables and downscaling models—A case study. *J. Earth Syst. Sci.* **2014**, *123*, 1603–1618. [CrossRef]
51. Chen, T.; Guestrin, C. XGBoost: A Scalable Tree Boosting System. In Proceedings of the 22nd ACM SIGKDD International Conference on Knowledge Discovery and Data Mining, New York, NY, USA, 13 August 2016; pp. 785–794.
52. Ma, M.; Zhao, G.; He, B.; Li, Q.; Dong, H.; Wang, S.; Wang, Z. XGBoost-based method for flash flood risk assessment. *J. Hydrol.* **2021**, *598*, 126382. [CrossRef]
53. Cortes, C.; Vapnik, V. Support-vector networks. *Mach. Learn.* **1995**, *20*, 273–297. [CrossRef]
54. Raghavendra, N.S.; Deka, P.C. Support vector machine applications in the field of hydrology: A review. *Appl. Soft Comput.* **2014**, *19*, 372–386. [CrossRef]
55. Ferreira, L.B.; da Cunha, F.F.; de Oliveira, R.A.; Filho, E.I.F. Estimation of reference evapotranspiration in Brazil with limited meteorological data using ANN and SVM—A new approach. *J. Hydrol.* **2019**, *572*, 556–570. [CrossRef]
56. Agatonovic-Kustrin, S.; Beresford, R. Basic concepts of artificial neural network (ANN) modeling and its application in pharmaceutical research. *J. Pharm. Biomed. Anal.* **2000**, *22*, 717–727. [CrossRef]
57. Ahmed, K.; Shahid, S.; Bin Haroon, S.; Xiao-Jun, W. Multilayer perceptron neural network for downscaling rainfall in arid region: A case study of Baluchistan, Pakistan. *J. Earth Syst. Sci.* **2015**, *124*, 1325–1341. [CrossRef]
58. Rumelhart, D.E.; Hinton, G.E.; Williams, R.J. Learning representations by back-propagating errors. *Nature* **1986**, *323*, 533–536. [CrossRef]
59. Frank, M.; Wolfe, P. An algorithm for quadratic programming. *Nav. Res. Logist. Q.* **1956**, *3*, 95–110. [CrossRef]
60. Markatou, M.; Tian, H.; Biswas, S.; Hripcsak, G.M. Analysis of Variance of Cross-Validation Estimators of the Generalization Error. *J. Mach. Learn. Res.* **2005**, *6*, 1127–1168. [CrossRef]
61. Lever, J.; Krzywinski, M.; Altman, N. Model selection and overfitting. *Nat. Methods* **2016**, *13*, 703–704. [CrossRef]
62. Fox, D.G. Judging Air Quality Model Performance: A Summary of the AMS Workshop on Dispersion Model Performance, Woods Hole, Mass., 8–11 September 1980. *Bull. Am. Meteorol. Soc.* **1981**, *62*, 599–609. [CrossRef]

63. McKee, T.B.; Doesken, N.J.; Kleist, J. The Relationship of Drought Frequency and Duration to Time Scales. In Proceedings of the 8th Conference on Applied Climatology, Anaheim, CA, USA, 17–22 January 1993; p. 6.
64. Angelidis, P.B.; Maris, F.; Kotsovinos, N.; Hrissanthou, V. Computation of Drought Index SPI with Alternative Distribution Functions. *Water Resour. Manag.* **2012**, *26*, 2453–2473. [CrossRef]
65. Willmott, C.; Matsuura, K. Advantages of the mean absolute error (MAE) over the root mean square error (RMSE) in assessing average model performance. *Clim. Res.* **2005**, *30*, 79–82. [CrossRef]
66. Yang, J.; Wang, B.; Bao, Q. Biweekly and 21–30-Day Variations of the Subtropical Summer Monsoon Rainfall over the Lower Reach of the Yangtze River Basin. *J. Clim.* **2010**, *23*, 1146–1159. [CrossRef]
67. Solomatine, D.P.; Ostfeld, A. Data-driven modelling: Some past experiences and new approaches. *J. Hydroinform.* **2008**, *10*, 3–22. [CrossRef]
68. Patel, D.; Canaday, D.; Girvan, M.; Pomerance, A.; Ott, E. Using machine learning to predict statistical properties of non-stationary dynamical processes: System climate, regime transitions, and the effect of stochasticity. *Chaos Interdiscip. J. Nonlinear Sci.* **2021**, *31*, 033149. [CrossRef]

*Article*

# Towards an Extension of the Model Conditional Processor: Predictive Uncertainty Quantification of Monthly Streamflow via Gaussian Mixture Models and Clusters

Jonathan Romero-Cuellar [1,2,*], Cristhian J. Gastulo-Tapia [1], Mario R. Hernández-López [1], Cristina Prieto Sierra [3] and Félix Francés [1]

1    Research Institute of Water and Environmental Engineering (IIAMA), Universitat Politècnica de València, Camino de Vera s/n, 46022 Valencia, Spain; cgastulo@unibagua.edu.pe (C.J.G.-T.); maherlo@alumni.upv.es (M.R.H.-L.); ffrances@upv.es (F.F.)
2    Agricultural Engineering Department, Universidad Surcolombiana, Avenida Pastrana—Cra 1, Huila 410001, Colombia
3    IHCantabria—Instituto de Hidráulica Ambiental, Universidad de Cantabria, 39011 Santander, Spain; prietoc@unican.es
*    Correspondence: jonathan.romero@usco.edu.co

**Citation:** Romero-Cuellar, J.; Gastulo-Tapia, C.J.; Hernández-López, M.R.; Prieto Sierra, C.; Francés, F. Towards an Extension of the Model Conditional Processor: Predictive Uncertainty Quantification of Monthly Streamflow via Gaussian Mixture Models and Clusters. *Water* **2022**, *14*, 1261. https://doi.org/10.3390/w14081261

Academic Editors: Binquan Li, Dong Wang, Dedi Liu, Yuanfang Chen and Ashish Sharma

Received: 28 February 2022
Accepted: 5 April 2022
Published: 13 April 2022

**Publisher's Note:** MDPI stays neutral with regard to jurisdictional claims in published maps and institutional affiliations.

**Abstract:** This research develops an extension of the Model Conditional Processor (MCP), which merges clusters with Gaussian mixture models to offer an alternative solution to manage heteroscedastic errors. The new method is called the Gaussian mixture clustering post-processor (GMCP). The results of the proposed post-processor were compared to the traditional MCP and MCP using a truncated Normal distribution (MCPt) by applying multiple deterministic and probabilistic verification indices. This research also assesses the GMCP's capacity to estimate the predictive uncertainty of the monthly streamflow under different climate conditions in the "Second Workshop on Model Parameter Estimation Experiment" (MOPEX) catchments distributed in the SE part of the USA. The results indicate that all three post-processors showed promising results. However, the GMCP post-processor has shown significant potential in generating more reliable, sharp, and accurate monthly streamflow predictions than the MCP and MCPt methods, especially in dry catchments. Moreover, the MCP and MCPt provided similar performances for monthly streamflow and better performances in wet catchments than in dry catchments. The GMCP constitutes a promising solution to handle heteroscedastic errors in monthly streamflow, therefore moving towards a more realistic monthly hydrological prediction to support effective decision-making in planning and managing water resources.

**Keywords:** uncertainty analysis; water resources; cluster analysis; Gaussian mixture model; probabilistic prediction

## 1. Introduction

Hydrological predictions are beneficial for water management and planning, such as arranging hydraulic infrastructure (irrigation and draining systems, aqueducts, reservoirs, among others), managing flood and drought risks, and estimating ecological flows, operations, and monitoring existing systems—among others [1]. In addition, estimating the predictive uncertainty of monthly streamflow plays a crucial role in supporting decision-making for water resources management, such as water supply, hydropower, and water balance [2]. Moreover, decision-making in the context of water resources is a complex practice due to the investments, the large scale, and the meaning of projects [3]. Furthermore, such hydrologic predictions are affected by various sources of uncertainty, mainly in observed data [4], the model's parameters [5,6], the model's structure [7,8], the initial conditions [9], the model's numerical solution [10], and the intrinsic non-deterministic performance of the systems [11]. Accordingly, Predictive Uncertainty Quantification (PUQ) is a fundamental tool for risk management and for supporting decision-making in an informed manner when administering water resources [12].

Predictive uncertainty is the probability of observations conditioned by all information and available knowledge (predictions) occurring until today [13]. Therefore, predictive uncertainty is conditioned on the model's structure, parameters, and input data [14]. Hence, PUQ is crucial for making reliable, sharp, and accurate hydrological predictions. It also characterises all the possible predictions and their respective occurrence probabilities [15]. This way of characterising uncertainty does not simplify the decision-making process but provides valuable information about what is not known in the system [16,17]. According to Prieto et al. [18], making predictions without quantifying uncertainty is not knowing reality. Hydrological post-processing methods are suitable for estimating the predictive uncertainty of deterministic hydrological predictions (point predictions).

Formally speaking, a hydrological post-processor is a statistical model employed to improve deterministic predictions by relating the hydrological model's outputs with the corresponding observations [19]. In practice, post-processors are used to characterise the hydrological model's uncertainty and to eliminate the systematic bias of predictions [20]. Post-processors are in charge of mitigating errors in the model's input and output data, parameters, initial conditions, boundary conditions, and structure. Hydrological post-processors have two main objectives: (i) estimating the predictive uncertainty of the hydrological model's deterministic outputs. In this context, hydrological post-processing can be understood as a simple method to convert deterministic predictions into probabilistic ones [21,22]; (ii) correcting the systematic bias of hydrological models to make more accurate predictions.

In recent years, different methods have been developed to estimate the predictive uncertainty of hydrological forecasts. To determine the structure of dependence between the predictions and observations, most methods are based on the meta-Gaussian model, owing to the statistical goodness and facilities that Gaussian variables present [13,14,23–26]. This procedure distributes bivariate probability distributions between deterministic predictions and observations. The errors of hydrological predictions are generally non-Gaussian, heteroscedastic, and autocorrelated [27–30]. To solve this problem, many post-processors apply normalisation methods, such as Normal Quantile Transform (NQT) [31], Box-Cox transformation [32], log-sinh transformation [33], etc.

The first work about predictive uncertainty and hydrological post-processing was conducted by Krzysztofowicz [13] in the context known as the Bayesian Forecasting Framework (BFS). This method developed a bivariate meta-Gaussian distribution function based on a Normal quantile transformation of two variables: observations and predictions according to Gaussian laws. This procedure is known as the Hydrological Uncertainty Processor (HUP) [34]. One of the disadvantages of the HUP is that it does not suitably represent the heteroscedasticity of the error variance. Todini [14] proposed the Model Conditional Processor (MCP), which employs a meta-Gaussian model to estimate the predictive uncertainty of one or a combination of many hydrological models. Coccia and Todini [35] extended the MCP by using Multivariate Truncated Normal distributions to model the joint distribution for many variables in the Normal space to solve the heteroscedastic error problem. Weerts et al. [36] applied quantile regression (QR) to deal with the heteroscedasticity of the hydrological variables' error. QR offers the advantage of analysing the relationship between the observations and predictions from different quantiles, which could be very important for understanding extreme data and managing data with heteroscedasticity [37]. Nonetheless, QR separately estimates one regression for each quantile, generating many parameters.

Similarly, Raftery et al. [38] introduced the Bayesian Model Average (BMA) method that uses many models. Uncertainty is estimated as the average weight of each model's predictive distribution [39]. BMA offers the disadvantage of uncertainty, being conditioned to the number of employed models and their diversification to represent the state variable's uncertainty.

Other hydrological post-processing methods have been implemented, and most employ Bayesian principles. For example, Wang et al. [25] presented the Bayesian Joint Probability (BJP) and Zhao et al. [26] introduced the General Linear Model Post-processor (GLMPP). There are also hydrological post-processors with different error models that had been evaluated under different climate conditions [40,41], such as post-processors that employ non-parametric methods [42], post-processors based on machine-learning principles [43–48], and post-processors based on the copula concept to establish the relation of the dependence among state variables [49–51]. This list of hydrological post-processors is not long, and readers can find more details in the work by Li et al. [52]. Likewise, for reviews of advances in uncertainty analysis, see Moges et al. [53] and Matott et al. [54].

The Model Conditional Processor (MCP) has been established as a hydrologic post-processor for quantifying predictive uncertainty in diverse applications. For instance, precipitation and temperature re-analyses [55], floods in real-time [56], ensemble predictions [57], and satellite rainfall information [58]. Although Coccia and Todini [35] provide insights to deal with the heteroscedastic error using multivariate truncated Normal distributions, the problem is still an open question, especially in monthly streamflow. This paper introduces the Gaussian mixture model and cluster method as a promising alternative to deal with the heteroscedasticity problem, namely that the forecast uncertainty increases with the magnitude of forecast variables.

Nowadays, the use of clusters has become popular in hydrology. For example, Parviz and Rasouli [59] made rain forecasts by artificial intelligence and cluster analysis; Yu et al. [60] implemented the regionalisation of hydroclimate variables with clustering; Basu et al. [61] worked with clusters to analyse floods; and Zhang et al. [62] used clusters and climate similarities to calibrate hydrological models, among others. Likewise, some studies use the Gaussian mixture to represent errors of hydrological variables. For example, Schaefli et al. [63] used a mixture of Normal distributions for quantifying hydrological modelling errors, Smith et al. [64] proposed a mixture of the likelihood for improved Bayesian inference of ephemeral catchments, and Li et al. [65] developed the Error reduction and representation in stages (ERRIS) in hydrological modelling for ensemble streamflow forecasting, which used a sequence of simple error models through four stages. Other authors employ the Gaussian mixture to estimate marginal probability distributions. Thus, Klein et al. [51] proposed a hydrological post-processor based on the bivariate Pair-copula concept and recommended the Gaussian mixture to estimate marginal distributions. Feng et al. [66] introduced a minor modification into the traditional HUP using the Gaussian mixture to estimate marginal distributions. Also, Yang et al. [67] proposed a Bayesian ensemble forecast method, comprising of a Gaussian mixture model (GMM), a hydrological uncertainty processer (HUP), and an autoregressive (AR) model. Finally, Kim et al. [68] used Gaussian mixture clustering to determine groundwater pollution by anthropic effects. It is important to notice that many Gaussian mixture applications were used to estimate marginal distributions.

The importance of estimating uncertainty and support for decision-making in water resources management and planning is stressed. When managing water resources, the monthly temporal discretisation scale is essential for planning the rules for operating in reservoirs, estimating the water balances of catchments, and administrating hydraulic infrastructure in the long term. To deal with these problems, monthly streamflow was employed to evaluate hydrological post-processing.

This paper develops an extension of the MCP [14], which merges clustering with the Gaussian mixture model to offer an alternative solution to manage heteroscedastic errors. The new method is called the Gaussian mixture clustering post-processor (GMCP). The results of the proposed post-processor were compared to the MCP [14] and the MCPt [35] by applying multiple deterministic and probabilistic verification indices. This research also assesses the GMCP's capacity to estimate the predictive uncertainty of the monthly streamflow under different climate conditions in the 12 MOPEX catchments [69] distributed in the SE part of the USA.

To achieve the above goals, this paper is structured as follows: the reference hydrological post-processing methods and the basis for the new post-processor are described; next, the origin of the hydrological predictions and the characteristics of the 12 MOPEX Project catchments are presented to prove the new post-processor's predictive performance; this is followed by the Results, Discussion, and Conclusions sections. All the analyses were carried out using the R Statistical software [70].

## 2. Materials and Methods

The GMCP post-processor is a statistical model to transform point predictions obtained by any deterministic model into probabilistic predictions, thus deriving the predictive uncertainty of the predictand. The GMCP computes the probability distribution of the observed data conditioned on the generic deterministic model's output (point predictions), along with its mode (median or mean) value and uncertainty band, which is asymptotically consistent in quantifying total uncertainty. In general, all MCP post-processors assessed are based on the following main assumptions:

1. Uncertainty of weather forecasts has been substantially reduced because past observations are employed as the hydrological model's input.
2. Predictions and observations correlating, and this system performance will continue in the future. Similarly, modelled variables are stationary during the calibration and application period. Non-stationarity can be accounted for using deterministic model non-stationarity [71,72]. Such extension is not considered in the present contribution, but a discussion is provided in Section 4.
3. A single deterministic model with a single parameter set is considered. Section 4 will discuss the possible extension of the GMCP post-processor to multi-model applications.
4. The calibration dataset is long enough to ensure sufficient information to upgrade the deterministic and post-processor models. The predictive capacity of the models is limited by proper calibration, which implies that sufficiently long records of observed data, guiding to a variety of hydrologic conditions, are available for model training.

As previously mentioned, this research aims to develop an extension of the MCP [14], which merges clusters with a Gaussian mixture model to offer an alternative solution to manage heteroscedastic errors. The method is identified with the acronym "GMCP" post-processor. This research also assesses the GMCP's capacity to estimate the predictive uncertainty of the monthly streamflow under different climate conditions in 12 catchments in the MOPEX Project [69]. The results of the proposed post-processor were compared to the MCP [14] and the MCPt [35].

### 2.1. Predictive Uncertainty

In hindcasting, predictive uncertainty describes the probability of any value conditioned to all the information and knowledge acquired by hydrological predictions [13,14,16]. Krzysztofowicz [13] and Todini [14] emphasise two basic theses. Firstly, the objective of hydrological predictions is to quantify the uncertainty of observations rather than the uncertainty of hydrological models. Secondly, the main aim to improve hydrological predictions is to estimate the actual streamflow and to reduce their predictive uncertainty. To better explain these ideas, and to keep in line with Todini [14], a joint probability distribution concept of observations $q_o$ and predictions $q_s$ is presented. Figure 1 shows the joint sample's frequency of $q_o$ and $q_s$ that can be used to quantify the joint probability density function. For any given hydrological model, predictions $q_s$ should be a function of the model parameters ($\theta$) and the input data ($x$) (precipitation, evapotranspiration, and others.) Therefore, joint density probability can be expressed as $f\left(q_o, \left(q_s | x, \hat{\theta}\right)\right)$. To predict $q_o$, the conditional predictive distribution must be derived from $q_o$ given $q_s$. This can be accomplished by conditioning the joint probability density to the $q_s$ predicted value (Figure 2) and renormalizing. This can be formally expressed as:

$$f(q_o|(q_s|x,\hat{\theta})) = \frac{f(q_{o'}(q_s|x,\hat{\theta}))}{\int f(q_{o'}(q_s|x,\hat{\theta}))\,dq_o} \tag{1}$$

It is stressed that the conditional predictive uncertainty of Equation (1) represents the predictive uncertainty given a hydrological model, input data, certain conditions, and some hydrological parameters. Accordingly, and for this paper, the term "predictive uncertainty" refers to "conditional predictive uncertainty". As Figure 1 shows, the conditional distribution $f(q_o|q_s)$ is not as dispersed as the marginal distribution is for $f(q_o)$ because uncertainty could be reduced by any further information provided by the hydrological model's predictions.

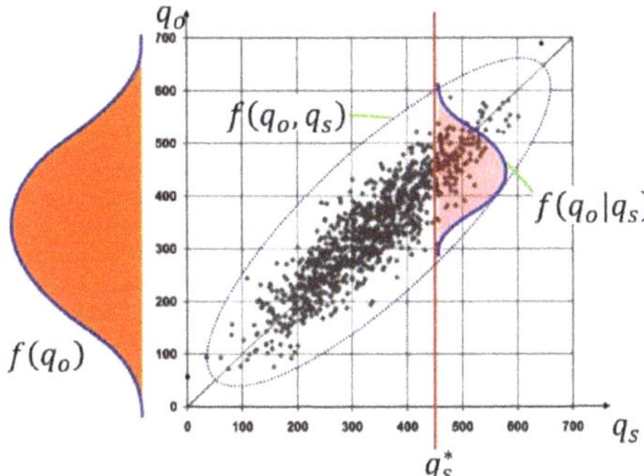


**Figure 1.** Predictive density is defined as the probability density of the observed variable $q_o$ that is conditional on the hydrological model's predictions, $q_s$, where $q_s$ is considered to be known in the prediction time (adapted from Todini [14]).

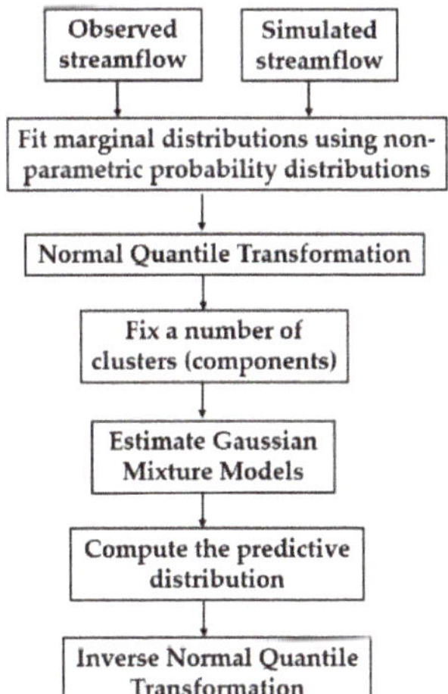


**Figure 2.** The flow chart of the proposed Gaussian Mixture Clustering Post-processor (GMCP).

### 2.2. Marginal Distribution and Normal Quantile Transformation

In general, the problem with Gaussian approaches in hydrology is that variables do not tend to be distributed as Gaussian. Therefore, some kind of statistical transformation must be applied to take the hydrological variables to the Gaussian space and to thus adjust the joint Gaussian probability density distribution (PDF) of both the observations and predictions. The present research applied Normal Quantile Transform (NQT) [31] to all the evaluated post-processing methods. Two auxiliary variables, $\eta_o$ and $\eta_s$, derive from NQT to replace $F(q_o)$ and $F(q_s)$ so that the probability distribution of the observations and predictions in the Gaussian space would, respectively, be:

$$\begin{aligned} \eta_o &= N^{-1}(F(q_o)), \\ \eta_s &= N^{-1}(F(q_s)), \end{aligned} \tag{2}$$

where $N$ represents the standardGaussian distribution with zero mean and unit variance, and $F()$ symbolises marginal distributions. The present research used non-parametric probability distributions to adjust marginal distributions because monthly streamflow is heterogeneous, and the data represent different hydrological situations that might not be easy to describe with the parametric distribution. The kernel density estimation method was applied to adjust the marginal distributions of the random variables [73]. For a random bivariate sample, $X_1, X_2, \ldots, X_n$, are obtained from a joint PDF, $f$, and the kernel density estimation is defined as:

$$\hat{f}(x; H) = n^{-1} \sum_{i=1}^{n} K_H(x - X_i), \tag{3}$$

where $x = (x_1, x_2)^{\mathrm{T}}$ and $X_i = (X_{i1}, X_{i2})^T, i = 1, 2, \ldots, n$. Here, $K(x)$ is the kernel, which is the asymmetric probability density function, $H$ is the symmetric and positive bandwidth matrix, and $K_H(x) = |H|^{-0.5} K(H^{-0.5}x)$. Selecting $K$ is not fundamental: the standard Gaussian distribution $K(x) = (2\pi)^{-1} e^{(-0.5 x^T x)}$ was used. Conversely, selecting $H$ is very important for $\hat{f}$ performance [73]. The most widely used parametrization for the bandwidth matrix is the diagonal $H = diag(h_1^2, h_2^2, \ldots, h_n^2)$ with no constraints in $H$, but it ensures that $H$ is positive and symmetric. For the present research, the kernel estimation was applied using the last square cross-validation method implemented in the ks library [74] of the R statistical software [70].

### 2.3. Hydrological Post-Processing Methods

The streamflow post-processing methods assessed in this research consist of implementing the Model Conditional Processor (MCP) [14] and some of its ramifications from the MCP using a truncated Normal distribution (MCPt) [35] to finish with the proposed extension of the MCP [14], which merges clustering with a Gaussian mixture model to offer an alternative solution to manage heteroscedastic errors. The new method is called the Gaussian mixture clustering post-processor (GMCP). Only a short overview of the theory of the methods is given here. For future details, we refer to cited publications.

#### 2.3.1. Model Conditional Processor (MCP)

Todini [14] proposed the Model Conditional Processor (MCP), a meta-Gaussian approach initially designed to estimate the predictive uncertainty of floods in real-time. The MCP can be used in several ways: bivariate (observed, simulated), multivariate (several prediction models), unique forecast horizon [35], and multiple lead-time [56]. The MCP is a well-accepted hydrological post-processing method by the hydrological community [55–58]. The MCP mainly establishes a joint probability distribution to describe the relationship between the deterministic hydrological predictions and the corresponding observations. The joint probability distribution is modelled as a bivariate Gaussian distribution, followed by adjusting the marginal distributions and transforming the Gaussian space variables. The MCP, herein employed, includes three steps. The first is the transformation of the predictions and observations to the Gaussian space by the NQT transformation method [31], as shown in Section 2.2. The second step is predictive distribution, which was calculated using Bayes' Theorem by assuming that both the predictions and observations are available simultaneously. In line with the notation of the present paper, observations $q_o$ were transformed to Gaussian space $\eta_o$ and predictions $q_s$ were transformed into $\eta_s$. Therefore, the relation between $\eta_o$ and $\eta_s$ was formulated using a bivariate Gaussian distribution:

$$\begin{bmatrix} \eta_o \\ \eta_s \end{bmatrix} \sim N(\boldsymbol{\mu}, \boldsymbol{\Sigma}), \tag{4}$$

where $\boldsymbol{\mu} = \begin{bmatrix} \mu_{\eta_o} \\ \mu_{\eta_s} \end{bmatrix}$ is the means vector and $\boldsymbol{\Sigma} = \begin{bmatrix} \sigma_{\eta_o}^2 & \rho_{\eta_o\eta_s}\sigma_{\eta_o}\sigma_{\eta_s} \\ \rho_{\eta_o\eta_s}\sigma_{\eta_o}\sigma_{\eta_s} & \sigma_{\eta_s}^2 \end{bmatrix}$ is the covariance matrix. In step three, the predictive uncertainty estimated in Gaussian space was reconverted into real space by the inverse of NQT. The series of observations was divided into two parts to identify the MCP parameters. The first half of the series was used for calibration purposes; that is, to identify marginal distributions and joint distribution thorough Bayes' Theorem. The second half of the series was employed to validate the MCP; the calibrated MCP was conditioned to new predictions to evaluate its performance for a group of parameters $\theta$ and the new predictions were transformed into Gaussian space $\eta_{s\_new}$:

$$\eta_{o\_new}|\eta_{s_{new}},\,\theta \sim N\left[\mu_{\eta_o} + \rho_{\eta_o\eta_s}\frac{\sigma_{\eta_o}}{\sigma_{\eta_s}}\left(\eta_{s_{new}} - \mu_{\eta_s}\right),\ \sigma_{\eta_o}^2\left(1 - \rho_{\eta_o\eta_s}^2\right)\right], \tag{5}$$

Interestingly, the MCP is simple to implement with a low computational cost because the bivariate Gaussian distribution is analytically processed. Likewise, the parameters are analytically identified, saving the total parametric inference cost. For further details about the MCP, we recommend that readers look at the work of Todini [14]. Next, an improved version of the traditional MCP is presented.

2.3.2. MCP Using Truncated Normal Distribution (MCPt)

To address the heteroscedasticity in the error variance, Coccia and Todini [35] extended the MCP [14] by joining two truncated Normal distributions (TND). The general recommendation is to use two TNDs to characterise the heteroscedasticity of the error variance properly. In line with our monthly streamflow research objective, two TNDs were used; that is, two variances were employed. The split of the Normal multivariate space into two parts is obtained by identifying an M-dimensional hyperplane:

$$H_p = \sum_{i=1}^{M} \eta_{s_i} = M \cdot a, \tag{6}$$

where $M$ is the number of models and $\eta_{s_i}$ is the prediction in Gaussian space. The threshold $a$ can be distinguished as the value of $\eta_{s_i}$ that minimizes the predictive variance of the upper sample. In other words, the value of $a$ is identified by minimizing the predictive variance of the upper sample. The predictive uncertainty for the sample above the truncation hyperplane is represented as:

$$f\left(\eta_o\middle|\eta_{s_i} = \eta_{s_i}^*,\, H_p^* > M*\right) \sim N\left(\mu_{\eta_o|\eta_{s_i}=\eta_{s_i}^*,H_p^*>M*a},\, \sigma_{\eta_o|\eta_{s_i}=\eta_{s_i}^*,H_p^*>M*a}^2\right) \tag{7}$$

Here, $\eta_{s_i}^*,\, H_p^*$ symbolize a new realization of predictions and a new hyperplane, respectively. Moreover, the predictive mean is represented by:

$$\mu_{\eta_o|\eta_{s_i}=\eta_{s_i}^*,H_p^*>M\cdot a} = \mu_{\eta_o} + \sum_{\eta_o\eta_s}\sum_{\eta_s\eta_s}^{-1}\left(\eta_{s_i}^* - \mu_{\eta_s}\right), \tag{8}$$

and the predictive variance:

$$\sigma_{\eta_o|\eta_{s_i}=\eta_{s_i}^*,H_p^*>M\cdot a}^2 = \sum_{\eta_o\eta_o} - \sum_{\eta_o\eta_s}\sum_{\eta_s\eta_s}^{1}\sum_{\eta_o\eta_s}^{T}, \tag{9}$$

In Equations (7)–(9), $\mu_{\eta_o}$ and $\mu_{\eta_s}$ are the sample means, while $\sigma^2_{\eta_o|\eta_{s_i}}$ is the conditional variance and $T$ is the transpose of a matrix. In addition, the model parameters, i.e., the mean, variance, and covariance matrices, are computed from the data of the upper sample. The predictive uncertainty distribution of the lower sample looks similar but is characterised by the values of the sample below the truncation hyperplane $H^*_p \leq M \cdot a$, for more details see [35]. The MCP and MCPt can work with a multi-model, but, for this research, we used only a single model for which we aimed to quantify the total predictive uncertainty. We did not include multiple models because of the ease of understanding and transparency of the procedure. This assumption is further discussed in Section 4.

### 2.3.3. Gaussian Mixture Clustering Post-Processor (GMCP)

The extension of the MCP method known as GMCP post-processor came about after merging the bivariate Gaussian outline and grouping it into clusters with the Gaussian mixture models (GMM). This means that the GMCP post-processor begins with MCP [14] for a single hydrological model in Section 2.3.1, but the GMCP post-processor offers a different way to deal with the heteroscedasticity of the error variance when the error variance is characterised by clustering with GMMs. The Gaussian mixture is well established in the literature to find homogeneous groups (clusters) in heterogeneous data. The idea of employing GMM to perform a cluster analysis is not new. Wolfe [75] was the first to test GMMs to find clusters. The GMMs offer the advantage of including a probability measure when assigning cluster data. This assignment is known as a soft cluster, where data have a probability of belonging to each cluster [76].

The basic idea behind mixture models of probability distributions to perform cluster analyses consists of assuming that data come from a mixture of underlying probability distributions. The most well-known approach is the Gaussian mixture model (GMM) [77], in which each observation is assumed to be distributed into $g$ Normal distributions and $g$ is the number of clusters (components). For more details, readers refer to the work of Fraley and Raftery [78]. Generally, when GMMs are employed to perform cluster analyses, the same model type is employed ($f_g(x|\theta_g)$) for all the components (clusters), which, in this case, is Gaussian, but with different means and covariance structures.

There are different automatic methods to select the number of mixture components and their parameters [79]. However, the number of mixture components can also be fixed by some prior knowledge about the modelled phenomenon. This research assumes that the joint probability distribution of the observed and simulated data (model error) can be grouped into three categories of variance, and thus choose a three-components Gaussian mixture model. We fixed the number of components a priori to three, thereby corresponding to the high, middle, and low flow period, which is typical of monthly streamflow. Using more than three components is possible, but we will show that three components are sufficient for monthly streamflow and water resources applications in our case studies.

The GMCP provides a semi-parametric outline to model unknown probability distributions, which are represented as the weighted Gaussian sum [80]. Specifically, GMMs possess the flexibility of non-parametric methods with the added advantage of a lower number of parameters, i.e., the dimension of the parameter's vector [81]. To express the mathematical basis of the cluster with GMMs, let us take $X$ as a random vector that stems from the $G$ in the Gaussian mixture distribution. For all $x \in X$, its probability density can be expressed as:

$$f(x|\vartheta) = \sum_{g=1}^{G} \pi_g \, f_g(x|\theta_g) = \sum_{g=1}^{G} \pi_g \, N_g(x|\mu_g, \Sigma_g), \tag{10}$$

where weights $\pi_g > 0$ and $\sum_{g=1}^{G} \pi_g = 1$, which are known as the mixture proportion or weighted weights; $f_g(x|\theta_g)$ is the $g$th component of probability density; $\theta = \theta_1, \ldots, \theta_G$ and $\pi = \pi_1, \ldots, \pi_G$ are the parameters; $N_g$ represents the Normal distribution; $\mu_g$ is the means vector; and $\Sigma_g$ is the covariance matrix for each component (cluster) $g$. This research employed two random variables (observed streamflow ($\eta_o$) and simulated streamflow ($\eta_s$) after normalisation). Then, each data pair ($\eta_o, \eta_s$) was modelled as if sampled from one of the $g$ probability distributions $\left(N_1(\mu_1, \Sigma_1),\ N_2(\mu_2, \Sigma_2),\ \ldots,\ N_g(\mu_g, \Sigma_g)\right)$. For example, assuming that three clusters were identified, the probability of belonging to a given cluster lowers as data points ($\eta_o, \eta_s$) move away from the cluster centre.

Now, let us assume that $z_i = (z_{i1}, \ldots, z_{iG})$ represents the membership of the component of observation $i$. Thus, $z_{ig} = 1$ if observation $i$ belongs to component $g$, and $z_{ig} = 0$ otherwise. Let us also assume that the $n$ vectors of data $x_1, \ldots, x_n$ are observations with no assigned component $g$. In this scenario, the likelihood function is:

$$\mathcal{L}(\vartheta|x) = \prod_{i=1}^{n} \sum_{g=1}^{G} \pi_g\, N(x_i|\theta_g), \tag{11}$$

where $N$ represents the Normal probability distribution. The parameters were estimated with the Expectation-Maximization (EM) algorithm [82]. This algorithm is an iterative procedure followed to estimate the maximum likelihood function. Having estimated the parameters, the predictive classification results are supplied by the *a posteriori* probability distribution:

$$\hat{z_{ig}} = \frac{\hat{\pi}_g\, f(x_i|\hat{\theta}_g)}{\sum_{h=1}^{G} \hat{\pi}_h\, f(x_i|\hat{\theta}_h)}, \tag{12}$$

for $i = 1, \cdots, n$. The complete cluster grouping analysis was implemented with GMMs using the *mclust* library [83] of the R statistics software [70]. Figure 2 displays the flow chart of the procedure for applying the GMCP post-processor.

*2.4. Case Studies*

The data, herein employed, were the observed and simulated monthly streamflow obtained from the "Second Workshop on Model Parameter Estimation Experiment (MOPEX)" [69]. The MOPEX project is a well-known reference database in the international hydrological community that has mainly been used to evaluate hydrological models and theories [8,10,84]. For example, and particular to this paper, Ye et al. [19] used the 12 catchments from the MOPEX database to compare the results from post-processing and calibrated the hydrological models. Thus, MOPEX offers a valuable opportunity to evaluate and compare the performance of new hydrological post-processing methods under different climate conditions. From the MOPEX database, 12 catchments were selected, which are distributed in the SE area of the USA. The Aridity Index (relation between potential evapotranspiration and precipitation) ranges from 0.43 to 2.22, and the Runoff Ratio (relation between surface run-off and precipitation) varies between 0.15 and 0.63 (Table 1). Thus, the 12 catchments selected from the MOPEX project represent different climate conditions (Figure 1). Basic information about them is supplied in Table 1. We selected the same 12 MOPEX catchments used by Ye et al. [19] to discuss the results.

Figure 3 depicts the Budyko curve for all 12 catchments from the MOPEX project. According to the Budyko hypothesis, if the energy available in a catchment suffices to evaporate humidity, then the catchment is limited by water availability (catchment B12 has the highest Aridity Index). Conversely, if the available energy does not suffice to evaporate humidity, the basin is limited by energy availability (catchment B2 is the exact opposite of B12 as it has the lowest Aridity Index). It is worth stressing that the 12 selected catchments were distributed all over the Budyko curve, as Figure 3 depicts, which ensures the critical evaluation of post-processors under different climate conditions.

**Table 1.** Hydrological information about the 12 catchments selected from the Mopex project.

| ID | Station Name | Elev. | Area (km²) | P | PET | Q | Run-Off Index (Q/P) | Aridity Index (PET/P) |
|---|---|---|---|---|---|---|---|---|
| B1 | Amite River Near Denham Springs, LA | 0 | 3315 | 1560 | 1068.5 | 612 | 0.39 | 0.67 |
| B2 | French Broad River at Asheville, NC | 594 | 2448 | 1378 | 588.9 | 795 | 0.58 | 0.43 |
| B3 | Tygart Valley River at Philippi, WV | 390 | 2372 | 1164 | 661.4 | 736 | 0.63 | 0.57 |
| B4 | Spring River Near Waco, MO | 254 | 3015 | 1075 | 1119.8 | 300 | 0.28 | 1.04 |
| B5 | S Branch Potomac River Nr Springfield, WV | 171 | 3810 | 1043 | 636 | 339 | 0.33 | 0.61 |
| B6 | Monocacy R At Jug Bridge Nr Frederick, MD | 71 | 2116 | 1042 | 906.1 | 421 | 0.4 | 0.87 |
| B7 | Rappahannock River Nr Fredericksburg, VA | 17 | 4134 | 1028 | 856.7 | 375 | 0.36 | 0.83 |
| B8 | Bluestone River Nr Pipestem, WV | 465 | 1020 | 1017 | 678 | 419 | 0.41 | 0.67 |
| B9 | East Fork White River at Columbus, IN | 184 | 4421 | 1014 | 838 | 377 | 0.37 | 0.83 |
| B10 | English River at Kalona, IA | 193 | 1484 | 881 | 989.9 | 261 | 0.3 | 1.12 |
| B11 | San Marcos River at Luling, TX | 98 | 2170 | 819 | 1462.5 | 170 | 0.21 | 1.79 |
| B12 | Guadalupe River Nr Spring Branch, TX | 289 | 3406 | 761 | 1691.1 | 116 | 0.15 | 2.22 |

Elev: elevation (m), P: mean areal precipitation (mm/year), PET: potential evapotranspiration (mm/year), Q: observed streamflow (mm/year).

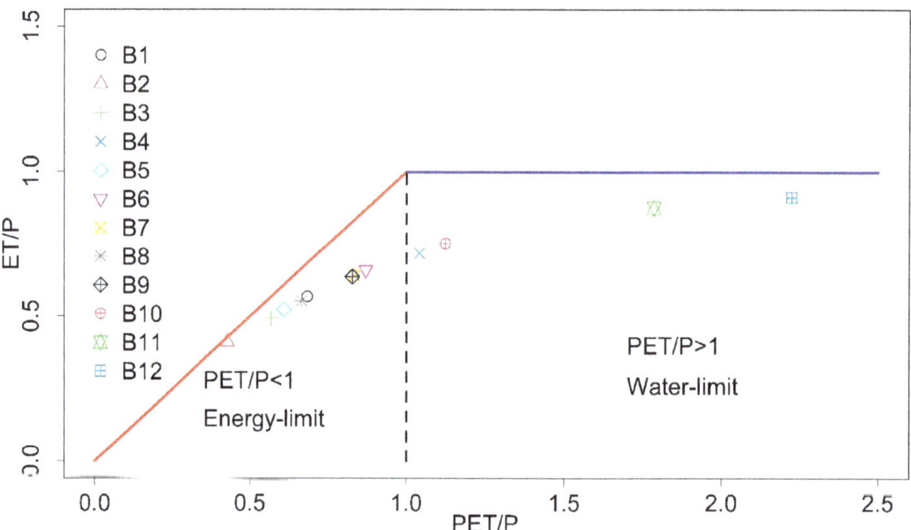


**Figure 3.** The Budyko curve for the 12 catchments selected from MOPEX. The values to reproduce the figure came directly from the MOPEX database.

## 2.5. Hydrological Model

The GR4J hydrological model predictions were employed [85], which are a well-known and widely used model in different parts of the world. GR4J is a lumped conceptual model with four calibration parameters: maximum capacity of the production store $x_1$(mm); groundwater exchange coefficient $x_2$(mm); 1-day-ahead maximum capacity of the routing store $x_3$(mm); and time base of unit hydrograph $x_4$(days). For further information about the model's description, readers refer to the work by Perrin et al. [85]. Daily predictions were aggregated on a monthly basis to evaluate the post-processors' performance for planning and managing water resources. We want to emphasise that the GR4J hydrological model predictions were not prioritised because they are input data. According to the aim of this paper, we focused on the performance of the post-processors.

*2.6. Verification Indices*

Assessment of the performance is vital to offer end-users an indication of the predictions´ reliability and uncertainty bands. Some verification indices exist that can be used to assess the performance of hydrological post-processors. This research employed deterministic and probabilistic verification indices, which evaluate the hydrological predictions´ accuracy, sharpness, and reliability. These indices were also recommended by Laio and Tamea [86], Renard et al. [6], and Thyer et al. [87]. The deterministic Nash–Sutcliffe efficiency index (NSE) [88] was applied to the predictive distribution mean. This index does not assess the complete predictive distribution but is a classic index and a general reference in hydrology. The NSE measures the squared differences between predictions $q_s$ and observations $q_o$, which are normalised by the variance of the observations:

$$\text{NSE} = 1 - \frac{\sum_{i=1}^{n}(q_s - q_o)^2}{\sum_{i=1}^{n}(q_o - \overline{q_o})^2}, \tag{13}$$

where $\overline{q_o}$ is the average of the observations. Probabilistic indices were employed to assess the predictive distributions. The predictive quantile–quantile (PQQ) plot [86] was applied. This diagram shows how probabilistic predictions represent the observations´ uncertainty [86,87]. If both predictive distribution and observations are consistent in the PQQ context, the value corresponding to the distribution $p$-value must be uniformly distributed throughout the interval $[0, 1]$. In other words, predictions are considered reliable when the relative frequency of the observations equals the frequency of predictions. This situation can be visually identified when the PQQ curve follows the bisector (line 1:1). Otherwise, predictive distribution deficiencies can be interpreted when the curve moves away from the bisector. Indeed, according to Laio and Tamea [86], the predictive distribution can display three patterns. If the PQQ plot follows the bisector, the predictive uncertainty is correctly estimated, and the observations are a random sample of the predictive distribution. Conversely, if the PQQ plot shows an "S"-shape, it means that the predictive distribution is underestimated (large bands) and an inverted "S"-shape implies an overestimated uncertainty (narrow bands). From the PQQ plot, we can deduce two indices: reliability and sharpness.

The reliability index quantifies the statistical consistency between the observations and predictive distribution:

$$\text{Reliability} = 1 - \frac{2}{n} \sum_{i=1}^{n} |F_U - F_{q_s}(q_o)|, \tag{14}$$

where $F_U$ is a uniform cumulative distribution function (CDF) and $F_{q_s}(q_o)$ is the predictive CDF. The reliability index ranges from 0 (the worst reliability) to 1 (perfect reliability).

The sharpness index is related to the predictive distribution concentration. In other words, it refers to the coverage provided by the distribution [6]:

$$\text{sharpness} = \frac{1}{n} \sum_{i=1}^{n} \frac{\text{E}[q_s]}{\sigma[q_s]}, \tag{15}$$

where $\text{E}[]$ and $\sigma[]$ are the operators of the expected value and standard deviation, respectively. The sharpness index range is $(0, \infty)$, and the predictive distributions with higher sharpness index (narrower) values are more accurate. Predictive distributions can be found with equal reliability indices but different degrees of sharpness, in which case the higher sharpness values are preferable because they denote more accurate predictive distributions.

Furthermore, the containing ratio (95%CR) was used. The 95%CR is the percentage of observations that fall within the 95% uncertainty band. In this research, the 95% band was estimated to be within the 2.5 and 97.5 percentiles. This allowed the quantification of the desired uncertainty to be achieved when the 95%CR came close to 95%. As the presented verification indices are well-known in the literature, no lengthy description is provided. However, readers are recommended to the works of Franz and Houge [84], Laio and Tamea [86], and Renard et al. [6] for further information.

### 2.7. Comparison Frame

It should be remembered that the main aim of the present paper is to develop an extension of the MCP [14], which merges clustering with a Gaussian mixture model to offer an alternative solution to manage heteroscedastic errors. In addition, comparing GMCP's performance to similar post-processing methods under different climate conditions is also needed as a benchmark. In order to perform the post-processing of monthly streamflow and to quantify predictive uncertainty, the following procedure was used.

First, daily streamflow predictions were obtained from the GR4J hydrological model [85], and were calibrated and validated by Ye et al. [19] for the 12 MOPEX catchments. Given this, the hydrological model outputs (previously calibrated and validated) become the inputs for the evaluated hydrological post-processors.

Second, the daily hydrological predictions were aggregated monthly because the post-processing methods were applied in the water resources management context.

Third, to evaluate the post-processing methods, the time series of both observations and predictions were divided into 20 years to calibrate the post-processors' parameters (1960–1980) and into 17 years for the validation (1981–1998).

Fourth, NQT [31] was applied to all the evaluated post-processors with non-parametric marginal distributions to map the observations and simulations to the Normal space. The three evaluated post-processors were separately implemented into the 12 MOPEX catchments to find the best performing post-processors. The 12 MOPEX catchments were selected because they were the same catchments employed in previous studies to compare hypotheses, which the hydrological community is very familiar with, e.g., [8,19,84,89].

Moreover, evaluating hydrological post-processors under different climate conditions allows for more general recommendations to be obtained [90].

Finally, evaluating the predictive uncertainty with only one verification index can lead to mistaken interpretations and wrong decision-making for managing water resources [41]. Consequently, many independent verification indices were used together instead of individual ones.

## 3. Results

The hydrological post-processing methods evaluated according to the framework described in the previous subsection are presented. The results correspond to the validation period, as it is the most critical period where the predictive uncertainty of the analysed methods is identified. Section 3.1 benchmarks the GCMP post-processor with the MCP [14] and MCPt [35] to quantify the predictive uncertainty of the monthly streamflow, which is conditional on deterministic model predictions. The case studies consider 12 MOPEX catchments with a diverse range of hydroclimatology. The Nash–Sutcliffe efficiency index (NSE), sharpness, and the containing ratio (95%CR) verification index are presented in Figure 4. Moreover, the PQQ plots, which assess the reliability, sharpness, and bias, are depicted in Figure 5.

An initial inspection of the results found considerable overlap in the performance verification indices achieved by the MCP and MCPt post-processors for monthly streamflow. The MCP and MCPt also showed poor performance in dry catchments. Conversely, the GMCP post-processor empirically made the most accurate, reliable, and sharpest predictions.

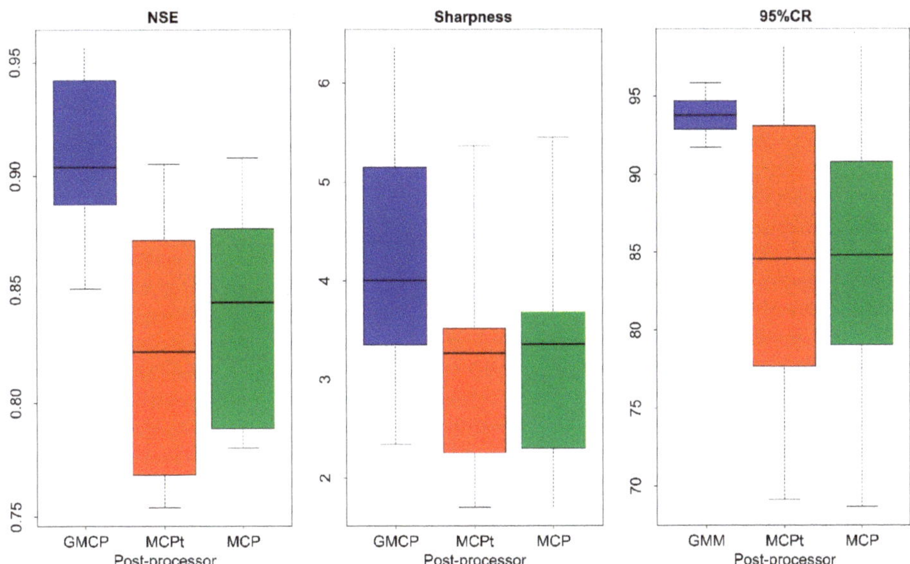

**Figure 4.** Performance of monthly predictions in terms of NSE, sharpness, and containing ratio (95%CR) for the three post-processors during the validation period (1980–1998) overall catchments.

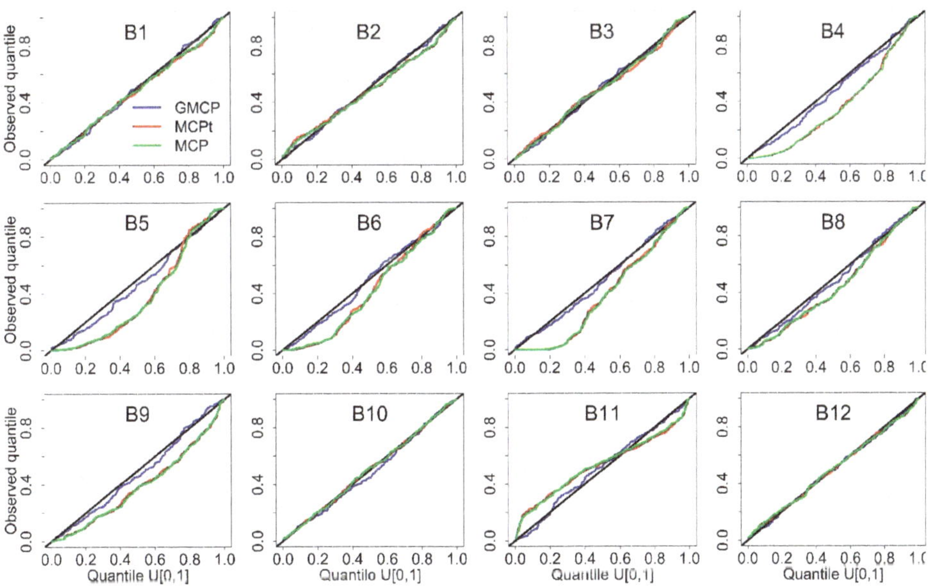

**Figure 5.** Predictive PQQ plot of the three evaluated post-processors and 12 MOPEX catchments during the validation period (1980–1998).

The streamflow forecast time series and corresponding skill for a single catchment, the San Marcos catchment (B11), are presented in Figure 6. Then, the relation of the Aridity index with the performance of post-processors is shown in Figure 7.

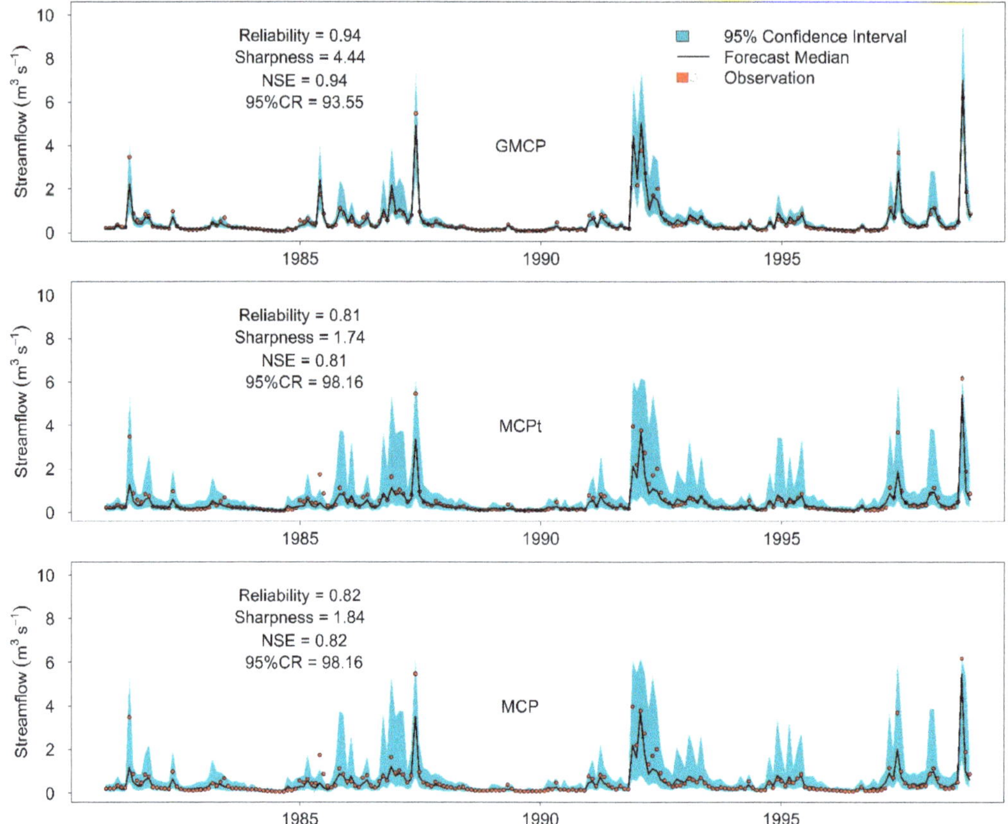



**Figure 6.** Time series of the median and 95% confidence interval of monthly streamflow predictions derived from the Gaussian mixture clustering post-processor (GMCP), model conditional processor (MCP), and MCP using the truncated Normal (MCPt), compared with observations from the validation period (1980–1998) in the San Marcos catchment (B11).

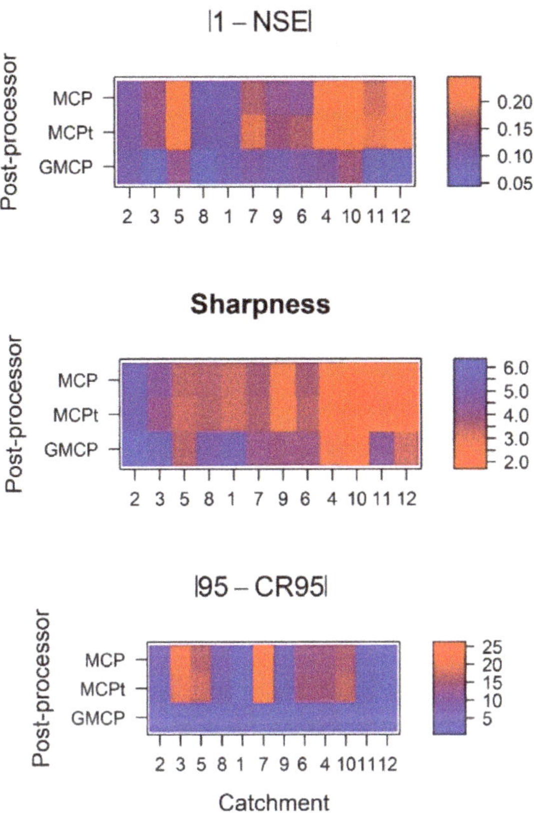

**Figure 7.** Comparison of the deterministic and probabilistic metrics computed for the three post-processors during the validation period (1980–1998) for 12 MOPEX catchments. MCP (model conditional processor), MCPt (MCP using the truncated Normal) and GMCP (Gaussian mixture clustering post-processor).

### 3.1. Comparison of Post-Processors: Individual Verification Indices

Figure 4 offers the average values for the verification indices for the 12 catchments in boxplot-type diagrams. In terms of the Nash–Sutcliffe efficiency index (NSE), Figure 4 shows considerable overlap in the boxplots corresponding to MCP and MCPt. This finding suggests little difference in the performance of these post-processors for monthly streamflow. Moreover, the NSE indices are generally suitable for all 12 catchments and assessed post-processors (Figure 4, left panel), and, according to the classification of Martinez and Gupta [91], NSE > 0.75 is considered a good result. Overall, these results suggest the GMCP is consistently better in terms of the NSE values because of its higher NSE indices and shows less dispersion in the boxplot.

Regarding the sharpness index, and in line with Figure 4 (middle panel), GMCP has the highest sharpness values. The sharpness index refers to the predictive distribution concentration [92]. High sharpness indices indicate that the predictive distribution is less dispersed or more concentrated, and therefore high sharpness indices are preferable [6]. The GMCP post-processor improves the sharpness index by 36.64% compared to the MCP and MCPt post-processors.

In terms of the containing ratio (95%CR), the GMCP post-processor outperforms the MCP and MCPt methods. The GMCP improves the 95%CR by 10.29% compared to the MCP and MCPt, which perform similarly. A proper predictive uncertainty estimation is achieved when the 95%CR comes close to 95%. According to Figure 4 (right panel), the 95%CR obtained by GMCP comes closer to 95%, and with lower variance. The average 95%CR is 93.82% for the GMCP post-processor compared with 85.06% for the MCP or MCPt.

These results show how the boxplots for MCP and MCPt methods overlap. These can indicate that the evaluated reference post-processors that are used with monthly streamflow perform the same in terms of accuracy, sharpness, and reliability. Conversely, the GMCP empirically made the most accurate, reliable, and sharpest predictions for the monthly streamflow of the 12 MOPEX catchments.

Regarding reliability, Figure 5 shows the predictive PQQ plots for the post-processors evaluated through the 12 MOPEX catchments. The PQQ plot indicates the predictive distribution's reliability. According to Figure 5, we stress that the predictive distribution of GMCP (blue line) in most of the evaluated catchments follows the diagonal line in the PQQ, which evidences a reliable predictive uncertainty estimation under different climate conditions. We can also note that the MCP (green curve) and MCPt (red curve) performance is similar for all the evaluated catchments. Furthermore, the PQQ plot for the MCP and MCPt deviates substantially from the 1:1 line in the B4–B9 catchments, indicating some bias. Also, the PQQ plot for the MCP and MCPt in the B11 catchment shows unreliable results, as the predictions are overconfident. In the following subsection, we provide the predictive uncertainty bands of the B11 catchment to explain the poor reliability issue.

### 3.2. Uncertainty Bands in San Marcos Catchment (B11)

The PQQ plots (Figure 5) evidence that the reference post-processors present reliability problems, while GMCP provides reliable results. The 95% confidence interval of predictive distribution is presented to illustrate these reliability difficulties better. We cannot present the 95% confidence interval for all the catchments and post-processors for space reasons. Given this, we only present the predictive distribution for the San Marcos catchment (B11), a dry catchment, because it clearly shows the reliability problems of the evaluated post-processors.

To illustrate these results, Figure 6 shows the time series of the median and 95% confidence interval of the monthly streamflow forecast at the San Marcos catchment (B11). The GMCP post-processor, which merges clustering with the Gaussian mixture model to deal with heteroscedastic errors, achieves the following verification indices: reliability index = 0.94, sharpness index = 4.44, NSE = 0.94, and the containing ratio (95%CR) = 93.55. Meanwhile, the MCP and MCPt, which perform similarly, have a worse reliability index (metric value = 0.82), sharpness index (metric value = 1.84), NSE (metric value = 0.82), and containing ratio (95%CR) (metric value = 98.16).

In terms of sharpness, the MCP and MCPt methods produce a wider 95% predictive range than the GMCP post-processor (Figures 4 and 6), which manifests as degradation in the sharpness index from 4.44 to 1.84. The widest uncertainty bands produced by the MCP and MCPt confirm the results obtained with the sharpness index (Figure 4, middle panel) and reliability (Figure 5).

Altogether, in the 12 MOPEX catchments, these results show that the GMCP post-processor achieves significant improvements in reliability, sharpness, NSE, and the containing ratio (95%CR). In addition, using the GMCP post-processor for monthly streamflow has an incremental impact on performance, as measured using deterministic and probabilistic verification indices. These results show the robust ability of the GMCP post-processor for better quantifying hydrologic uncertainty and producing enhanced probabilistic streamflow forecasts.

### 3.3. Influence of the Aridity Index

Figure 7 shows the comparison between the deterministic and probabilistic verification indices for the new GMCP and the two reference post-processors during the validation period (1980–1998) in the 12 MOPEX catchments. Note that the horizontal axis in Figure 7 sorts the catchments from wettest to driest, whereas the vertical axis denotes the post-processor name.

In NSE index terms, and given the purpose of unifying the Figure 7 legend, $|1 - \text{NSE}|$ is shown, where the values close to 0 are the most optimum ones (blue colors in Figure 7). Figure 7 (upper panel) shows the differences in the performance of post-processors in the catchments. For example, the GMCP shows the best performance (blue colors) in most of the evaluated catchments (Figure 7, upper panel). Moreover, the performance in NSE terms for the reference post-processors was generally similar, while the worst performances were obtained in drier catchments (B4, B10, B11, and B12), except for the B5 catchment, which is humid.

In terms of the sharpness index, Figure 7 (middle panel) shows that the lowest sharpness values, which were for the driest catchments (B4, B10, B11 and B12), and the highest values were for the wettest (B2 and B3). In most catchments, the GMCP achieves higher sharpness values than the reference post-processors (MCP and MCPt) (Figure 7, middle panel).

For the 95%CR index, the statistics $|95 - 95\%\text{CR}|$ were calculated, where values close to 0 are preferable and interpreted as the best performance. Similarly, the other verification indices—and as shown in Figure 7 (lower panel)—the GMCP best performed in all the evaluated catchments. Unlike the other verification indices, the 95%CR did not indicate a worsened performance for post-processors in the driest catchments (B4, B10, B11 and B12).

Overall, the results suggest that streamflow forecasts using the GMCP post-processor are better (i.e., NSE and sharpness) than that of the MCP and MCPt methods, particularly in dry catchments. For example, in dry catchments (B4, B10, B11 and B12), the GMCP processor improves the NSE index by 16.66 % compared to the MCP and MCPt methods.

## 4. Discussion

Predictive uncertainty quantification (PUQ) is essential for supporting effective decision-making and planning for water resources management [93]. In recent years, PUQ has become essential in hydrological predictions [16]. A wide range of methods has been developed to evaluate the predictive uncertainty of the variables of interest. This paper develops an extension of the MCP [14], which merges clusters with Gaussian mixture models to offer an alternative solution to manage heteroscedastic errors. The new method is called the Gaussian mixture clustering post-processor (GMCP). The results of the proposed post-processor were compared to the MCP [14] and the MCPt [35] by applying multiple deterministic and probabilistic verification indices. This research also assesses the GMCP's capacity to estimate the predictive uncertainty of the monthly streamflow under different climate conditions in the 12 MOPEX catchments [70] that are distributed in the SE part of the USA.

Overall, GMCP has shown significant potential in generating more reliable, sharp, and accurate monthly streamflow predictions, especially for dry catchments. Compared to the benchmark methods, GMCP shows more consistency in the validation period than MCP and MCPt (Figure 4). The improvement in the GMCP compared to the MCP and MCPt can be attributed to the procedure used by GMCP to model the dependence structure between observation and forecast (residual error model). GMCP joins the variables via Gaussian mixture models and clusters. Therefore, the Gaussian mixture distribution treats model residuals as three clusters with different means and variances. The Gaussian mixture distribution can capture the peak and the tails of the underlying residual density for all catchments, indicating reliable, sharp, and accurate forecasts. Consequently, this dependence structure of the residual error model faces the assumption of homoscedastic error variance, which provides poor probabilistic predictions. In addition, note that the only difference between the MCP (or MCPt) and GMCP post-processors is the use of clusters and Gaussian mixture models in the GMCP. Hence, the performance improvement must be due to this difference.

Moreover, the MCP and MCPt methods provide similar performances for monthly streamflow predictions regarding the NSE index, reliability, sharpness, and containing ratio (95%CR) (Figures 4–6). The MCPt was designed by Coccia and Todini [35] to improve the reliability and sharpness of predictions, particularly for high flows, and has worked well for flood applications. The MCPt used the truncated Normal distribution (TND) to deal with heteroscedastic errors. In theory, the TDN reduces the standard error of high flows when there is a significant difference between low and high flows, which is the case in flood applications. However, these differences are minor for monthly streamflow, so using the standard Normal distribution or TDN provided similar results.

Figure 5 shows that the PQQ plot for the MCP and MCPt deviates substantially from the 1:1 line in the B4–B9 catchments, indicating some bias. However, the proposed GMCP can obtain unbiased results in the same catchments. One possible explanation is that the MCP uses a linear regression in the Normal transformed space for bias correction, while the GMCP uses three Gaussian mixture models with different means and variances corresponding to the high, middle, and low flow period. Therefore, the bias corrector of GMCP is more robust and flexible than the MCP. As well, Figure 5 depicts that the PQQ plot for the MCP and MCPt in the San Marcos (B11) catchment shows unreliable results, as predictions are overconfident, while GMCP provides reliable results. San Marcos (B11) is a dry catchment with complex residual errors [19]. It is possible that the residual error model of MCP is not enough to represent the complex errors of the San Marcos (B11) catchment. The residual error model of MCP has two assumptions that are undoubtedly inappropriate for the San Marcos (B11) catchment. The MCP assumes homoscedastic errors and a linear relationship between observed and simulated Normal transformed variables. Conversely, the residual error model of GMCP is more complex because of using clustering and Gaussian mixture models.

Our findings in this study confirm the insights of Schaefli et al. [63], namely that using a finite mixture model constitutes a promising solution to residual model errors and to estimate the total modelling uncertainty in hydrological model calibration studies. However, there are two differences between this research and the previous work of Schaefli et al. [63]. First, we used the "post-processing" strategy, where the hydrological model parameters were estimated first using an objective function, followed by a separate estimation of the residual error model parameters. In contrast, Schaefli et al. [63] used the more classical "joint" strategy to estimate all parameters simultaneously using a single likelihood function. Second, we merged the Gaussian mixture model with clusters and used them in the framework of the Model Conditional Processor (MCP) [14]. Likewise, Li et al. [65], who developed the ERRIS post-processor, used a mixture of two Gaussian distributions to represent the residual error model. GMCP and ERRIS have some similarities: (1) both are post-processors of deterministic hydrological models for hydrological uncertainty quantification, (2) both apply a transformation to normalised data, and (3) both use a Gaussian mixture distribution to model residual errors. However, GMCP and ERRIS have some differences. For example, ERRIS uses a linear regression in the transformed space for bias correction, uses an autoregressive model to update hydrological simulation, and is implemented in stages.

We want to discuss some assumptions mentioned in the Materials and Methods in Section 2. First, although GMCP has been conceived to be applied to one single model (point prediction), a multi-model application would be possible. An extension of the GMCP consists of a matrix of predictions and various deterministic models (one column for each model), yet here we study the simpler scalar version of the model. Second, there are possible ways to advance towards the application of GMCP in a non-stationary context. The simplest option is using a deterministic model for non-stationarity [71,72]. We also suggest considering a deterministic model with time-varying (perhaps seasonal) parameters, under the assumption that the uncertainty of the model for non-stationarity is represented by a stationary distribution. In addition, we recommend the use of data assimilation to update hydrological predictions [94]. Third, in the GMCP, we used Gaussian mixture distributions and fixed the number of mixture components (clusters) to three—corresponding to the high, middle, and low flow period, which are typical of monthly streamflow. This practical choice is based on a priori information about the sources and behaviour of the residual error model. Therefore, identifying the number of clusters is purely heuristic accounting for a priori knowledge about the total error model. Fourth, in the GMCP, any probabilistic prediction is primarily based on the conditions monitored during the considered observation period only, and thus particular care should be used when extrapolation to out-of-sample conditions.

This research confirms the importance of using multiple independent verification indices to assess hydrological post-processors. For example, if one considers the containing ratio (95%CR) verification index alone, all post-processors yield comparable performances, and there is no argument for selecting any of them. Nonetheless, once the sharpness index and reliability index are considered explicitly, the GMCP post-processor can be recommended for significantly better sharpness and reliability than the MCP and MCPt. These results align with Woldemeskel et al. [41], who showed that evaluating the predictive uncertainty with a single metric can lead to suboptimum conclusions.

Moreover, examining and evaluating hydrological post-processors in catchments with different climate and hydrological conditions ensures suitable comparisons and helps to generalise the obtained results [47,95]. Furthermore, the diverse climate conditions of catchments analysed allow us to deduce functional relationships between climatic indices and the post-processors' performance. This research attempted to establish a relation between the Aridity index and the post-processors' performance (Figure 7). In most dry catchments, the MCP and MCPt perform relatively worse, especially in terms of the sharpness and NSE index. This result is because streamflow data for dry catchments contain too many days with low flow (defined as flow below 2% of the mean flow [12]). Thus, dry catchments require more complex residual error modelling methods [64]. Our findings agree with Ye et al. [19], who found that the GLMPP post-processor [26] could not improve the predictions or reduce uncertainty in the same dry MOPEX catchments.

In this study, all post-processors provide a clear improvement in hydrological predictions. Post-processing usually leads to better performance verification indices than deterministic hydrological predictions alone because post-processing works directly to correct the errors in the model outputs [19]. Accordingly, Farmer and Vogel [22] stated that the prudent management of environmental resources requires probabilistic predictions, which offer the potential to quantify predictive uncertainty, and can avoid the false sense of security associated with point predictions [16]. Generally speaking, predictive distribution explicitly represents the system's uncertainty, and it can, therefore, perform risk management in a more informed manner. In addition, the probabilistic approach can be put to further use for process-based deterministic hydrological modelling and by coupling it with a hydrological post-processor to convert deterministic predictions into probabilistic predictions. Probabilistic predictions offer an opportunity to improve the operational planning and management of water resources.

A promising improvement is to extend the GMCP post-processor in future work using a multi-model or a chain of hydrological models. Also, it can be interesting to validate the GMCP using daily and hourly data and couple the GMCP method with data assimilation to update the states of hydrological models. Bourgin et al. [94] recommended using data assimilation and post-processing in forecasting because data assimilation strongly impacts forecast accuracy, while post-processing strongly impacts forecast reliability. Besides, evaluating the GMCP post-processor in many catchments is beneficial for establishing its robustness.

Another area for further investigation is overcoming the data transformation. In hydrology, data transformation is a popular approach to reduce the heteroscedasticity of the error model because these approaches are simple to implement and can give satisfactory results in hydrological modelling [14,41,51,65]. However, Schaefli et al. [63], Brown and Seo [42], and others indicated that this approach is questionable. A detailed discussion of the implications of data transformation is beyond the context of this paper. Nevertheless, we recommend reading the work of Hunter et al. [96], which established the detrimental impact of calibrating hydrological parameters in the real space and calibrating the error model parameters in the transformed space using post-processing methods on the quality of probabilistic predictions. Moreover, a future possible study could be to extend the GMCP post-processor using a link function to avoid the Normal quantile transformation, especially the link function, which has provided promising results in the context of the Generalized Linear Model. Finally, another improvement can be the selection of the number of GMCP clusters using unsupervised learning, for example, by using cluster indicators [97].

## 5. Conclusions and Summary

Considering that predictive uncertainty is crucial for providing reliable, sharp, and accurate probabilistic streamflow predictions, the Model Conditional Processor (MCP) [14] is a well-known method for quantifying predictive uncertainty by providing a posterior distribution conditioned on the deterministic model forecast. This study develops an extension of the MCP [14], which merges clustering with the Gaussian mixture model to offer an alternative solution to manage heteroscedastic errors. The new method is called the Gaussian mixture clustering post-processor (GMCP). The results of the proposed post-processor were compared to the MCP [14] and the MCPt [35] by applying multiple deterministic and probabilistic verification indices. This research also assesses the GMCP's capacity to estimate the predictive uncertainty of the monthly streamflow under different climate conditions in the 12 MOPEX catchments [70] distributed in the SE part of the USA. The summary of the most important empirical findings based on the detailed analysis of the results are as follows:

1.  In general, all three post-processors showed promising results. However, the GMCP post-processor has shown significant potential in generating more reliable, sharp, and accurate monthly streamflow predictions than the MCP and MCPt methods, especially in dry catchments.
2.  The MCP and MCPt methods provided similar performances for monthly streamflow predictions regarding the NSE index, reliability, sharpness, and containing ratio (95%CR).
3.  The MCP and MCPt showed a better performance in wet catchments than in dry catchments.

Overall, when used for post-processing monthly predictions, the GMCP method provides an opportunity to improve forecast performance further than is possible using the MCP and MCPt methods, especially in dry catchments. In addition, it is worth mentioning that incorporating clusters and Gaussian mixture models into the Model Conditional Processor framework constitutes a promising solution to handle heteroscedastic errors in monthly streamflow, therefore moving towards a more realistic monthly hydrological prediction to support effective decision-making in planning and managing water resources.

**Author Contributions:** Conceptualization, F.F., M.R.H.-L. and J.R.-C.; methodology and software, M.R.H.-L. and J.R.-C.; formal analysis, F.F. and J.R.-C.; data curation and visualization, C.J.G.-T., M.R.H.-L. and J.R.-C.; writing—original draft preparation, J.R.-C.; writing—review and editing, C.P.S., F.F. and J.R.-C.; supervision and funding acquisition, F.F. All authors have read and agreed to the published version of the manuscript.

**Funding:** This research was funded by the department of Huila Scholarship Program No. 677 (Colombia) and Colciencias, the Vice-Presidents Research and Social Work office of the Universidad Surcolombiana, the Spanish Ministry of Science and Innovation through research project TETIS-CHANGE (ref. RTI2018-093717-B-I00). Cristina Prieto acknowledges the financial support from the Government of Cantabria through the Fénix Program.

**Data Availability Statement:** All data for the US catchments are taken from MOPEX dataset, which are available at http://www.nws.noaa.gov/ohd/mopex/mo_datasets.htm (accessed on 15 May 2018). The code used to generate the results presented in this study can be found on GitHub (https://github.com/jhonrom/GMCP.git) (accessed on 25 January 2022).

**Acknowledgments:** We are grateful to Qingyun Duan for information of the MOPEX experiment. We also are grateful to the editor and two anonymous reviewers for their thoughtful comments on this manuscript.

**Conflicts of Interest:** The authors declare no conflict of interest.

# References


1. Stakhiv, E.; Stewart, B. Needs for Climate Information in Support of Decision-Making in the Water Sector. *Procedia Environ. Sci.* **2010**, *1*, 102–119. [CrossRef]
2. Hamilton, S.H.; Fu, B.; Guillaume, J.H.; Badham, J.; Elsawah, S.; Gober, P.; Hunt, R.J.; Iwanaga, T.; Jakeman, A.J.; Ames, D.P.; et al. A framework for characterising and evaluating the effectiveness of environmental modelling. *Environ. Model. Softw.* **2019**, *118*, 83–98. [CrossRef]
3. Chang, F.J.; Guo, S. Advances in Hydrologic Forecasts and Water Resources Management. *Water* **2020**, *12*, 1819. [CrossRef]
4. Kavetski, D.; Kuczera, G.; Franks, S.W. Bayesian analysis of input uncertainty in hydrological modeling: 1. Theory. *Water Resour. Res.* **2006**, *42*, W03408. [CrossRef]
5. Gan, Y.; Liang, X.-Z.; Duan, Q.; Ye, A.; Di, Z.; Hong, Y.; Li, J. A systematic assessment and reduction of parametric uncertainties for a distributed hydrological model. *J. Hydrol.* **2018**, *564*, 697–711. [CrossRef]
6. Renard, B.; Kavetski, D.; Kuczera, G.; Thyer, M.; Franks, S.W. Understanding predictive uncertainty in hydrologic modeling: The challenge of identifying input and structural errors. *Water Resour. Res.* **2010**, *46*, W05521. [CrossRef]
7. Bulygina, N.; Gupta, H. Estimating the uncertain mathematical structure of a water balance model via Bayesian data assimilation. *Water Resour. Res.* **2009**, *45*, W00B13. [CrossRef]
8. Clark, M.P.; Slater, A.G.; Rupp, D.E.; Woods, R.A.; Vrugt, J.A.; Gupta, H.V.; Wagener, T.; Hay, L.E. Framework for Understanding Structural Errors (FUSE): A modular framework to diagnose differences between hydrological models. *Water Resour. Res.* **2008**, *44*, W00B02. [CrossRef]
9. Reichle, R.H.; Mclaughlin, D.B.; Entekhabi, D. Hydrologic Data Assimilation with the Ensemble Kalman Filter. *American Meteorol. Soc.* **2002**, *130*, 103–114. [CrossRef]
10. Clark, M.P.; Kavetski, D. Ancient numerical daemons of conceptual hydrological modeling: 1. Fidelity and efficiency of time stepping schemes. *Water Resour. Res.* **2010**, *46*, W10510. [CrossRef]
11. Reichert, P. Conceptual and Practical Aspects of Quantifying Uncertainty in Environmental Modelling and Decision Support. 2012. Available online: https://www.semanticscholar.org/paper/Conceptual-and-Practical-Aspects-of-Quantifying-in-Reichert/4dbd0397c9cb925cff1eea445e8d18428ef4a95a (accessed on 16 July 2019).
12. McInerney, D.; Thyer, M.; Kavetski, D.; Bennett, B.; Lerat, J.; Gibbs, M.; Kuczera, G. A simplified approach to produce probabilistic hydrological model predictions. *Environ. Model. Softw.* **2018**, *109*, 306–314. [CrossRef]
13. Krzysztofowicz, R. Bayesian theory of probabilistic forecasting via deterministic hydrologic model. *Water Resour. Res.* **1999**, *35*, 2739–2750. [CrossRef]
14. Todini, E. A model conditional processor to assess predictive uncertainty in flood forecasting. *Int. J. River Basin Manag.* **2008**, *6*, 123–137. [CrossRef]
15. Loucks, D.P.; van Beek, E. *Water Resource Systems Planning and Management*; Springer International Publishing: Cham, Switzerland, 2017.
16. Todini, E. Paradigmatic changes required in water resources management to benefit from probabilistic forecasts. *Water Secur.* **2018**, *3*, 9–17. [CrossRef]
17. Refsgaard, J.C.; van der Sluijs, J.P.; Højberg, A.L.; Vanrolleghem, P.A. Uncertainty in the environmental modelling process—A framework and guidance. *Environ. Model. Softw.* **2007**, *22*, 1543–1556. [CrossRef]
18. Prieto, C.; le Vine, N.; Kavetski, D.; García, E.; Medina, R. Flow Prediction in Ungauged Catchments Using Probabilistic Random Forests Regionalization and New Statistical Adequacy Tests. *Water Resour. Res.* **2019**, *55*, 4364–4392. [CrossRef]
19. Ye, A.; Duan, Q.; Yuan, X.; Wood, E.F.; Schaake, J. Hydrologic post-processing of MOPEX streamflow simulations. *J. Hydrol.* **2014**, *508*, 147–156. [CrossRef]
20. Hopson, T.M.; Wood, A.; Weerts, A.H. Motivation and overview of hydrological ensemble post-processing. In *Handbook of Hydrometeorological Ensemble Forecasting*; Springer: Berlin/Heidelberg, Germany, 2019; pp. 783–793.
21. Montanari, A.; Koutsoyiannis, D. A blueprint for process-based modeling of uncertain hydrological systems. *Water Resour. Res.* **2012**, *48*, W09555. [CrossRef]
22. Farmer, W.H.; Vogel, R.M. On the deterministic and stochastic use of hydrologic models. *Water Resour. Res.* **2016**, *52*, 5619–5633. [CrossRef]
23. Montanari, A.; Brath, A. A stochastic approach for assessing the uncertainty of rainfall-runoff simulations. *Water Resour. Res.* **2004**, *40*, W01106. [CrossRef]
24. Montanari, A.; Grossi, G. Estimating the uncertainty of hydrological forecasts: A statistical approach. *Water Resour. Res.* **2008**, *44*, W00B08. [CrossRef]
25. Wang, Q.J.; Robertson, D.E.; Chiew, F.H.S. A Bayesian joint probability modeling approach for seasonal forecasting of streamflows at multiple sites. *Water Resour. Res.* **2009**, *45*, W05407. [CrossRef]

26. Zhao, L.; Duan, Q.; Schaake, J.; Ye, A.; Xia, J. A hydrologic post-processor for ensemble streamflow predictions. *Adv. Geosci.* **2011**, *29*, 51–59. [CrossRef]

27. McInerney, D.; Thyer, M.; Kavetski, D.; Lerat, J.; Kuczera, G. Improving probabilistic prediction of daily streamflow by identifying Pareto optimal approaches for modeling heteroscedastic residual errors. *Water Resour. Res.* **2017**, *53*, 2199–2239. [CrossRef]

28. Schoups, G.; Vrugt, J.A. A formal likelihood function for parameter and predictive inference of hydrologic models with correlated, heteroscedastic, and non-Gaussian errors. *Water Resour. Res.* **2010**, *46*, W10531. [CrossRef]

29. Smith, T.; Marshall, L.; Sharma, A. Modeling residual hydrologic errors with Bayesian inference. *J. Hydrol.* **2015**, *528*, 29–37. Available online: https://www.sciencedirect.com/science/article/pii/S0022169415004011 (accessed on 3 May 2017). [CrossRef]

30. Sorooshian, S.; Dracup, J.A. Stochastic parameter estimation procedures for hydrologie rainfall-runoff models: Correlated and heteroscedastic error cases. *Water Resour. Res.* **1980**, *16*, 430–442. [CrossRef]

31. Van Der Waerden, B. Order tests for two-sample problem and their power I. *Indag. Math.* **1952**, *55*, 453–458. [CrossRef]

32. Box, G.E.P.; Cox, D.R. An Analysis of Transformations. *J. R. Stat. Soc.* **1964**, *26*, 211–252. Available online: https://www.semanticscholar.org/paper/An-Analysis-of-Transformations-Box-Cox/6e820cf11712b9041bb625634612a535476f0960 (accessed on 24 May 2019). [CrossRef]

33. Wang, Q.J.; Shrestha, D.L.; Robertson, D.E.; Pokhrel, P. A log-sinh transformation for data normalization and variance stabilization. *Water Resour. Res.* **2012**, *48*, W05514. [CrossRef]

34. Krzysztofowicz, R.; Kelly, K.S. Hydrologic uncertainty processor for probabilistic river stage forecasting. *Water Resour. Res.* **2000**, *36*, 3265–3277. [CrossRef]

35. Coccia, G.; Todini, E. Recent developments in predictive uncertainty assessment based on the model conditional processor approach. *Hydrol. Earth Syst. Sci.* **2011**, *15*, 3253–3274. [CrossRef]

36. Weerts, A.H.; Winsemius, H.C.; Verkade, J.S. Estimation of predictive hydrological uncertainty using quantile regression: Examples from the National Flood Forecasting System (England and Wales). *Hydrol. Earth Syst. Sci.* **2011**, *15*, 255–265. [CrossRef]

37. Tyralis, H.; Papacharalampous, G. Quantile-Based Hydrological Modelling. *Water* **2021**, *13*, 3420. [CrossRef]

38. Raftery, A.E.; Gneiting, T.; Balabdaoui, F.; Polakowski, M. Using Bayesian Model Averaging to Calibrate Forecast Ensembles. *Mon. Weather Rev.* **2005**, *133*, 1155–1174. [CrossRef]

39. Darbandsari, P.; Coulibaly, P. Inter-Comparison of Different Bayesian Model Averaging Modifications in Streamflow Simulation. *Water* **2019**, *11*, 1707. [CrossRef]

40. Evin, G.; Thyer, M.; Kavetski, D.; McInerney, D.; Kuczera, G. Comparison of joint versus postprocessor approaches for hydrological uncertainty estimation accounting for error autocorrelation and heteroscedasticity. *Water Resour. Res.* **2014**, *50*, 2350–2375. [CrossRef]

41. Woldemeskel, F.; McInerney, D.; Lerat, J.; Thyer, M.; Kavetski, D.; Shin, D.; Tuteja, N.; Kuczera, G. Evaluating post-processing approaches for monthly and seasonal streamflow forecasts. *Hydrol. Earth Syst. Sci.* **2018**, *22*, 6257–6278. [CrossRef]

42. Brown, J.D.; Seo, D.-J. Evaluation of a nonparametric post-processor for bias correction and uncertainty estimation of hydrologic predictions. *Hydrol. Process.* **2013**, *27*, 83–105. [CrossRef]

43. Solomatine, D.P.; Shrestha, D.L. A novel method to estimate model uncertainty using machine learning techniques. *Water Resour. Res.* **2009**, *45*, W00B11. [CrossRef]

44. López, P.L.; Verkade, J.S.; Weerts, A.H.; Solomatine, D.P. Alternative configurations of quantile regression for estimating predictive uncertainty in water level forecasts for the upper Severn River: A comparison. *Hydrol. Earth Syst. Sci.* **2014**, *18*, 3411–3428. [CrossRef]

45. Sikorska, A.E.; Montanari, A.; Koutsoyiannis, D. Estimating the Uncertainty of Hydrological Predictions through Data-Driven Resampling Techniques. *J. Hydrol. Eng.* **2015**, *20*, A4014009. [CrossRef]

46. Ehlers, L.B.; Wani, O.; Koch, J.; Sonnenborg, T.O.; Refsgaard, J.C. Using a simple post-processor to predict residual uncertainty for multiple hydrological model outputs. *Adv. Water Resour.* **2019**, *129*, 16–30. [CrossRef]

47. Tyralis, H.; Papacharalampous, G.; Burnetas, A.; Langousis, A. Hydrological post-processing using stacked generalization of quantile regression algorithms: Large-scale application over CONUS. *J. Hydrol.* **2019**, *577*, 123957. [CrossRef]

48. Papacharalampous, G.; Tyralis, H.; Langousis, A.; Jayawardena, A.W.; Sivakumar, B.; Mamassis, N.; Montanari, A.; Koutsoyiannis, D. Probabilistic Hydrological Post-Processing at Scale: Why and How to Apply Machine-Learning Quantile Regression Algorithms. *Water* **2019**, *11*, 2126. [CrossRef]

49. Schefzik, R.; Thorarinsdottir, T.L.; Gneiting, T. Uncertainty Quantification in Complex Simulation Models Using Ensemble Copula Coupling. *Stat. Sci.* **2013**, *28*, 616–640. [CrossRef]

50. Madadgar, S.; Moradkhani, H. Improved Bayesian multimodeling: Integration of copulas and Bayesian model averaging. *Water Resour. Res.* **2014**, *50*, 9586–9603. [CrossRef]

51. Klein, B.; Meissner, D.; Kobialka, H.-U.; Reggiani, P. Predictive Uncertainty Estimation of Hydrological Multi-Model Ensembles Using Pair-Copula Construction. *Water* **2016**, *8*, 125. [CrossRef]

52. Li, W.; Duan, Q.; Miao, C.; Ye, A.; Gong, W.; Di, Z. A review on statistical postprocessing methods for hydrometeorological ensemble forecasting. *Wiley Interdiscip. Rev. Water* **2017**, *4*, e1246. [CrossRef]

53. Moges, E.; Demissie, Y.; Larsen, L.; Yassin, F. Review: Sources of Hydrological Model Uncertainties and Advances in Their Analysis. *Water* **2020**, *13*, 28. [CrossRef]

54. Matott, L.S.; Babendreier, J.E.; Purucker, S.T. Evaluating uncertainty in integrated environmental models: A review of concepts and tools. *Water Resour. Res.* **2009**, *45*, 6421. [CrossRef]

55. Reggiani, P.; Coccia, G.; Mukhopadhyay, B. Predictive Uncertainty Estimation on a Precipitation and Temperature Reanalysis Ensemble for Shigar Basin, Central Karakoram. *Water* **2016**, *8*, 263. [CrossRef]

56. Barbetta, S.; Coccia, G.; Moramarco, T.; Todini, E. Case Study: A Real-Time Flood Forecasting System with Predictive Uncertainty Estimation for the Godavari River, India. *Water* **2016**, *8*, 463. [CrossRef]

57. Biondi, D.; Todini, E. Comparing Hydrological Postprocessors Including Ensemble Predictions into Full Predictive Probability Distribution of Streamflow. *Water Resour. Res.* **2018**, *54*, 9860–9882. [CrossRef]

58. Massari, C.; Maggioni, V.; Barbetta, S.; Brocca, L.; Ciabatta, L.; Camici, S.; Moramarco, T.; Coccia, G.; Todini, E. Complementing near-real time satellite rainfall products with satellite soil moisture-derived rainfall through a Bayesian Inversion approach. *J. Hydrol.* **2019**, *573*, 341–351. [CrossRef]

59. Parviz, L.; Rasouli, K. Development of Precipitation Forecast Model Based on Artificial Intelligence and Subseasonal Clustering. *J. Hydrol. Eng.* **2019**, *24*, 04019053. [CrossRef]

60. Yu, Y.; Shao, Q.; Lin, Z. Regionalization study of maximum daily temperature based on grid data by an objective hybrid clustering approach. *J. Hydrol.* **2018**, *564*, 149–163. [CrossRef]

61. Basu, B.; Srinivas, V.V. Regional flood frequency analysis using kernel-based fuzzy clustering approach. *Water Resour. Res.* **2014**, *50*, 3295–3316. [CrossRef]

62. Zhang, H.; Huang, G.H.; Wang, D.; Zhang, X. Multi-period calibration of a semi-distributed hydrological model based on hydroclimatic clustering. *Adv. Water Resour.* **2011**, *34*, 1292–1303. [CrossRef]

63. Schaefli, B.; Talamba, D.B.; Musy, A. Quantifying hydrological modeling errors through a mixture of normal distributions. *J. Hydrol.* **2007**, *332*, 303–315. [CrossRef]

64. Smith, T.; Sharma, A.; Marshall, L.; Mehrotra, R.; Sisson, S. Development of a formal likelihood function for improved Bayesian inference of ephemeral catchments. *Water Resour. Res.* **2010**, *46*, W12551. [CrossRef]

65. Li, M.; Wang, Q.J.; Bennett, J.C.; Robertson, D.E. Error reduction and representation in stages (ERRIS) in hydrological modelling for ensemble streamflow forecasting. *Hydrol. Earth Syst. Sci.* **2016**, *20*, 3561–3579. [CrossRef]

66. Feng, K.; Zhou, J.; Liu, Y.; Lu, C.; He, Z. Hydrological Uncertainty Processor (HUP) with Estimation of the Marginal Distribution by a Gaussian Mixture Model. *Water Resour. Manag.* **2019**, *33*, 2975–2990. [CrossRef]

67. Yang, X.; Zhou, J.; Fang, W.; Wang, Y. An Ensemble Flow Forecast Method Based on Autoregressive Model and Hydrological Uncertainty Processer. *Water* **2020**, *12*, 3138. [CrossRef]

68. Kim, K.H.; Yun, S.T.; Park, S.S.; Joo, Y.; Kim, T.S. Model-based clustering of hydrochemical data to demarcate natural versus human impacts on bedrock groundwater quality in rural areas, South Korea. *J. Hydrol.* **2014**, *519*, 626–636. [CrossRef]

69. Duan, Q.; Schaake, J.; Andréassian, V.; Franks, S.; Goteti, G.; Gupta, H.; Gusev, Y.; Habets, F.; Hall, A.; Hay, L.; et al. Model Parameter Estimation Experiment (MOPEX): An overview of science strategy and major results from the second and third workshops. *J. Hydrol.* **2006**, *320*, 3 17. [CrossRef]

70. R Core Team. R: A Language and Environment for Statistical Computing. Vienna, Austria. 2013. Available online: Ftp://ftp.uvigo.es/CRAN/web/packages/dplR/vignettes/intro-dplR.pdf (accessed on 16 April 2019).

71. Pathiraja, S.; Marshall, L.; Sharma, A.; Moradkhani, H. Detecting non-stationary hydrologic model parameters in a paired catchment system using data assimilation. *Adv. Water Resour.* **2016**, *94*, 103–119. [CrossRef]

72. Deb, P.; Kiem, A.S. Evaluation of rainfall–runoff model performance under non-stationary hydroclimatic conditions. *Hydrol. Sci. J.* **2020**, *65*, 1667–1684. [CrossRef]

73. Simonoff, J.S. *Smoothing Methods in Statistics*; Springer: New York, NY, USA, 1996.

74. Duong, T. Ks: Kernel density estimation and kernel discriminant analysis for multivariate data in R. *J. Stat. Softw.* **2007**, *21*, 1–16. [CrossRef]

75. Wolfe, J.H. Pattern clustering by multivariate mixture analysis. *Multivariate Behav. Res.* **1970**, *5*, 329–350. [CrossRef]

76. Boehmke, B.; Greenwell, B.M. *Hands-On Machine Learning with R*, 1st ed.; Chapman and Hall: London, UK; CRC: Boca Raton, FL, USA, 2019.

77. Banfield, J.D.; Raftery, A.E. Model-Based Gaussian and Non-Gaussian Clustering. *Biometrics* **1993**, *49*, 821. [CrossRef]

78. Fraley, C.; Raftery, A.E. Model-based clustering, discriminant analysis, and density estimation. *J. Am. Stat. Assoc.* **2002**, *97*, 611–631. [CrossRef]

79. James, L.F.; Priebe, C.E.; Marchette, D.J. Consistent estimation of mixture complexity. *Ann. Stat.* **2001**, *29*, 1281–1296. [CrossRef]

80. McLachlan, G.J.; Peel, D. *Finite Mixture Models*; John Wiley & Sons, Inc.: Hoboken, NJ, USA, 2000.

81. Melnykov, V.; Maitra, R. Finite mixture models and model-based clustering. *Stat. Surv.* **2010**, *4*, 80–116. [CrossRef]

82. McLachlan, G.J.; Krishnan, T. *The EM Algorithm and Extensions*; Wiley-Interscience: Hoboken, NJ, USA, 2008.

83. Zhang, W.; Di, Y. Model-based clustering with measurement or estimation errors. *Genes* **2020**, *11*, 185. [CrossRef] [PubMed]

84. Franz, K.J.; Hogue, T.S. Evaluating uncertainty estimates in hydrologic models: Borrowing measures from the forecast verification community. *Hydrol. Earth Syst. Sci.* **2011**, *15*, 3367–3382. [CrossRef]

85. Perrin, C.; Michel, C.; Andréassian, V. Improvement of a parsimonious model for streamflow simulation. *J. Hydrol.* **2003**, *279*, 275–289. [CrossRef]

86. Laio, F.; Tamea, S. Verification tools for probabilistic forecasts of continuous hydrological variables. *Hydrol. Earth Syst. Sci.* **2007**, *11*, 1267–1277. [CrossRef]

87. Thyer, M.; Renard, B.; Kavetski, D.; Kuczera, G.; Franks, S.W.; Srikanthan, S. Critical evaluation of parameter consistency and predictive uncertainty in hydrological modeling: A case study using Bayesian total error analysis. *Water Resour. Res.* **2009**, *45*, W00B14. [CrossRef]

88. Nash, J.E.; Sutcliffe, J.V. River flow forecasting through conceptual models part I—A discussion of principles. *J. Hydrol.* **1970**, *10*, 282–290. [CrossRef]

89. Kavetski, D.; Clark, M.P. Ancient numerical daemons of conceptual hydrological modeling: 2. Impact of time stepping schemes on model analysis and prediction. *Water Resour. Res.* **2010**, *46*, 2009WR008896. [CrossRef]

90. Gupta, H.V.; Perrin, C.; Blöschl, G.; Montanari, A.; Kumar, R.; Clark, M.; Andréassian, V. Large-sample hydrology: A need to balance depth with breadth. *Hydrol. Earth Syst. Sci.* **2014**, *18*, 463–477. [CrossRef]

91. Martinez, G.F.; Gupta, H.V. Toward improved identification of hydrological models: A diagnostic evaluation of the 'abcd' monthly water balance model for the conterminous United States. *Water Resour. Res.* **2010**, *46*, W08507. [CrossRef]

92. Gneiting, T.; Balabdaoui, F.; Raftery, A.E. Probabilistic forecasts, calibration and sharpness. *J. R. Stat. Soc. Ser. B Stat. Methodol.* **2007**, *69*, 243–268. [CrossRef]

93. Tolson, B.A.; Shoemaker, C.A. Efficient prediction uncertainty approximation in the calibration of environmental simulation models. *Water Resour. Res.* **2008**, *44*, 4411. [CrossRef]

94. Bourgin, F.; Ramos, M.H.; Thirel, G.; Andréassian, V. Investigating the interactions between data assimilation and post-processing in hydrological ensemble forecasting. *J. Hydrol.* **2014**, *519*, 2775–2784. [CrossRef]

95. Papacharalampous, G.; Tyralis, H.; Koutsoyiannis, D.; Montanari, A. Quantification of predictive uncertainty in hydrological modelling by harnessing the wisdom of the crowd: A large-sample experiment at monthly timescale. *Adv. Water Resour.* **2020**, *136*, 103470. [CrossRef]

96. Hunter, J.; Thyer, M.; McInerney, D.; Kavetski, D. Achieving high-quality probabilistic predictions from hydrological models calibrated with a wide range of objective functions. *J. Hydrol.* **2021**, *603*, 126578. [CrossRef]

97. Breaban, M.; Luchian, H. A unifying criterion for unsupervised clustering and feature selection. *Pattern Recognit.* **2011**, *44*, 854–865. [CrossRef]

*Article*

# Revision of Frequency Estimates of Extreme Precipitation Based on the Annual Maximum Series in the Jiangsu Province in China

Yuehong Shao [1,*], Jun Zhao [1,2], Jinchao Xu [1,2], Aolin Fu [1] and Junmei Wu [3]

[1] School of Hydrology and Water Resources, Nanjing University of Information Science and Technology (NUIST), Nanjing 210044, China; zsmzyq@126.com (J.Z.); xujinchao301@foxmail.com (J.X.); -fual177@126.com (A.F.)
[2] Nanjing Hydraulic Research Institute, Nanjing 210029, China
[3] Kunshan Meteorology Bereaus, Kunshan 215300, China; wjmqingkong@126.com
[*] Correspondence: syh@nuist.edu.cn; Tel.: +86-138-1589-7365

**Abstract:** Frequency estimates of extreme precipitation are revised using a regional L-moments method based on the annual maximum series and Chow's equation at lower return periods for the Jiangsu area in China. First, the study area is divided into five homogeneous regions, and the optimum distribution for each region is determined by an integrative assessment. Second, underestimation of quantiles and the applicability of Chow's equation are verified. The results show that quantiles are underestimated based on the annual maximum series, and that Chow's formula is applicable for the study area. Next, two methods are used to correct the underestimation of frequency estimation. A set of rational and reliable frequency estimations is obtained using the regional L-moments method and the two revised methods, which can indirectly provide a robust basis for flood control and water resource management. This study extends previous works by verifying underestimation of the quantiles and the provision of two improved methods for obtaining reliable quantile estimations of extreme precipitation at lower recurrence intervals, especially in solving reliable estimates for a 1-year return period from the integral lower limit of the frequency distribution.

**Keywords:** regional l-moments; revision of frequency estimation of extreme precipitation; chow's equation; annual maximum series; annual exceedance series

**Citation:** Shao, Y.; Zhao, J.; Xu, J.; Fu, A.; Wu, J. Revision of Frequency Estimates of Extreme Precipitation Based on the Annual Maximum Series in the Jiangsu Province in China. *Water* **2021**, *13*, 1832. https://doi.org/10.3390/w13131832

Academic Editors: Yuanfang Chen, Dong Wang, Dedi Liu, Binquan Li and Ashish Sharma

Received: 24 May 2021
Accepted: 26 June 2021
Published: 30 June 2021

**Publisher's Note:** MDPI stays neutral with regard to jurisdictional claims in published maps and institutional affiliations.

## 1. Introduction

Natural flood disasters occur frequently in China. As a consequence, flood control is an important topic relevant to the preservation of human life, property, and society [1,2]. Scientific and robust flood control standards are critical to engineering and urban flood control design, for which an important theoretical basis of estimation is hydrological frequency calculation [3]. Rapid economic development and enhanced environmental consciousness have led to increased attention on extreme hydrometeorological events and growing concern for events occurring at lower return periods, fueled by the increasing seriousness of urban waterlogging disasters [1,4]. However, the sampling method and the choice of probability distribution can influence frequency estimations at low recurrence intervals [5]. Therefore, knowledge regarding sampling and the optimum distribution is a key element of frequency analysis.

The determination of an appropriate distribution is an important step in frequency analysis. Selection of the optimum distribution has been extensively researched because the theoretical distribution curve is unknown [6,7]. For example, the person type III (PE3) distribution has been selected as the appropriate fit in the United States, and the generalized extreme value (GEV) distribution has been recommended in more than ten countries [8,9]. However, the adoption of a "one-size-fits-all" scenario may lead to poor accuracy in quantile estimation due to heterogeneity and discordancy associated with different sites in the region [10]. Therefore, some researchers have recommended selecting an optimum distribution for each homogeneous region based on practical grounds, showing that the accuracy of quantile estimates is significantly improved [8,11,12]. In this study, Monte Carlo (MC) simulations and a diagram of L-moments ratios are used to determine goodness-of-fit. However, the criterion from MC simulation can be unreliable when data are serially linked or cross-correlated among sites [13]. Therefore, a summary assessment is performed using different statistical criteria to determine the optimum distribution, in order to avoid obtaining an arbitrary result from any single test.

For sampling, the annual maximum series (AMS) and partial duration series (PDS) can be used to select extreme values from a long hydrological time series. In the AMS, the largest event in each year is extracted and recorded in a series that contains critical information such as extreme precipitation or peak flow amount. These data are easily obtained and widely used in hydrological statistical analysis [14–17]. However, the AMS extracts only the largest event, and secondary events occurring in one year may exceed the annual maximum of other years. In addition, annual maximum events observed in dry years may be very small, and interpretations based on these events can lead to significant bias with respect to the outcome of an extreme value analysis [18–20]. Extensive research has shown that quantiles based on AMS data are underestimated to a certain extent at low return periods [1,12,21]. For example, Lin et al. showed underestimation of extreme precipitation based on AMS data from daily precipitation data obtained from 1438 stations in the southwestern United States [20]. However, less research has been done to assess the underestimation of quantiles in China, or to verify the results in a specific area [22].

The PDS method extracts all of the extreme events above a truncation or threshold level for the analysis and therefore does not suffer from the drawbacks inherent to the AMS data. If a descending sort of PDS is selected, such that the number of values in the series equals the number of years on record, the series is called the annual exceedance series (AES). The AES data not only simplifies sample selection and subsequent statistical analysis, but also gives similar results to that obtained with PDS data and can be regarded as a special case of the PDS [20]. A complete description and solid theoretical basis of precipitation and flood processes exists in the PDS. Previous research has shown that the PDS is more efficient for quantile estimation than the AMS because it is more suitable for a heavy-tailed distribution, which is common in hydrological applications [23–25]. The accuracy of estimation based on the PDS is closely related to the selection of an appropriate threshold level and independence of the sample data [26–28]. Construction of a PDS model can be hampered by several difficulties and is less commonly used in hydrologic research than AMS methods. First, events should be independent; hence, criteria explicitly identifying independent events must be defined. Second, the selection of an appropriate threshold is important to the result and should ensure that a maximum amount of relevant information is included in the analysis without violating basic statistical assumptions. Third, the return period of the PDS in sampling units is not consistent with the return period of the AMS in years. Conversion and verification of recurrence intervals are difficult with PDS data [19,29,30].

Although it has a solid theoretical base, difficulties such as data availability make the construction of a PDS model difficult. At most sites in China, only AMS data are available for a variety of reasons. Therefore, the development of a simple and feasible method for calculating reliable quantile estimates on the basis of AMS data is critical. Chow [21] derived a relation for AMS and AES between two recurrence intervals corresponding to the same event that has been widely accepted and used in engineering practice [31]. Lin et al. subsequently verified the applicability of Chow's equation in the southwestern United States [20]. Takeuchi noted that the precision of estimations can satisfy the requirement using the return period in sampling units of Chow's equation if the size of the PDS is in accordance with a Poisson distribution [32]. Ghahraman noted that the relationship between the recurrence intervals of AMS and PDS should be a function of rainfall duration and the number of samples [33]. Is the frequency conversion related to local hydrological characteristics and other factors? Is Chow's formula applicable to hydrological frequency analysis in China? These questions require in-depth analysis and research. If Chow's formula is appropriate for sites in China, calculation of reliable frequency estimates based on the AMS data and Chow's formula can be performed. Because only AMS data exist for most sites in China, this highlights the purpose and importance of this study.

The objectives of this study are to verify the underestimation of quantiles, provide two revised methods for reliably estimating the frequency of extreme precipitation, and to solve the problem of the distribution integral lower limit. To achieve these objectives, different homogeneous regions are first identified and the optimum distribution for each homogeneous region in the study area is determined. Second, we compare exceedance frequencies with the exceedance probabilities in order to verify whether quantiles are underestimated based on AMS data in the study area. Third, frequency estimates are computed using real AMS data, real AES data, and generated AES data based on the Chow's equation and AMS data, to verify the applicability of Chow's equation in this study. Last, a set of reliable frequency estimates is obtained using a regional L-moments method based on AMS data and Chow's equation. We revise the estimation of quantiles at each site at low return periods for the AMS data and also solve the quantiles for a 1-year return period; the latter is a major merit of this research that extends previous work conducted to date.

## 2. Materials and Methods

### 2.1. Study Area

An important part of Yangtze River Delta, Jiangsu Province ($116°18'–121°57'$ E, $30°45'–35°20'$ N) covers an area of about $1.07 \times 105$ km$^2$ and is located downstream of Yangtze River and Huaihe River basins. The terrain is dominated by plains, accounting for more than 70% of the area. Hills are concentrated in the southwest, accounting for 14.3% of the total area. The terrain slopes from west to east. The river network is intricate and includes the three major river systems of the Yishusi River drainage: Downstream of the Huaihe River, the Yangtze River, and Taihu Lake stream. Jiangsu is located in a transitional subtropical to warm temperate climate zone. The area is characterized by four distinct seasons, which are cold and dry in winter, and warm and humid with plum rains in the late spring and early summer, and typhoons in summer and autumn. The annual average rainfall is 996 mm. Precipitation gradually increases from south to north and is greater on the coast than inland. Rainstorm zones are mainly located in the south of Yimeng Mountain. The elevation, stream network, and meteorological stations of the study area are shown in Figure 1.

### 2.2. Data

Daily precipitation from meteorological stations was obtained for this study from the National Meteorological Information Centre of the China Meteorological Administration (http://cdc.cma.gov.cn/shuju). Data from 63 representative stations in the Jiangsu area obtained between 1961 and 2011 were used for analysis. The AMS was extracted from the daily precipitation data using a bubble sort method. The annual maximum series $x_1$, $x_2$, $x_3$, … … , $x_N$ is a collection of the maximum data for each year, where $N$ is the number of years in the observed time series. The partial duration series $y_1$, $y_2$, $y_3$, … … , $y_M$ is a collection of exceedance over a certain truncation level, the $M^{th}$ largest in the whole time series of $N_{yr}$. In this study, the threshold value was equal to three. That is, the three largest daily rainfalls were selected from each year and form the PDS. The PDS was sorted and intercepted the largest $N$ events in descending order, which includes the AES. Therefore, the AES may be regarded as a special case of the PDS. The frequency estimations for extreme precipitation based on AMS and AES were assessed and compared.

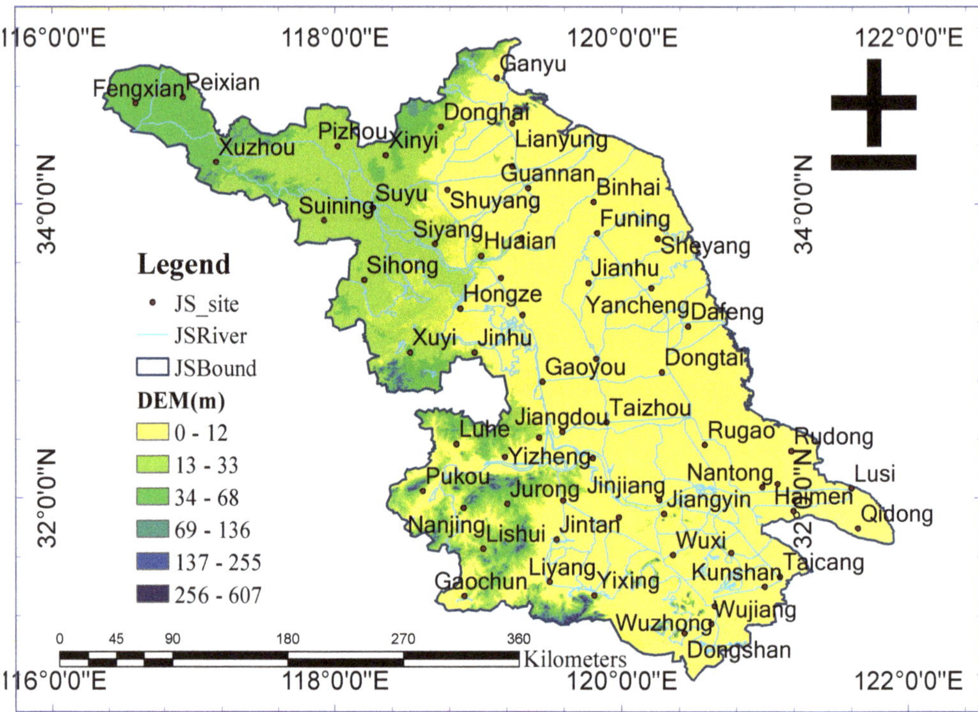



**Figure 1.** Map showing the locations of meteorological stations in Jiangsu province.

### 2.3. Methodology

#### 2.3.1. Regional L-Moments Method

The L-moments method is aimed at the issue of the robust parameter estimation. Regional analysis provides a solution to reduce the uncertainties that exist in at-site statistical analysis. Accordingly, many studies have shown that a regional L-moments method is a reasonable and reliable method to improve the precision and accuracy of frequency estimation [12,20].

Regional frequency analysis employs data from several sites in a region to estimate the frequency distribution of the underlying population at each site. The approach makes the assumption that the shape of the probability distribution function is shared among a group of sites. An index-flood procedure was used in the estimation of precipitation frequency. It assumes that the frequency distribution at each of the $N$ sites in a region is identical apart from a site-specific scaling factor, the index-flood, and that the region is homogeneous. That is, the quantile estimates at site $i$, $Q_{T,j,i}$ can be computed by a regional component that reflects the common precipitation character and a local component that reflects the site-specific scaling factor. The formula can be written as:

$$Q_{T,j,i} = q_{T,j} \times \quad \overline{x}_{i,j} \tag{1}$$

where $\overline{x}_{i,j}$ is commonly the at-site sample mean used for the location estimator, $j = 1, 2, \ldots,$ $N$; $q_{T,j}$, namely the regional growth factor (RGF), is defined as the dimensionless regional frequency distribution common to the N sites in the region at multiple desired return periods, $T_j$. It can be determined by a set of regional parameters that are weighted average values over $N$ sites for a selected distribution. For example, the regional Linear coefficient of deviation ($L$-$C_v$) can be written as follows:

$$\hat{L}_{Cv}^{(R)} = \sum_{i}^{N} n_i \hat{L}_{Cv}^{(i)} / \sum_{i}^{N} n_i, \ i = 1, 2, \ldots, N \tag{2}$$

where $\hat{L}_{Cv}^{(R)}$ and $\hat{L}_{Cv}^{(i)}$ are respectively denoted as the regional $L$-$C_v$ and the single station $L$-$C_v$ at site $i$.

### 2.3.2. Identification of Homogeneous Regions

The identification of homogeneous regions is an important task. First, cluster analysis is used to identify homogeneous regions on the basis of four variables: Longitude, latitude, elevation, and the mean annual precipitation. This analysis is conducted using Ward's method based on Euclidean distance by Statistical Analysis System (SAS) hierarchical clustering software [34] More details on cluster analysis can be found in reference [13]. Second, a measurement of heterogeneity ($H$) is used to assess hydrological similarity and determine regional homogeneity. H is denoted as:

$$H_i = \left( V_i - \mu_{V_i} \right) / \sigma_{V_i} \ i = 1, 2, 3 \tag{3}$$

where $\mu_{V_i}$ and $\sigma_{V_i}$ are the expectation and standard deviation of $V_i$, which can be defined as follows:

$$V_1 = \left\{ \sum_{i=1}^{N} n_i (t^{(i)} - t^R)^2 / \sum_{i=1}^{N} n_i \right\}^{1/2},$$

$$V_2 = \sum_{i=1}^{N} n_i \left\{ (t^{(i)} - t^R)^2 + (t_3^{(i)} - t_3^R)^2 \right\}^{1/2} / \sum_{i=1}^{N} n_i, \tag{4}$$

$$V_3 = \sum_{i=1}^{N} n_i \left\{ (t_3^{(i)} - t_3^R)^2 + (t_4^{(i)} - t_4^R)^2 \right\}^{1/2} / \sum_{i=1}^{N} n_i.$$

where $t^{(i)}$, $t_3^{(i)}$, and $t_4^{(i)}$ are separately the coefficient of sample L-moments. $t^R$, $t_3^R$, and $t_4^R$ denoted regional average L-moments coefficient weighted the site's record lengths, which are defined as:

$$t^R = \sum_{i=1}^{N} n_i t^{(i)} / \sum_{i=1}^{N} n_i, \quad t_3^R = \sum_{i=1}^{N} n_i t_3^{(i)} / \sum_{i=1}^{N} n_i, \quad t_4^R = \sum_{i=1}^{N} n_i t_4^{(i)} / \sum_{i=1}^{N} n_i \tag{5}$$

where $N$ is the number of sites, $n_i$ is the site's record lengths. Hosking and Wallis [13] suggested that a region may be considered "acceptable homogeneous" if $H < 1$, "possibly heterogeneous" if $1 \leq H < 2$, "definitely heterogeneous" if $H \geq 2$, and "possibly correlated" if $H < 0$.

Finally, a measurement of discordancy ($D_i$) is used to identify data that are grossly discordant with the region as a whole [13]. The critical values for discordancy experiments are dependent on the number of sites in the region [13]. More detailed information on these procedures can be found in [12,13].

### 2.3.3. The Goodness-of-Fit

The MC simulation and the Root Mean Square Error (*RMSE*) of the sample L-moments are used to determine the appropriate distribution according to an arbitrary result of any one test. Due to the relative stability and flexibility of 3 parameters, five kinds of commonly used 3-parameter distributions are investigated for goodness-of-fit as follows [12,13]: Generalized logistic (GLO), GEV, generalized normal (GNO), generalized pareto (GPA), and PE3.

A large number of synthetic datasets are generated by MC simulation and used to access the deviation from the mean point to the distribution in $L$-$C_k$ scale. For each distribution, the goodness-of-fit measure is defined as follows:

$$Z^{DIST} = \left( \tau_4^{DIST} - t_4^R + B_4 \right) / \sigma_4 \tag{6}$$

where $t_4^R$ is the regional average L-kurtosis, weighted proportionally to the site's record length; $\tau_4^{DIST}$ is the L-kurtosis of the fitted distribution, where DIST can be any of GLO, GEV, GNO, GPA, and PE3. For the $m^{th}$ simulated region, after the regional average L-kurtosis $t_4^{[m]}$ obtained, the bias ($B_4$) and standard deviation ($\sigma_4$) of $t_4^R$ can be calculated as follows:

$$B_4 = \left[ \sum_{m=1}^{N_{sim}} \left( t_4^{[m]} - t_4^R \right) \right] / N_{sim} \tag{7}$$

$$\sigma_4 = \left\{ \left[ \sum_{m=1}^{N_{sim}} \left( t_4^{[m]} - t_4^R \right)^2 - N_{sim} B_4^2 \right] / (N_{sim} - 1) \right\}^{1/2} \tag{8}$$

Assuming that $Z$ takes the form of a standard Gaussian distribution, the criterion $|Z| \leq 1.64$ is chosen as the cutoff threshold, and the smallest $|Z|$ value, the best distribution.

The $MC$ simulation emphasizes the effect of the regional average. The $RMSE$ is used to assess the variability of the sample $L\text{-}C_k$ of the real data at $N$ sites to accurately evaluate the distribution pattern. The $RMSE$ is calculated for each of the plausible distributions as follows:

$$RMSE = \left\{ \sum_{i=1}^{N} n_i (S_{i,L-Ck} - D_{i,L-Ck})^2 / \sum_{i=1}^{N} n_i \right\}^{1/2} \tag{9}$$

where $S_{i,L\text{-}Ck}$ is the sample $L\text{-}C_k$ at site $i$ and $D_{i,L\text{-}Ck}$ is the distribution's $L\text{-}C_k$ at sample $L\text{-}Cs$ of site $i$. The distribution with the smallest $RMSE$ is selected as the most appropriate distribution based on this experiment. More details of the $MC$ and $RMSE$ methods can be found in the literature [12,13,20].

### 2.3.4. Conversion of AES-AMS

Chow [21] derived a relation between the two recurrence intervals $T_{AMS}$ and $T_{AES}$ corresponding to the same event, as follows:

$$T_{AES} = \left[ \ln\left( \frac{T_{AMS}}{T_{AMS} - 1} \right) \right]^{-1} \text{ or } T_{AMS} = \frac{1}{1 - e^{-\frac{1}{T_{AES}}}} \tag{10}$$

where $T_{AMS}$ and $T_{AES}$ are, respectively, the return period of AMS and AES.

Chow's equation is a frequency conversion relation that has been widely adopted for use in engineering research. Table 1 gives the return periods based on AES data. Frequency estimation can be computed by using non-exceedance probability ($P_{NON}$). However, the computer program cannot be computed if $P_{NON}$ equals zero. From Equation (10), it is clear that it is not computable for a 1-year event under AMS data. If Chow's equation is applicable to this study area, we can not only correct the frequency estimation at low return periods based on AMS data, but can also compute quantiles for a 1-year recurrence interval based on Chow's equation.

**Table 1.** Return periods based on AMS data.

| $T_{AES}$ (-year) | $T_{AMS}$ (-year) | $P = 1/T_{AMS}$ | $P_{NON} = 1 - 1/T_{AMS}$ |
|:---:|:---:|:---:|:---:|
| N/A | 1 | 1.0 | 0.0 * |
| 1.44 | 2 | 0.50 | 0.50 |
| 4.48 | 5 | 0.20 | 0.80 |
| 9.49 | 10 | 0.10 | 0.90 |
| 24.50 | 25 | 0.04 | 0.96 |
| 49.50 | 50 | 0.02 | 0.98 |

* Note: It is incomputable for 1-year event based on AMS data.

## 3. Results

### 3.1. Results and Analysis of the Goodness-of-Fit

The study area was divided into five homogeneous regions according to the above-mentioned procedures in methods section (Figure 2). After identification of homogeneous regions, the optimum distribution is determined based on the regional L-moments analysis.

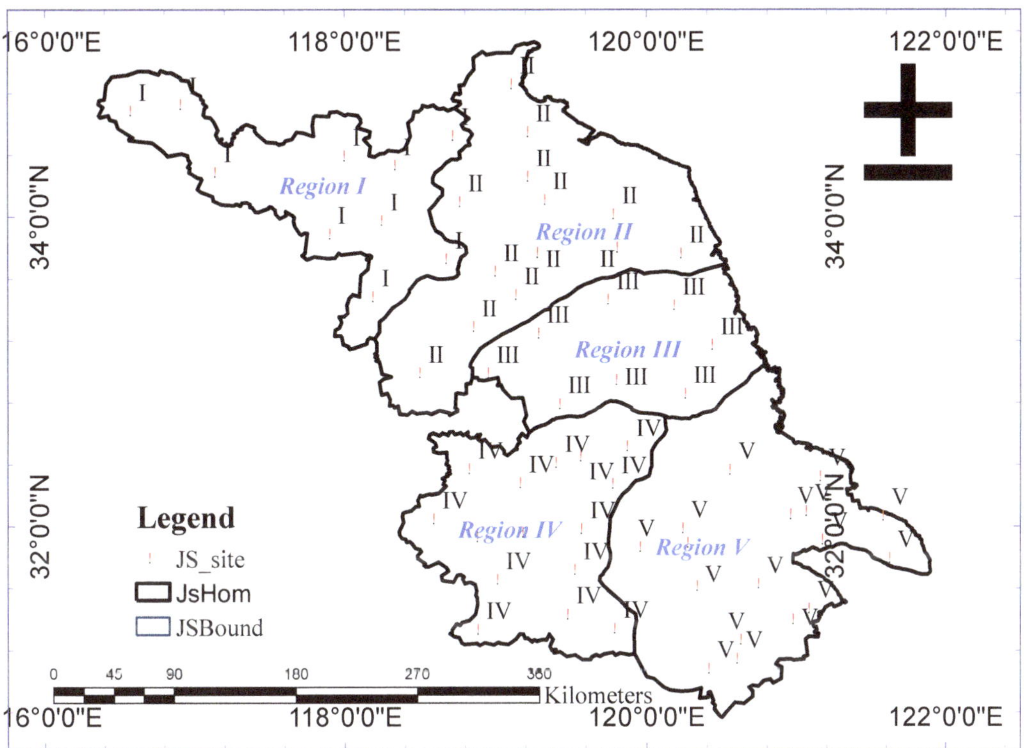


**Figure 2.** Spatial map of five homogeneous regions.

The results of the goodness-of-fit simulation experiments for the five regions are shown in Table 2. It can be seen from Table 2 that GLO and GEV are, respectively, the optimum distributions based on the two indices of $|Z|$ and the *RMSE* for regions I and III, IV, and V. For region II, GNO is the best distribution from the $|Z|$ value, and GEV is the best based on the *RMSE*. However, the difference between the $|Z|$ and *RMSE* estimations is small. Therefore, GNO and GEV can be considered as the best-fitting distributions based on the two tests. However, abrupt changes in frequency estimations at the borders of adjacent homogeneous regions should be avoided. The frequency estimation has a good correlation with the tail thickness of distribution and decreases in the order of GLO, GEV, GNO, GPA, and PE3. The regions adjacent to region II have an appropriate fit with GLO. Therefore, GEV is the best distribution based on the two tests and the change in adjacent regions. Similarly, the optimum distributions of the five regions for the AES data are GNO, GPA, GPA, GPA, and GPA.

**Table 2.** Results of *MC* and *RMSE* measures for a 1-day duration.

| Test Index | Distribution | Homogeneous Region | | | | |
|---|---|---|---|---|---|---|
| | | **I** | **II** | **III** | **IV** | **V** |
| | GLO | −0.17 | 2.63 | −0.28 | 2.00 | 2.47 |
| | GEV | −1.39 | 0.26 | −1.66 | 0.05 | 0.37 |
| Z | GNO | −2.05 | −0.25 | −2.05 | −0.78 | −0.56 |
| | GPA | −4.46 | −5.11 | −4.87 | −4.71 | −4.79 |
| | PE3 | −3.21 | −1.35 | −2.83 | −2.27 | −2.24 |
| $Z_{min}$ | | GLO | GNO | GLO | GEV | GEV |
| | GLO | 0.0404 | 0.0650 | 0.0401 | 0.0591 | 0.0655 |
| | GEV | 0.0424 | 0.0445 | 0.0509 | 0.0395 | 0.0447 |
| *RMSE* | GNO | 0.0834 | 0.0466 | 0.0604 | 0.0452 | 0.0470 |
| | GPA | 0.0555 | 0.0885 | 0.1239 | 0.0811 | 0.0773 |
| | PE3 | 0.1057 | 0.0546 | 0.0797 | 0.0632 | 0.0628 |
| $RMSE_{min}$ | | GLO | GEV | GLO | GEV | GEV |

*3.2. Comparison between Exceedance Frequency and Exceedance Probability*

The regional L-moments analysis is applied to obtain the quantiles at each station on a region-by-region basis. The exceedance frequencies from 2-year to 100-year return periods are calculated station-by-station and averaged first over the region and then over the study area. The data exceedance frequencies at each station for study area are found from Table S1 in Supplementary Material. The average region-by-region exceedance frequencies are shown in Table 3 over the entire study area. It can be seen from Table 3 that the average exceedance frequencies are higher than the corresponding theoretical exceedance probabilities for 2-year to 100-year return periods over the study area, which are 0.507, 0.206, 0.111, 0.045, 0.021, and 0.011 for 2-year, 5-year, 10-year, 25-year, 50-year, and 100-year return periods, respectively. The corresponding real return periods are calculated to be 1.97 years, 4.85 years, 8.99 years, 22.25 years, 47.49 years, and 91.87 years, which indicates that extreme precipitation events occur more frequently. These data indicate that current quantile estimates based on AMS data are underestimated for frequent events in the study area.

**Table 3.** Average exceedance frequencies for the study area.

| Region | Return Period (R.P.)/Exceedance Probability (E.P.) | | | | | |
|---|---|---|---|---|---|---|
| | **2-yr** | **5-yr** | **10-yr** | **25-yr** | **50-yr** | **100-yr** |
| | **0.50** | **0.20** | **0.10** | **0.04** | **0.02** | **0.01** |
| I | 0.518 | 0.210 | 0.124 | 0.051 | 0.022 | 0.012 |
| II | 0.505 | 0.201 | 0.106 | 0.045 | 0.021 | 0.011 |
| III | 0.500 | 0.201 | 0.120 | 0.044 | 0.020 | 0.010 |
| IV | 0.508 | 0.217 | 0.101 | 0.041 | 0.022 | 0.013 |
| V | 0.502 | 0.202 | 0.106 | 0.044 | 0.021 | 0.009 |
| *Average E.P.* | 0.507 | 0.206 | 0.111 | 0.045 | 0.021 | 0.011 |
| *Real R.P.* | 1.97-yr | 4.85-yr | 8.99-yr | 22.25-yr | 47.49-yr | 91.87-yr |

*3.3. Verification of the Applicability of Chow's Equation in the Study*

The underestimated frequencies from the AMS data can be revised for low-return periods if Chow's equation is applicable to this study area. The procedure for verification is as follows: First, the quantiles based on real AES and AMS data are independently estimated for the study area (Table 4). Then the AES–AMS ratios are obtained based on frequency estimates from 2-year to 100-year return periods. Second, frequency estimations and their ratios are calculated based on real AMS data, where AES is obtained based on

AMS data and Chow's equation. The best-fit distribution of each homogeneous region is used to calculate the frequency estimates. The verification results are shown in Figure 3.

**Table 4.** Quantile estimates of extreme precipitation for a 1-day duration with different return periods in homogeneous region I.

| Site Name | Quantile Estimates Based on AES Data | | | | | |
|---|---|---|---|---|---|---|
| | 2-yr | 5-yr | 10-yr | 25-yr | 50-yr | 100-yr |
| Fengxian | 92.2 | 117.3 | 139.7 | 175.1 | 206.8 | 243.3 |
| Peixian | 98.4 | 125.2 | 149.1 | 186.9 | 220.7 | 259.7 |
| Pizhou | 106.6 | 135.6 | 161.5 | 202.4 | 239.0 | 281.2 |
| Xuzhou | 103.2 | 131.3 | 156.4 | 196.0 | 231.5 | 272.4 |
| Xinyi | 98.2 | 124.9 | 148.8 | 186.4 | 220.2 | 259.1 |
| Donghai | 98.9 | 125.9 | 149.9 | 187.9 | 221.9 | 261.1 |
| Suining | 115.4 | 146.8 | 174.9 | 219.2 | 258.9 | 304.6 |
| Suyu | 116.5 | 148.2 | 176.5 | 221.2 | 261.3 | 307.5 |
| Siyang | 104.9 | 133.4 | 158.9 | 199.1 | 235.2 | 276.7 |
| Sihong | 104.5 | 132.7 | 154.9 | 185.5 | 209.6 | 234.5 |
| Quantile estimates based on AMS data and Chow's equation | | | | | | |
| Fengxian | 91.0 | 116.6 | 139.1 | 174.9 | 207.9 | 246.9 |
| Peixian | 99.4 | 127.4 | 152.0 | 191.1 | 227.2 | 269.8 |
| Pizhou | 106.0 | 135.9 | 162.1 | 203.8 | 242.2 | 287.6 |
| Xuzhou | 103.9 | 133.2 | 158.8 | 199.7 | 237.3 | 281.9 |
| Xinyi | 96.8 | 124.1 | 148.0 | 186.1 | 221.2 | 262.7 |
| Donghai | 99.0 | 126.9 | 151.4 | 190.3 | 226.2 | 268.6 |
| Suining | 111.7 | 143.2 | 170.8 | 214.7 | 255.2 | 303.1 |
| Suyu | 111.9 | 143.4 | 171.0 | 215.0 | 255.5 | 303.5 |
| Siyang | 105.6 | 135.3 | 161.4 | 202.9 | 241.1 | 286.4 |
| Sihong | 100.0 | 128.2 | 152.9 | 192.2 | 228.5 | 271.4 |
| *Mean RE (%)* | 1.72 | 1.60 | 1.41 | 1.68 | 2.49 | 3.64 |

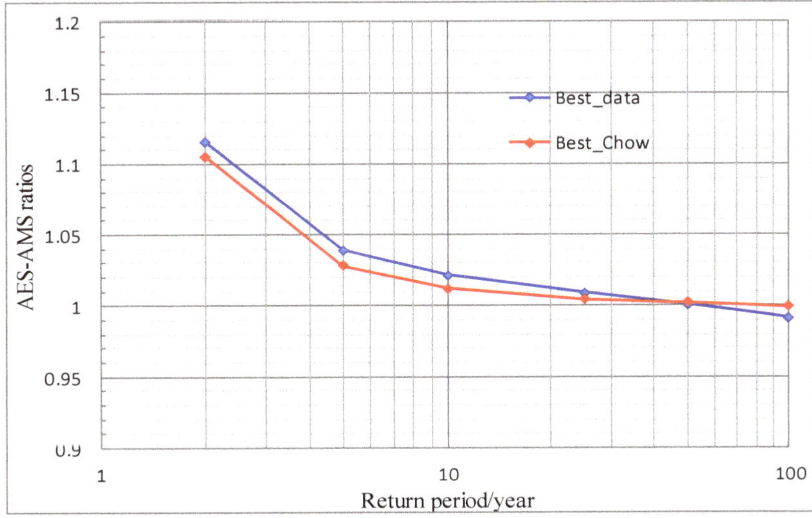


**Figure 3.** Comparison of AES–AMS ratios obtained using Chow's equation and real data.

Taking Region I as an example, quantile estimates based on AES data, AMS data, and Chow's equation are shown in Table 4. The results from region II, III, IV, and V can be found in Table S2 in Supplementary Material. The frequency estimates obtained from the real AES data have good consistency with that obtained from computed AES data. The mean relative error from 2-year to 100-year return periods is 2.09%, indicating that the AMS data and Chow's equation may be used as an alternative method when only AMS data are available. Figure 3 indicates that the general trend is consistent between Chow's equation and the data. The AES–AMS ratio gradually decreases with increasing length of the return period; ratios are >1 when the recurrence intervals are less than 25 years, near 1 between 25-year and 50-year return periods, and <1 for return periods longer than 50 years. However, the magnitude of the decrease is largest when the recurrence interval is less than 25 years. It can also be seen from the curve trend that different sampling methods can have a substantial effect on the low-return-period interval. Taken overall, the best-fit Chow's case is consistent with the best-fit real data, indicating that Chow's equation can be used as a simple method to obtain reasonable AES-AMS ratios consistent with those obtained from real AES and AMS data.

### 3.4. Reliable Frequency Estimation and Spatiotemporal Analysis

By applying Chow's equation, quantiles derived from AMS data can be revised at low return periods. A set of rational and reliable frequency estimations can be obtained using a regional L-moments method based on the AMS data and Chow's equation. Solving the quantile for a 1-year return period is most important, which means solving the integral lower limits of the frequency distribution curve. For example, frequency estimates for different return periods in region I are shown in Table 4. It can be seen that the estimates increase incrementally with the length of the recurrence interval as a whole. The maximum estimation occurs at the Suyu station. We compared the quantile estimates with the maximum of the 24-h observation series for each station, which can indirectly reflect estimation accuracy to some extent due to the unknown true value of the frequency estimate. Considering the 24-h record length (51 years) of the Suyu station, frequency estimates for the 50-year recurrence interval were selected to assess consistency. The quantile estimate is 255.5 mm, which is consistent with the maximum observed 24-h value (253.9 mm). The frequency estimates also agree with observations at other sites and provide a scientific basis for flood disaster warnings and urban construction, among other uses.

Figure 4 shows the spatial mapping of frequency estimates for a 24-h duration for 1-year, 10-year, 25-year, and 50-year return periods, which have similar patterns. The estimated values at the northern end of the study area are greater than those at the southern area, and all of the estimates increase with increased length of the return period. The highest frequency estimates are observed near the Suqian and Lianyungang stations in the northern part of Jiangsu, and low values are observed in the southern Taihu lake basin. These data suggest that Xuzhou and Lianyungang are in a high-risk area of extreme precipitation that may be subject to flash floods. Therefore, decision makers should pay heightened attention to the risk of flooding and water resource management in these areas.

### 3.5. Validation of Frequency Estimations of Extreme Precipitation

Because the true value of the frequency estimation is unknown, the accuracy of the estimated value cannot be evaluated using the error of the estimated value and the true value. However, the accuracy of quantiles is indirectly reflected by a comparison of the estimation and observation at the same frequency. The plotting-position estimator is used to compute the experience frequency. The experience frequency is defined as follows:

$$P = (i + A)/(n + B) \tag{11}$$

where $i$ is the sequence number from ascending series, $n$ is the number of sequence length for each site, $A$ and $B$ are the parameters, $A$ is equal to $-0.35$, and $B$ is zero [13].

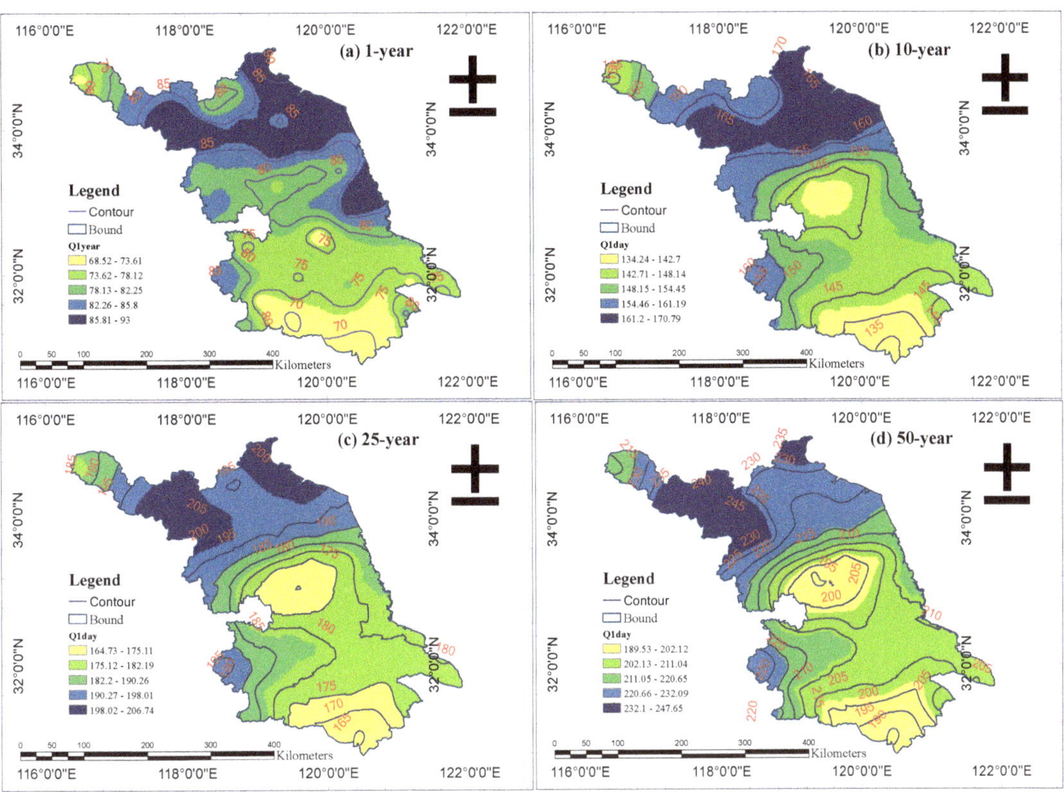


**Figure 4.** Map of quantile estimates for (**a**) 1-year, (**b**) 10-year, (**c**) 25-year, and (**d**) 50-year return periods.

Three statistical criteria, *RE*, *RMSE*, and the correlation coefficient (*r*) are used to judge the precision of frequency estimates. The results can be found in Table 5. The average *RE*, *RMSE*, and *r* of all stations in the study area are 5.56%, 0.107 mm, and 0.969, respectively. This indicates that the estimation is in good agreement with the observation. Figure 5 shows a scatterplot of measured and observed quantiles at the same frequency for each site in each homogeneous region. It can be seen from Figure 5 and Table 5 that the simulation is consistent with the set of observations as a whole, with a value of r greater than 0.96 in each homogeneous region. From the above, it may be concluded that the frequency estimation is reasonable and reliable for low return periods. As a whole, frequency estimations based on the regional L-moments method are in good agreement with observations.

**Table 5.** Comparisons between estimation and observation.

| Homogeneous Regions | RE (%) | RMSE (mm) | r |
|---|---|---|---|
| Region I | 5.47 | 0.101 | 0.975 |
| Region II | 5.31 | 0.093 | 0.975 |
| Region III | 4.77 | 0.09 | 0.971 |
| Region IV | 5.88 | 0.118 | 0.965 |
| Region V | 5.96 | 0.119 | 0.961 |
| All | 5.56 | 0.107 | 0.969 |

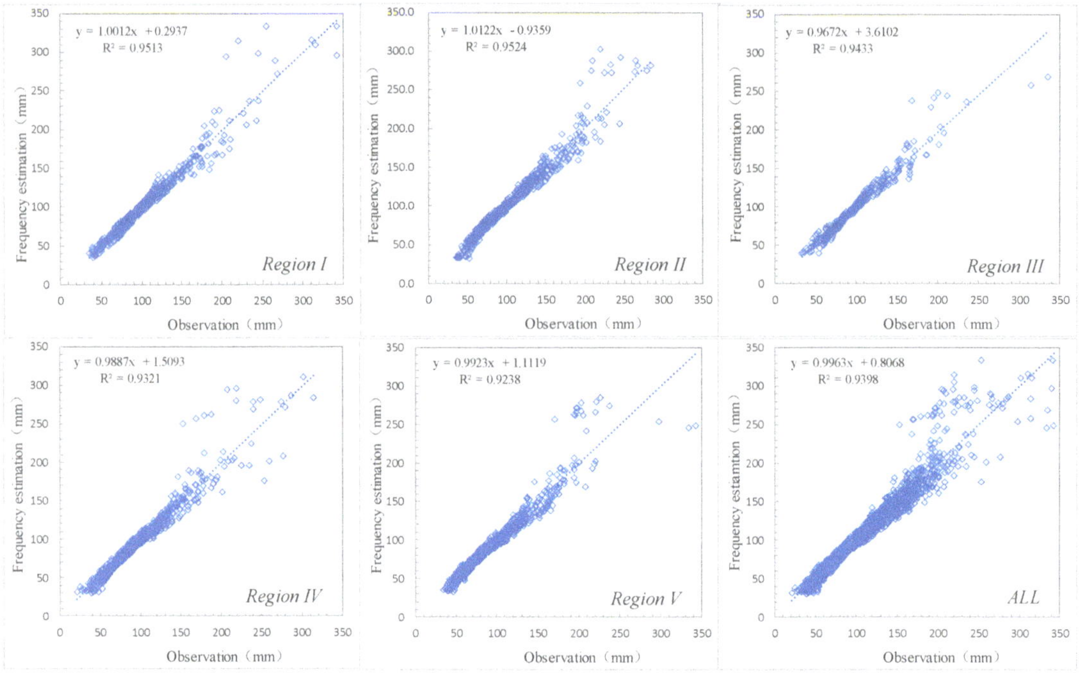

**Figure 5.** Scatterplot of estimations and observations at the same frequency for each homogeneous region.

## 4. Discussion

The results of this study are valuable for revising the underestimation of quantiles and obtaining a set of reliable quantile estimates in the study area. However, some issues may benefit from a more in-depth analysis in future research.

The determination of homogeneous region is an important step in regional frequency analysis. The optimum distributions are, respectively, GEV and GPA distribution based on the AMS data and AES data for most regions of the study area, which is consistent with many previous studies. Many research studies have shown that GEV is the most commonly used for AMS analysis, and GPA is frequently proposed for PDS analysis [35–39]. However, the national guidelines and regulation for calculation design storm and flood in China recommend the use of PE3 distribution, which is inconsistent with the research in this paper [40]. The main attribution includes that PE3 is a recommended choice based on the conventional moment method, which is very different from when sample size is small, or the skewness of the sample is considerable. Many research studies have proved that L-moments are less subject to bias in estimation and enable more reliable inferences to be made from small samples than conventional moment method [22,41–43]. Therefore, the optimum distribution of this paper is rational and reliable based on the summary of judgement. At the same time, identification of homogeneous region may cause the discontinuity around the boundary of adjacent regions. Very few papers dealing with the discontinuity are found in the literature. So, spatial consistency should be considered and further research in the next study.

Research has shown that the quantiles based on AMS data are underestimated at low return periods in this paper, which is in accordance with previous research and theories [1,18,20,22]. Lin et al. verified the result that exists a significant underestimation based on 1438 stations data in southwestern United States [1,20]. Frequency estimates are underestimated but the magnitude of underestimation is not obvious in this study. Some possible reasons are analyzed and discussed as follows. By analyzing the AMS data of the site, it is found that a negative correlation exists between the exceedance frequency and skewness coefficient of the station; that is, the larger the positive skewness coefficient, the smaller the exceedance frequency. The frequency distribution diagram based on the AMS data is used to analyze the causes (Figure 6). Taking Region I and Region V for example, the stations with the largest L-Cs (58013 and 58,349 sites) and with the smallest L-Cs (58,131 and 58,345 sites) are selected to analyze the underestimated reasons in this study. It can be seen from Figure 6 that the station with the largest L-Cs have the maximum rainfall value in the corresponding region, and the AMS sequence with the smallest L-Cs is approximate to normal distribution, which has uniform and continuous characteristics and no extra-large value. We may come to the conclusion that the distribution of sparse and discontinuous extreme precipitation data at large value intervals is the main factor resulting in low exceedance frequency values. Second, factors including small sample sizes and data series of inadequate length can also affect the calculation of the exceedance frequency. Therefore, a larger range and longer sequence of data should be collected and analyzed in future research.

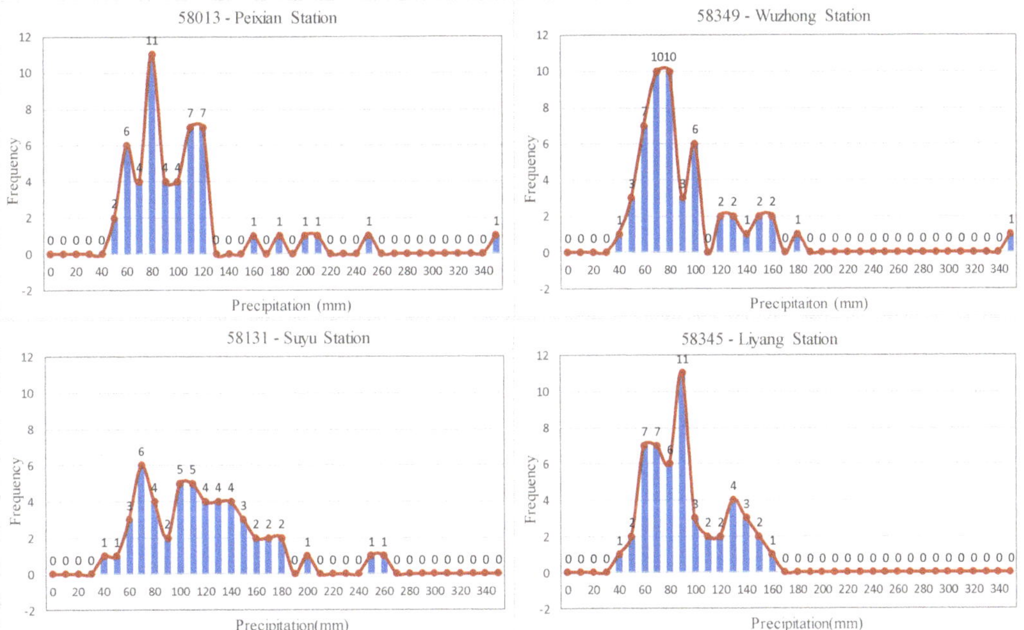

**Figure 6.** Frequency diagram of AMS data at representative site.

A set of rational and reliable frequency estimations can be obtained based on the abovementioned two methods, especially quantiles for the 1-year return period in this study. It is the main innovation of this paper that was missing from previous works. Many researchers have solved the quantile estimates for greater than or equal to a 2-year recurrence interval in regional frequency analysis [16,44–46]. Only few papers solving the quantiles of a 1-year return period are found in the literature [47–49]. However, these studies assume that the non-exceedance probability of a 1-year return period is equal to 0.1 because it is not computable if the non-exceedance probability equals zero. The assumption has no theoretical basis, and the quantile results are inaccurate at lower recurrence intervals due to the underestimation of the AMS analysis. On the basis of verification of Chow's equation applicability, the quantiles can be corrected based on AMS data at low-return periods. It is a very valuable practical element in China because only

AMS data is available from China Meteorological Administration since the 1960s [50]. As a whole, frequency estimations based on the regional L-moments method are in good agreement with observations. The findings validate the reliability of frequency estimations at low return periods. However, the quantiles are slightly overestimated or underestimated in cases of large extreme precipitation. Potential reasons for the deviation of quantiles at large return periods include spatial inconsistencies around the boundary of adjacent regions, short sampling series, and other factors. Future research should conduct a more in-depth analysis of these possibilities.

It is a pity that hourly extreme precipitation data were not available for this study. Ideally, complete quantile estimates from 1-h to 30-day durations would be carried out in the future in the study region. Thus, a complete set of spatiotemporal frequency estimates from multi-duration and multi-return periods can be obtained in the region, which can provide more of a quantitative and scientific basis for decision making. Such data would provide a reference criterion of different duration for comparison, and also provide a stronger scientific basis for issuing storm disaster and flash flood warnings.

## 5. Conclusions

In this paper, a regional frequency analysis of extreme precipitation in the province of Jiangsu in the Yangtze River Delta was studied using regional L-moments methods. A set of rational and reliable frequency estimation was obtained based on AMS data and Chow's equation. Some of the main findings obtained from the research are as follows:

The study area is categorized into five homogeneous regions using cluster analysis. Five distributions (GLO, GEV, GNO, GPA, and PE3) are investigated, and MC simulations and RMSE tests are used to identify the optimum distribution in each homogeneous region. The best-fit distributions based on AMS data are GLO, GEV, GLO, GEV, and GEV for the five homogeneous regions, respectively. The best-fit distributions based on AES data are GNO, GPA, GPA, GPA, and GPA, respectively. By comparing exceedance frequencies with exceedance probabilities it can be seen that extreme precipitation events occur more frequently, and that current quantile estimates based on AMS data are underestimated for frequent events in the study area.

Verification of Chow's equation in this study area shows that there is generally good consistency between real AES data, and AES data generated using Chow's equation and real AMS data. As a whole, the results indicate that Chow's equation can be used as a simple method to obtain reasonable AES-AMS ratios, similar to those obtained from real AES and AMS data. This finding also means that frequency estimations can be revised at lower return periods based on real AMS data and Chow's equation. Two methods can be used to correct for underestimation of frequency estimates. The first method is to use AES data in combination with theoretical exceedance probabilities, such as 0.5, 0.2, 0.1, 0.04, and 0.02 for the corresponding return periods of 2 years, 5 years, 10 years, 25 years, and 50 years. The second way is to use AMS data in combination with the correction of return periods based on Chow's conversion equation. The two methods are equivalent in quantile estimation. However, the second method is strongly recommended due to its simple data processing requirements and reliable results, especially when only AMS data are available for the study area.

A set of rational and reliable frequency estimations can be obtained using the regional L-moments method based on AMS data and Chow's equation. Solving the quantile for a 1-year return period is most important, which means the integral lower limits of the frequency distribution curve. The results show that the estimates increase incrementally with the recurrence interval, and that the estimates agree with observations as a whole. The spatial mapping of quantiles shows that similar patterns exist for 1-year, 10-year, 25-year, and 50-year return periods, and that quantiles in the northern part of the study are greater than in the southern area. The highest frequency estimates are observed near the Suqian and Lianyungang stations in the northern part of Jiangsu, and low values are observed in the southern Taihu lake basin. This suggests that the Xuzhou and Lianyungang areas are likely at high risk of flash floods due to extreme precipitation. Decision makers should pay heightened attention to flood risk and water resource management in these areas.

Based on the three criteria of RE, RMSE, and r, the accuracy of estimations can be evaluated by comparing estimations and observations at the same frequency. The results show that frequency estimations are in good agreement with observations, with the average RE, RMSE, and r of all stations being 5.56%, 0.107 mm, and 0.969, respectively, especially a r > 0.96 was found in each homogeneous region. Frequency estimations based on the regional L-moments method are in good agreement with observations. The findings validate the reliability of frequency estimations at low return periods. A set of reliable quantile estimates are obtained based on two revised ways, which provide a new perspective in regional frequency analysis.

**Supplementary Materials:** The following are available online at https://www.mdpi.com/article/10.3390/w13131832/s1, Table S1: Exceedance frequencies of stations based on AMS data at each station for each region. Table S2: Quantile estimates of extreme precipitation for a 1-day duration from 2-yr to 100-yr return periods in homogeneous region II, III, IV and V.

**Author Contributions:** For research articles with several authors, Conceptualization, Y.S. and J.Z.; methodology, Y.S.; validation, Y.S., J.Z. and J.X.; formal analysis, A.F.; investigation, J.W.; data curation, J.Z.; writing—original draft preparation, Y.S.; writing—review and editing. All authors have read and agreed to the published version of the manuscript.

**Funding:** This work is financially supported by the Meteorological Open Research Fund in Huaihe River Basin (HRM201702), by the Special Fund for Natural Science Foundation of Jiangsu province (BK20141001), by the China Postdoctoral Science Foundation ( No. 2020T130309, No. 2019M651892), by the Jiangsu Water Resources Science and Technology Project (No. 2020022).

**Informed Consent Statement:** It is not applicable for studies not involving humans.

**Data Availability Statement:** The data used to support the findings of this study are available from the corresponding author upon request.

**Acknowledgments:** The authors are also grateful to Lin Bingzhang for providing some constructive suggestions. The authors would like to acknowledge the anonymous reviewers for their thoughtful comments and suggestions

**Conflicts of Interest:** The authors declare that they have no conflict of interest.

## References

1. Lin, B.Z.; Shao, Y.H.; Yan, G.X.; Zhang, Y.H. The core research on the development of engineering hydrology calculation promoted by hydrometeorology. New development of hydrological science and technology. In *China Hydrology Symposium Proceedings*; Hohai University Press: Nanjing, China, 2012; pp. 50–63.
2. Sun, R.C.; Yuan, H.L.; Liu, X.L.; Jiang, X.M. Evaluation of the latest satellite–gauge precipitation products and their hydrologic applications over the Huaihe River basin. *J. Hydrol.* **2016**, *536*, 302–319. [CrossRef]
3. Ba, H.T.; Cholette, M.E.; Borghesani, P.; Zhou, Y.; Ma, L. Opportunistic maintenance considering non-homogenous opportunity arrivals and stochastic opportunity durations. *Reliab. Eng. Syst. Safe* **2017**, *160*, 151–161.
4. Hailegeorgis, T.T.; Alfredsen, K. Analyses of extreme precipitation and runoff events including uncertainties and reliability in design and management of urban water infrastructure. *J. Hydrol.* **2017**, *544*, 290–305. [CrossRef]
5. Nguyen, T.H.; Outayek, S.E.; Lim, S.H.; Nguyen, V.T.V. A systematic approach to selecting the best probability models for annual maximum rainfalls—A case study using data in Ontario (Canada). *J. Hydrol.* **2017**, *553*, 49–58. [CrossRef]
6. Haddad, K.; Rahman, A. Selection of the best fit flood frequency distribution and parameter estimation procedure—A case study for Tasmania in Australia. *Stoch. Environ. Res. Risk Assess.* **2011**, *25*, 415–428. [CrossRef]
7. Merz, B.; Thieken, A.H. Flood risk curves and uncertainty bounds. *Nat. Hazards* **2009**, *51*, 437–458. [CrossRef]
8. Cunnane, C. *Statistical Distributions for Flood Frequency Analysis*; Operational Hydrological Report No. 5/33; World Meteorological Organization (WMO): Geneva, Switzerland, 1989.
9. England, J.F. Flood frequency and design flood estimation procedures in the United States: Progress and challenges. *Austral. J. Water Resour.* **2011**, *15*, 33–46. [CrossRef]
10. Nagy, B.K.; Mohssen, M.; Hughey, K.F.D. Flood frequency analysis for a braided river catchment in New Zealand: Comparing annual maximum and partial duration series with varying record lengths. *J. Hydrol.* **2017**, *547*, 365–374. [CrossRef]
11. Mohssen, M. Partial duration series in the annual domain. In Proceedings of the 18th World IMACS and MODSIM International Congress, Cairns, Australia, 13–17 July 2009; pp. 2694–2700.

12. Shao, Y.H.; Wu, J.M.; Li, M. Study on quantile estimates of extreme precipitation and their spatiotemporal consistency adjustment over the Huaihe River basin. *Theor. Appl. Climatol.* **2017**, *127*, 495–511. [CrossRef]

13. Hosking, J.R.M.; Wallis, J.R. *Regional Frequency Analysis: An Approach Based on L-Moments*; Cambridge University Press: Cambridge, UK, 1997.

14. Neves, M.; Gomes, D.P. Geo-statistics for spatial extremes: A case study of maximum annual rainfall in Portugal. *Procedia Environ. Sci.* **2011**, *7*, 246–251. [CrossRef]

15. Kuo, Y.M.; Chu, H.J.; Pan, T.Y.; Yu, H.L. Investigating common trends of annual maximum rainfalls during heavy rainfall events in southern Taiwan. *J. Hydrol.* **2011**, *409*, 749–758. [CrossRef]

16. Shao, Y.H.; Wu, J.M.; Ye, J.Y.; Liu, Y.H. Frequency analysis and its spatiotemporal characteristics of precipitation extreme events in China during 1951-2010. *Theor. Appl. Climatol.* **2015**, *121*, 775–787. [CrossRef]

17. Tiwari, H.; Rai, S.P.; Sharma, N.; Kumar, D. Computational approaches for annual maximum river flow series. *Ain Shams Eng. J.* **2017**, *8*, 51–58. [CrossRef]

18. Chow, V.T.; Maidment, D.R.; Mays, L.W. *Applied Hydrology*; McGraw-Hill: New York, NY, USA, 1988; pp. 380–385.

19. Madsen, H.; Pearson, C.P.; Rosbjerg, D. Comparison of annual maximum series and partial duration series methods for modeling extreme hydrologic events 1. At-site modeling. *Water Resour. Res.* **1997**, *33*, 747–757. [CrossRef]

20. Lin, B.Z.; Bonnin, G.M.; Martin, D.; Parzybok, T.M.; Riley, D. Regional frequency studies of annual extreme precipitation in the United States based on regional L-moments analysis. In Proceedings of the World Environmental and Water Resource Congress, Omaha, NE, USA, 21–25 May 2006; pp. 1–11.

21. Chow, V.T. *Handbook of Applied Hydrology*; McGraw-Hill: New York, NY, USA, 1964.

22. Wu, J.M.; Lin, B.Z.; Pu, J. Underestimation of precipitation quantile estimates based on AMS data. *J. Nanjing Univ. Inf. Sci. Technol. Nat. Sci. Ed.* **2016**, *8*, 374–379.

23. Bhunya, P.K.; Berndtsson, R.; Jain, S.K.; Kumar, R. Flood analysis using negative binomial and Generalized Pareto models in partial duration series (PDS). *J. Hydrol.* **2013**, *497*, 121–132. [CrossRef]

24. Karim, F.; Hasan, M.; Marvanek, S. Evaluating Annual Maximum and Partial Duration Series for Estimating frequency of Small Magnitude Floods. *Water* **2017**, *9*, 481. [CrossRef]

25. Agilan1, V.; Umamahesh, N.V. Non-Stationary Rainfall Intensity-Duration-Frequency Relationship: A comparison between Annual Maximum and Partial Duration Series. *Water Resour. Manag.* **2017**, *31*, 1825–1841. [CrossRef]

26. Begueria, S. Uncertainties in partial duration series modeling of extremes related to the choice of threshold value. *J. Hydrol.* **2005**, *303*, 215–230. [CrossRef]

27. Nibedita, G.; Ramakar, J. Flood Frequency Analysis of Tel Basin of Mahanadi River System, India using Annual Maximum and POT Flood Data. *Aquat. Procedia* **2015**, *4*, 427–434.

28. Ahmadi, F.; Radmaneh, F.; Parham, G.A.; Mirabbasi, R. Comparison of the performance of power law and probability distributions in the frequency analysis of flood in Dez Basin, Iran. *Nat. Hazards* **2017**, *87*, 1313–1331. [CrossRef]

29. United States Water Resources Council. *Guidelines for Determining Flood Flow Frequency*; Bull. 17B; Hydrology Community, Water Resources Council: Washington, DC, USA, 1982.

30. Franchini, M.; Galeati, G.; Lolli, M. Analytical derivation of the flood frequency curve through partial duration series analysis and a probabilistic representation of the runoff coefficient. *J. Hydrol.* **2005**, *303*, 1–15. [CrossRef]

31. Claps, P.; Laio, F. Can continuous stream flow data support flood frequency analysis? An alternative to the partial duration series approach. *Water Resour. Res.* **2003**, *39*, 12–16. [CrossRef]

32. Takeuchi, K. Annual maximum series and partial duration series-evaluation of Langbein's formula and Chow's discussion. *J. Hydrol.* **1984**, *68*, 275–284. [CrossRef]

33. Ghahraman, B.; Khalili, D. A revisit to partial duration series of short duration rainfalls. *Iran. J. Sci. Technol.* **2004**, *28*, 547–558.

34. SAS. *SAS/STAT User's Guide, Release 6.03*; SAS Institute: Cary, NC, USA, 1988.

35. Pilon, P.J.; Condie, R.; Harvey, K.D. *Consolidated Frequency Analysis Package (CFA), User Manual for Version 1-DEC Pro Series*; Water Resources Branch, Inland Waters Directorate, Environment Canada: Ottawa, ON, USA, 1985.

36. Adamowski, K.; Liang, G.C.; Patry, G.G. Annual maxima and partial duration flood series analysis by parametric and non-parametric methods. *Hydrol. Process* **1998**, *12*, 1685–1699. [CrossRef]

37. Adamowski, K. Regional analysis of annual maximum and partial duration flood data by nonparametric and L-moment methods. *J. Hydrol.* **2000**, *229*, 219–231. [CrossRef]

38. Salinas, J.L.; Castellarin, A.; Kohnová, S.; Kjeldsen, T.R. Regional parent flood frequency distributions in Europe-Part 2: Climate and scale controls. *Hydrol. Earth Syst. Sci.* **2014**, *18*, 4391–4401. [CrossRef]

39. Ball, J.; Babister, M.; Nathan, R.; Weeks, W.; Weinmann, E.; Retallick, M.; Testoni, I. *Australian Rainfall and Runoff: A Guide to Flood Estimation*; Geoscience Australia: Canberra, Australia, 2016.

40. SL44. *Regulation for Calculating Design Flood of Water Resources and Hydropower Projects in China*; China WaterPower Press: Beijing, China, 2006.

41. Vogel, R.M.; Fennessey, N.M. L-Moment diagrams should replace product moment diagrams. *Water Resour. Res.* **1993**, *29*, 1745-1752. [CrossRef]

42. Delicadoa, P.; Goriab, M.N. A small sample comparison of maximum likelihood, moments and L-moments methods for the asymmetric exponential power distribution. *Comput. Stat. Data Anal.* **2008**, *52*, 1661–1673. [CrossRef]

43.  Gubareva, T.S.; Gartsman, B.I. Estimating distribution parameters of extreme hydrometeorological characteristics by L-moments method. *Water Resour.* **2010**, *37*, 437–445. [CrossRef]

44.  Norbiato, D.; Borga, M.; Sangati, M.; Zanon, F. Regional Frequency Analysis of Extreme Precipitation in the eastern Italian Alps and the August 29, 2003 Flash Flood. *J. Hydrol.* **2007**, *345*, 149–166. [CrossRef]

45.  Zakaria, Z.A.; Shabri, A. Regional frequency analysis of extreme rainfalls using partial L-moments method. *Theor. Appl. Climatol.* **2013**, *113*, 83–94. [CrossRef]

46.  Du, H.; Xia, J.; Zeng, S.D. Regional frequency analysis of extreme precipitation and its spatio-temporal characteristics in the Huai River Basin, China. *Nat. Hazards* **2014**, *70*, 195–215. [CrossRef]

47.  Neykov, N.M.; Neytchev, P.N.; Gelder, P.H.A.J.M.V.; Todorov, V.K. Robust detection of discordant sites in regional frequency analysis. *Water Resour. Res.* **2007**, *43*, 1–10. [CrossRef]

48.  Yang, T.; Shao, Q.X.; Hao, Z.C.; Chen, X.; Zhang, Z.X.; Xu, C.Y.; Sun, L.M. Regional frequency analysis and spatio-temporal pattern characterization of rainfall extremes in the Pearl River Basin, China. *J. Hydrol.* **2010**, *380*, 386–405. [CrossRef]

49.  Hu, C.; Xia, J.; She, D.X.; Xu, C.Y.; Zhang, L.P.; Song, Z.H.; Zhao, L. A modified regional L-moment method for regional extreme precipitation frequency analysis in the Songliao River Basin of China. *Atmos. Res.* **2019**, *230*, 104629. [CrossRef]

50.  Deng, P.D. Review on probability and application of two sampling methods for urban storm. *Water Wastewater Eng.* **2006**, *32*, 39–42.

*Article*

# MLE-Based Parameter Estimation for Four-Parameter Exponential Gamma Distribution and Asymptotic Variance of Its Quantiles

Songbai Song [1,*], Yan Kang [1], Xiaoyan Song [1] and Vijay P. Singh [2,3]

[1] College of Water Resources and Architectural Engineering & Key Laboratory of Agricultural Water and Soil Engineering, Ministry of Education, Northwest A & F University, Shaanxi 712100, China; kangyan@nwsuaf.edu.cn (Y.K.); xiaoyansong@nwsuaf.edu.cn (X.S.)

[2] Department of Biological and Agricultural Engineering, Zachry Department of Civil Engineering, Texas A&M University, College Station, TX 77843-2117, USA; vsingh@tamu.edu

[3] National Water Center, UAE University, Al Ain 15551, United Arab Emirates

* Correspondence: ssb6533@nwafu.edu.cn

**Abstract:** The choice of a probability distribution function and confidence interval of estimated design values have long been of interest in flood frequency analysis. Although the four-parameter exponential gamma (FPEG) distribution has been developed for application in hydrology, its maximum likelihood estimation (MLE)-based parameter estimation method and asymptotic variance of its quantiles have not been well documented. In this study, the MLE method was used to estimate the parameters and confidence intervals of quantiles of the FPEG distribution. This method entails parameter estimation and asymptotic variances of quantile estimators. The parameter estimation consisted of a set of four equations which, after algebraic simplification, were solved using a three dimensional Levenberg-Marquardt algorithm. Based on sample information matrix and Fisher's expected information matrix, derivatives of the design quantile with respect to the parameters were derived. The method of estimation was applied to annual precipitation data from the Weihe watershed, China and confidence intervals for quantiles were determined. Results showed that the FPEG was a good candidate to model annual precipitation data and can provide guidance for estimating design values.

**Keywords:** four-parameter exponential gamma distribution; levenberg-marquardt algorithm; maximum likelihood estimation; variance and covariance matrix; Weihe watershed

**Citation:** Song, S.; Kang, Y.; Song, X.; Singh, V.P. MLE-Based Parameter Estimation for Four-Parameter Exponential Gamma Distribution and Asymptotic Variance of Its Quantiles. *Water* **2021**, *13*, 2092. https://doi.org/10.3390/w13152092

Academic Editor: Giuseppe Pezzinga

Received: 9 June 2021
Accepted: 29 July 2021
Published: 30 July 2021

**Publisher's Note:** MDPI stays neutral with regard to jurisdictional claims in published maps and institutional affiliations.

## 1. Introduction

Hydrological frequency analysis is important for planning, designing and managing water resources projects. The design values (e.g., design flood, design rainfall) computed by frequency analysis involve uncertainties due to the sampling method, sample length, empirical frequency formula, cumulative distribution function (CDF) or probability density function (PDF), parameter estimation method, goodness-of-fit test, and extent of data extrapolation [1,2]. Among these uncertainty sources, there has been a considerable interest in the choice of CDF for a given sample, because the true CDF of a hydrological variable is unknown. Rao and Hamed summarized commonly used distributions: normal and related, gamma family, extreme value, Wakeby, and logistic as well as their application [3].

Further, quantifying the uncertainty of the estimated design values is important in the planning, design, and management of water resources projects [1,4]. In practice, standard error and confidence interval (confidence limits) are employed to measure the uncertainty of a statistical quantity [4]. Rao and Hamed provided confidence intervals of some common distributions, with parameters estimated by the method of moments (MOM), the probability weighted moments (PWM), and maximum likelihood estimation (MLE) [3]. Shin [5] and Shin et al. [6] summarized methods for computing confidence intervals. Methods of estimating the confidence intervals of design quantities include Monte Carlo simulation [7], approximate method [8], analytical method [5,9–11], asymptotic variance of the expected moments algorithm [12], bootstrap method [13–16], and standard error of regional population index flood (RPIF) [17]. Studies show that confidence intervals mainly depend on the method of estimation of the parameters of the probability distribution function.

The four-parameter exponential gamma (FPEG) distribution has been applied in hydrology in China and be specialized into 10 kinds of probability distribution functions: gamma, Pearson type III (P-III), K–M, Weibull, Chi-square, exponential, normal, Pearson type V (P-V), log-normal, and Gumbel. The properties of FPEG distribution and relations between this distribution and others, and its potential for application, have been investigated [18,19]. However, the MLE-based parameter estimation method and algorithms for computing confidence intervals of the design values for the FPEG distribution have received little attention.

The objective of this paper, therefore, is to present the MLE method for estimating the FPEG distribution parameters, and derive confidence intervals of quantiles using asymptotic variances. The method of parameter estimation involved a set of four equations which are solved by a three dimensional Levenberg–Marquardt algorithm. Following Kendall and Stuart [20], the expected values of the second-order partial derivatives of the log-likelihood function with respect to the parameters, and the explicit formulae for the variances and covariances are analytically derived. The proposed estimation procedure is illustrated by using observed annual precipitation data.

The paper is organized as follows. Describing the FPEG distribution and estimation of its quantiles. A set of four equations of the MLE method for parameters and confidence intervals of quantiles are derived in Section 2, followed by an application to annual precipitation from the Weihe watershed in China in Section 3. Conclusions, along with a summary of the main features of the proposed method, are given in Section 4.

## 2. Theory and Methodology

### 2.1. Probability Density Function and Cumulative Distribution Function

The FPEG distribution has the probability density function (PDF), $f(x)$, expressed as [18,19]:

$$f(x) = \begin{cases} \frac{\beta^\alpha}{b\Gamma(\alpha)}(x-\delta)^{\frac{\alpha}{b}-1}e^{-\beta(x-\delta)^{\frac{1}{b}}} & ; \quad \delta \le x < \infty \\ 0 & ; \quad \text{otherwise} \end{cases} \tag{1}$$

where $\alpha > 0$ is the shape parameter; $\beta > 0$ is the scale parameter; $\delta > 0$ is the location parameter; $b > 0$ is the transformation parameter; $\Gamma(\alpha)$ is the complete gamma function; $x$ is the value of the random variable $X$. Figure 1 shows some typical shapes of the PDF. The CDF can be expressed as

$$p = F(X \le x) = \int_\delta^x f(t)dt = \int_\delta^x \frac{\beta^\alpha}{b\Gamma(\alpha)}(t-\delta)^{\frac{\alpha}{b}-1}e^{-\beta(t-\delta)^{\frac{1}{b}}}dt$$

When the design frequency $p$ is given, its correspondance to design value $x_p$ (quantile) can be expressed as

$$p = F(X \le x_p) = \int_\delta^{x_p} f(x)dx = \int_\delta^{x_p} \frac{\beta^\alpha}{b\Gamma(\alpha)}(x-\delta)^{\frac{\alpha}{b}-1}e^{-\beta(x-\delta)^{\frac{1}{b}}}dx \tag{2}$$

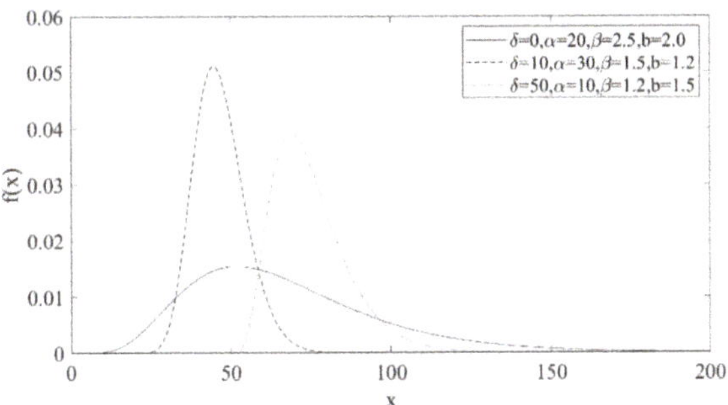

**Figure 1.** Examples of the probability density function of the four-parameter exponential gamma distribution.

Equation (2) can be transformed to a one-parameter gamma using the following transformation. Let $t = \beta(x - \delta)^{\frac{1}{b}}$, then $x = \delta + \frac{1}{\beta^b}t^b$, $x - \delta = \frac{1}{\beta^b}t^b$ and $dx = \frac{b}{\beta^b}t^{b-1}dt$, when $x = \delta, t = 0$; $x = x_p, t_p = \beta(x_p - \delta)^{\frac{1}{b}}$ substitution of these quantities into Equation (2) results in [18,19]:

$$p = \int_0^{t_p} \frac{\beta^\alpha}{b\Gamma(\alpha)} \left(\frac{1}{\beta^b}t^b\right)^{\frac{\alpha}{b}-1} e^{-t}\frac{b}{\beta^b}t^{b-1}dt = \int_0^{t_p} \frac{\beta^\alpha}{b\Gamma(\alpha)}\frac{1}{\beta^{\alpha-b}}t^{\alpha-b}e^{-t}\frac{b}{\beta^b}t^{b-1}dt = \int_0^{t_p} \frac{1}{\Gamma(\alpha)}t^{\alpha-1}e^{-t}dt \tag{3}$$

where $t_p = \beta(x_p - \delta)^{\frac{1}{b}}$, which can be determined by the incomplete gamma function.

### 2.2. Estimation of Quantiles

The quantile corresponding to the probability of exceedance $p$, $x_p$, is obtained as

$$x_p = \delta + \frac{1}{\beta^b}t_p^b \tag{4}$$

Also, the estimator $x_p$ may be generally written in terms of the mathematical expectation $E(X)$, the coefficient of variation $C_v$, and the frequency factor $\Phi_p$ as

$$x_p = E(X)\left(1 + \Phi_p C_v\right) \tag{5}$$

Given the probability of exceedance $p$, the frequency factor $\Phi_p$ can be written in the following form [18,19]:

$$\Phi_p = \frac{t_p^b\Gamma(\alpha) - \Gamma(\alpha + b)}{\sqrt{\Gamma(\alpha)\Gamma(\alpha + 2b) - \Gamma^2(\alpha + b)}} \tag{6}$$

Note that from Equations (3) and (6), the frequency factor $\Phi_p$ is a function of the probability of exceedance and parameters $\alpha$ and $b$. Some numerical values for such a function are shown in Table 1. For large $\alpha$ (e.g., $\alpha > 100$) it is seen that the differences among the $\Phi_p$ values for a given $p$ are subtle.

**Table 1.** Frequency factors for the four-parameter exponential gamma distribution.

### $b = 0.50$

| $\alpha$ | Exceedance Probability $p$ | | | | | | | | | | | |
|---|---|---|---|---|---|---|---|---|---|---|---|---|
| | 0.001 | 0.01 | 0.02 | 0.10 | 0.30 | 0.50 | 0.70 | 0.90 | 0.95 | 0.99 | 0.995 | 0.999 |
| 0.10 | 6.1052 | 4.0173 | 3.2803 | 1.3079 | −0.0892 | −0.4810 | −0.5628 | −0.5697 | −0.5697 | −0.5697 | −0.5697 | −0.5697 |
| 0.20 | 5.0293 | 3.4670 | 2.9186 | 1.4286 | 0.1880 | −0.4003 | −0.7010 | −0.8087 | −0.8147 | −0.8160 | −0.8160 | −0.8160 |
| 0.30 | 4.5665 | 3.2055 | 2.7291 | 1.4308 | 0.3045 | −0.3081 | −0.7170 | −0.9631 | −0.9950 | −1.0087 | −1.0094 | −1.0097 |
| 0.40 | 4.3045 | 3.0513 | 2.6134 | 1.4183 | 0.3625 | −0.2462 | −0.7029 | −1.0574 | −1.1260 | −1.1691 | −1.1729 | −1.1753 |
| 0.50 | 4.1350 | 2.9494 | 2.5356 | 1.4050 | 0.3957 | −0.2047 | −0.6844 | −1.1151 | −1.2196 | −1.3028 | −1.3132 | −1.3215 |
| 0.60 | 4.0161 | 2.8770 | 2.4796 | 1.3934 | 0.4168 | −0.1757 | −0.6676 | −1.1516 | −1.2869 | −1.4136 | −1.4332 | −1.4518 |
| 0.70 | 3.9278 | 2.8228 | 2.4375 | 1.3835 | 0.4312 | −0.1545 | −0.6534 | −1.1755 | −1.3362 | −1.5050 | −1.5355 | −1.5679 |
| 0.80 | 3.8596 | 2.7807 | 2.4046 | 1.3753 | 0.4416 | −0.1385 | −0.6417 | −1.1920 | −1.3732 | −1.5807 | −1.6225 | −1.6713 |
| 0.90 | 3.8051 | 2.7470 | 2.3782 | 1.3684 | 0.4494 | −0.1259 | −0.6320 | −1.2037 | −1.4017 | −1.6438 | −1.6967 | −1.7632 |
| 1.00 | 3.7605 | 2.7193 | 2.3565 | 1.3625 | 0.4555 | −0.1159 | −0.6239 | −1.2124 | −1.4242 | −1.6967 | −1.7602 | −1.8448 |
| 2.00 | 3.5423 | 2.5845 | 2.2506 | 1.3323 | 0.4817 | −0.0701 | −0.5840 | −1.2436 | −1.5195 | −1.9561 | −2.0881 | −2.3133 |
| 4.00 | 3.4078 | 2.5032 | 2.1871 | 1.3141 | 0.4955 | −0.0455 | −0.5617 | −1.2564 | −1.5655 | −2.0975 | −2.2754 | −2.6114 |
| 6.00 | 3.3513 | 2.4699 | 2.1614 | 1.3070 | 0.5010 | −0.0361 | −0.5534 | −1.2611 | −1.5823 | −2.1492 | −2.3442 | −2.7233 |
| 8.00 | 3.3179 | 2.4506 | 2.1466 | 1.3031 | 0.5041 | −0.0308 | −0.5488 | −1.2638 | −1.5917 | −2.1773 | −2.3815 | −2.7840 |
| 10.00 | 3.2951 | 2.4375 | 2.1366 | 1.3005 | 0.5062 | −0.0273 | −0.5459 | −1.2656 | −1.5978 | −2.1955 | −2.4056 | −2.8229 |
| 20.00 | 3.2380 | 2.4054 | 2.1123 | 1.2945 | 0.5115 | −0.0190 | −0.5390 | −1.2702 | −1.6123 | −2.2377 | −2.4610 | −2.9117 |
| 40.00 | 3.1966 | 2.3827 | 2.0953 | 1.2905 | 0.5152 | −0.0133 | −0.5344 | −1.2734 | −1.6221 | −2.2653 | −2.4971 | −2.9686 |
| 60.00 | 3.1779 | 2.3726 | 2.0878 | 1.2887 | 0.5168 | −0.0108 | −0.5325 | −1.2749 | −1.6263 | −2.2770 | −2.5123 | −2.9924 |
| 80.00 | 3.1666 | 2.3665 | 2.0833 | 1.2877 | 0.5178 | −0.0094 | −0.5314 | −1.2757 | −1.6288 | −2.2839 | −2.5212 | −3.0063 |
| 100.00 | 3.1588 | 2.3623 | 2.0802 | 1.2871 | 0.5185 | −0.0084 | −0.5306 | −1.2763 | −1.6305 | −2.2885 | −2.5272 | −3.0156 |
| 200.00 | 3.1392 | 2.3519 | 2.0725 | 1.2854 | 0.5202 | −0.0059 | −0.5288 | −1.2778 | −1.6348 | −2.2998 | −2.5418 | −3.0382 |
| 400.00 | 3.1251 | 2.3445 | 2.0670 | 1.2843 | 0.5214 | −0.0042 | −0.5275 | −1.2789 | −1.6377 | −2.3077 | −2.5520 | −3.0538 |
| 600.00 | 3.1188 | 2.3412 | 2.0646 | 1.2838 | 0.5220 | −0.0034 | −0.5269 | −1.2794 | −1.6390 | −2.3112 | −2.5564 | −3.0606 |
| 800.00 | 3.1151 | 2.3392 | 2.0632 | 1.2835 | 0.5223 | −0.0029 | −0.5266 | −1.2797 | −1.6398 | −2.3132 | −2.5590 | −3.0646 |
| 1000.00 | 3.1125 | 2.3379 | 2.0622 | 1.2833 | 0.5225 | −0.0026 | −0.5263 | −1.2799 | −1.6404 | −2.3146 | −2.5608 | −3.0674 |

### $b = 1.00$

| $\alpha$ | Exceedance Probability $p$ | | | | | | | | | | | |
|---|---|---|---|---|---|---|---|---|---|---|---|---|
| | 0.001 | 0.01 | 0.02 | 0.10 | 0.30 | 0.50 | 0.70 | 0.90 | 0.95 | 0.99 | 0.995 | 0.999 |
| 0.10 | 10.3207 | 4.7070 | 3.2225 | 0.5254 | −0.2611 | −0.3144 | −0.3162 | −0.3162 | −0.3162 | −0.3162 | −0.3162 | −0.3162 |
| 0.20 | 8.7251 | 4.4773 | 3.2969 | 0.9054 | −0.1766 | −0.4008 | −0.4437 | −0.4472 | −0.4472 | −0.4472 | −0.4472 | −0.4472 |
| 0.30 | 7.8853 | 4.2712 | 3.2435 | 1.0677 | −0.0793 | −0.4142 | −0.5245 | −0.5471 | −0.5477 | −0.5477 | −0.5477 | −0.5477 |
| 0.40 | 7.3406 | 4.1111 | 3.1792 | 1.1540 | −0.0043 | −0.4031 | −0.5731 | −0.6287 | −0.6318 | −0.6324 | −0.6325 | −0.6325 |
| 0.50 | 6.9491 | 3.9845 | 3.1197 | 1.2060 | 0.0525 | −0.3854 | −0.6021 | −0.6959 | −0.7043 | −0.7070 | −0.7071 | −0.7071 |
| 0.60 | 6.6497 | 3.8815 | 3.0671 | 1.2400 | 0.0965 | −0.3670 | −0.6197 | −0.7513 | −0.7673 | −0.7741 | −0.7744 | −0.7746 |
| 0.70 | 6.4109 | 3.7957 | 3.0209 | 1.2635 | 0.1316 | −0.3497 | −0.6305 | −0.7970 | −0.8221 | −0.8352 | −0.8361 | −0.8366 |
| 0.80 | 6.2145 | 3.7229 | 2.9802 | 1.2804 | 0.1601 | −0.3339 | −0.6371 | −0.8352 | −0.8699 | −0.8912 | −0.8931 | −0.8942 |
| 0.90 | 6.0493 | 3.6601 | 2.9442 | 1.2930 | 0.1839 | −0.3197 | −0.6410 | −0.8673 | −0.9118 | −0.9426 | −0.9459 | −0.9482 |
| 1.00 | 5.9078 | 3.6052 | 2.9120 | 1.3026 | 0.2040 | −0.3069 | −0.6433 | −0.8946 | −0.9487 | −0.9899 | −0.9950 | −0.9990 |
| 2.00 | 5.1148 | 3.2798 | 2.7110 | 1.3362 | 0.3106 | −0.2274 | −0.6383 | −1.0382 | −1.1629 | −1.3092 | −1.3410 | −1.3821 |
| 4.00 | 4.5311 | 3.0226 | 2.5421 | 1.3404 | 0.3811 | −0.1640 | −0.6181 | −1.1276 | −1.3168 | −1.5884 | −1.6639 | −1.7857 |
| 6.00 | 4.2681 | 2.9020 | 2.4605 | 1.3369 | 0.4105 | −0.1347 | −0.6054 | −1.1627 | −1.3827 | −1.7207 | −1.8220 | −1.9975 |
| 8.00 | 4.1105 | 2.8284 | 2.4100 | 1.3332 | 0.4274 | −0.1169 | −0.5967 | −1.1822 | −1.4210 | −1.8010 | −1.9194 | −2.1316 |
| 10.00 | 4.0026 | 2.7775 | 2.3748 | 1.3301 | 0.4387 | −0.1048 | −0.5904 | −1.1949 | −1.4466 | −1.8562 | −1.9869 | −2.2261 |
| 20.00 | 3.7345 | 2.6487 | 2.2848 | 1.3198 | 0.4656 | −0.0743 | −0.5733 | −1.2242 | −1.5083 | −1.9941 | −2.1571 | −2.4690 |
| 40.00 | 3.5449 | 2.5558 | 2.2191 | 1.3106 | 0.4838 | −0.0526 | −0.5601 | −1.2429 | −1.5502 | −2.0918 | −2.2791 | −2.6468 |
| 60.00 | 3.4610 | 2.5142 | 2.1894 | 1.3060 | 0.4916 | −0.0430 | −0.5539 | −1.2507 | −1.5683 | −2.1351 | −2.3334 | −2.7269 |
| 80.00 | 3.4111 | 2.4893 | 2.1716 | 1.3031 | 0.4962 | −0.0372 | −0.5502 | −1.2552 | −1.5789 | −2.1608 | −2.3658 | −2.7749 |
| 100.00 | 3.3770 | 2.4723 | 2.1593 | 1.3011 | 0.4993 | −0.0333 | −0.5476 | −1.2582 | −1.5861 | −2.1784 | −2.3880 | −2.8079 |
| 200.00 | 3.2927 | 2.4298 | 2.1288 | 1.2957 | 0.5068 | −0.0236 | −0.5410 | −1.2655 | −1.6037 | −2.2219 | −2.4430 | −2.8900 |
| 400.00 | 3.2332 | 2.3996 | 2.1070 | 1.2918 | 0.5121 | −0.0167 | −0.5362 | −1.2704 | −1.6159 | −2.2526 | −2.4819 | −2.9483 |
| 600.00 | 3.2069 | 2.3862 | 2.0973 | 1.2900 | 0.5144 | −0.0136 | −0.5341 | −1.2725 | −1.6213 | −2.2661 | −2.4991 | −2.9743 |
| 800.00 | 3.1912 | 2.3782 | 2.0915 | 1.2889 | 0.5157 | −0.0118 | −0.5328 | −1.2737 | −1.6245 | −2.2742 | −2.5094 | −2.9898 |
| 1000.00 | 3.1806 | 2.3727 | 2.0875 | 1.2881 | 0.5167 | −0.0105 | −0.5319 | −1.2746 | −1.6267 | −2.2797 | −2.5164 | −3.0003 |

**Table 1.** *Cont.*

| | $b = 1.50$ | | | | | | | | | | | |
| | Exceedance Probability $p$ | | | | | | | | | | | |
| $\alpha$ | 0.001 | 0.01 | 0.02 | 0.10 | 0.30 | 0.50 | 0.70 | 0.90 | 0.95 | 0.99 | 0.995 | 0.999 |
|---|---|---|---|---|---|---|---|---|---|---|---|---|
| 0.10 | 12.8886 | 4.0481 | 2.3121 | 0.0921 | −0.1944 | −0.1992 | −0.1993 | −0.1993 | −0.1993 | −0.1993 | −0.1993 | −0.1993 |
| 0.20 | 11.5995 | 4.3915 | 2.8159 | 0.3898 | −0.2229 | −0.2788 | −0.2830 | −0.2831 | −0.2831 | −0.2831 | −0.2831 | −0.2831 |
| 0.30 | 10.7500 | 4.4459 | 2.9974 | 0.5824 | −0.2028 | −0.3260 | −0.3465 | −0.3481 | −0.3481 | −0.3481 | −0.3481 | −0.3481 |
| 0.40 | 10.1286 | 4.4298 | 3.0780 | 0.7137 | −0.1704 | −0.3519 | −0.3965 | −0.4032 | −0.4033 | −0.4033 | −0.4033 | −0.4033 |
| 0.50 | 9.6438 | 4.3907 | 3.1154 | 0.8089 | −0.1367 | −0.3652 | −0.4360 | −0.4516 | −0.4521 | −0.4522 | −0.4522 | −0.4522 |
| 0.60 | 9.2498 | 4.3439 | 3.1311 | 0.8810 | −0.1049 | −0.3713 | −0.4673 | −0.4949 | −0.4963 | −0.4966 | −0.4966 | −0.4966 |
| 0.70 | 8.9202 | 4.2953 | 3.1349 | 0.9377 | −0.0758 | −0.3729 | −0.4923 | −0.5338 | −0.5368 | −0.5376 | −0.5376 | −0.5376 |
| 0.80 | 8.6385 | 4.2474 | 3.1318 | 0.9833 | −0.0495 | −0.3719 | −0.5125 | −0.5689 | −0.5741 | −0.5758 | −0.5759 | −0.5759 |
| 0.90 | 8.3938 | 4.2013 | 3.1245 | 1.0208 | −0.0256 | −0.3693 | −0.5290 | −0.6007 | −0.6085 | −0.6117 | −0.6119 | −0.6120 |
| 1.00 | 8.1783 | 4.1573 | 3.1147 | 1.0521 | −0.0040 | −0.3656 | −0.5426 | −0.6295 | −0.6405 | −0.6456 | −0.6460 | −0.6461 |
| 2.00 | 6.8717 | 3.8286 | 2.9915 | 1.2080 | 0.1351 | −0.3192 | −0.6040 | −0.8156 | −0.8645 | −0.9074 | −0.9141 | −0.9206 |
| 4.00 | 5.8088 | 3.4886 | 2.8158 | 1.2896 | 0.2518 | −0.2547 | −0.6233 | −0.9689 | −1.0757 | −1.2040 | −1.2335 | −1.2744 |
| 6.00 | 5.3092 | 3.3073 | 2.7105 | 1.3131 | 0.3060 | −0.2171 | −0.6220 | −1.0377 | −1.1801 | −1.3709 | −1.4207 | −1.4975 |
| 8.00 | 5.0061 | 3.1907 | 2.6395 | 1.3224 | 0.3383 | −0.1923 | −0.6179 | −1.0777 | −1.2442 | −1.4806 | −1.5464 | −1.6537 |
| 10.00 | 4.7981 | 3.1076 | 2.5876 | 1.3266 | 0.3602 | −0.1743 | −0.6136 | −1.1043 | −1.2884 | −1.5596 | −1.6381 | −1.7706 |
| 20.00 | 4.2817 | 2.8908 | 2.4471 | 1.3286 | 0.4131 | −0.1268 | −0.5971 | −1.1664 | −1.3981 | −1.7689 | −1.8857 | −2.0978 |
| 40.00 | 3.9208 | 2.7297 | 2.3388 | 1.3226 | 0.4487 | −0.0909 | −0.5806 | −1.2060 | −1.4743 | −1.9263 | −2.0762 | −2.3604 |
| 60.00 | 3.7632 | 2.6569 | 2.2887 | 1.3178 | 0.4638 | −0.0746 | −0.5719 | −1.2221 | −1.5072 | −1.9981 | −2.1641 | −2.4847 |
| 80.00 | 3.6702 | 2.6131 | 2.2583 | 1.3144 | 0.4725 | −0.0647 | −0.5664 | −1.2313 | −1.5266 | −2.0413 | −2.2175 | −2.5610 |
| 100.00 | 3.6071 | 2.5830 | 2.2373 | 1.3118 | 0.4784 | −0.0580 | −0.5625 | −1.2373 | −1.5397 | −2.0710 | −2.2543 | −2.6140 |
| 200.00 | 3.4523 | 2.5082 | 2.1845 | 1.3045 | 0.4926 | −0.0411 | −0.5522 | −1.2516 | −1.5716 | −2.1452 | −2.3467 | −2.7485 |
| 400.00 | 3.3444 | 2.4550 | 2.1467 | 1.2985 | 0.5023 | −0.0291 | −0.5445 | −1.2611 | −1.5936 | −2.1980 | −2.4131 | −2.8462 |
| 600.00 | 3.2971 | 2.4314 | 2.1298 | 1.2957 | 0.5065 | −0.0238 | −0.5410 | −1.2651 | −1.6032 | −2.2215 | −2.4427 | −2.8902 |
| 800.00 | 3.2690 | 2.4174 | 2.1197 | 1.2939 | 0.5090 | −0.0206 | −0.5388 | −1.2674 | −1.6089 | −2.2355 | −2.4604 | −2.9166 |
| 1000.00 | 3.2499 | 2.4078 | 2.1128 | 1.2927 | 0.5106 | −0.0184 | −0.5374 | −1.2690 | −1.6128 | −2.2451 | −2.4725 | −2.9346 |

| | $b = 2.00$ | | | | | | | | | | | |
| | Exceedance Probability $p$ | | | | | | | | | | | |
| $\alpha$ | 0.001 | 0.01 | 0.02 | 0.10 | 0.30 | 0.50 | 0.70 | 0.90 | 0.95 | 0.99 | 0.995 | 0.999 |
|---|---|---|---|---|---|---|---|---|---|---|---|---|
| 0.10 | 13.3536 | 2.8762 | 1.3613 | −0.0467 | −0.1307 | −0.1311 | −0.1311 | −0.1311 | −0.1311 | −0.1311 | −0.1311 | −0.1311 |
| 0.20 | 12.9833 | 3.6087 | 2.0068 | 0.0986 | −0.1764 | −0.1875 | −0.1879 | −0.1879 | −0.1879 | −0.1879 | −0.1879 | −0.1879 |
| 0.30 | 12.4989 | 3.9246 | 2.3406 | 0.2345 | −0.1935 | −0.2295 | −0.2326 | −0.2327 | −0.2327 | −0.2327 | −0.2327 | −0.2327 |
| 0.40 | 12.0542 | 4.0914 | 2.5456 | 0.3473 | −0.1950 | −0.2612 | −0.2708 | −0.2714 | −0.2714 | −0.2714 | −0.2714 | −0.2714 |
| 0.50 | 11.6592 | 4.1868 | 2.6831 | 0.4409 | −0.1884 | −0.2851 | −0.3039 | −0.3062 | −0.3062 | −0.3062 | −0.3062 | −0.3062 |
| 0.60 | 11.3083 | 4.2425 | 2.7802 | 0.5195 | −0.1777 | −0.3030 | −0.3330 | −0.3379 | −0.3381 | −0.3381 | −0.3381 | −0.3381 |
| 0.70 | 10.9946 | 4.2742 | 2.8512 | 0.5864 | −0.1650 | −0.3164 | −0.3585 | −0.3674 | −0.3677 | −0.3677 | −0.3677 | −0.3677 |
| 0.80 | 10.7121 | 4.2903 | 2.9041 | 0.6440 | −0.1512 | −0.3266 | −0.3811 | −0.3949 | −0.3955 | −0.3956 | −0.3956 | −0.3956 |
| 0.90 | 10.4561 | 4.2962 | 2.9442 | 0.6942 | −0.1371 | −0.3342 | −0.4010 | −0.4206 | −0.4217 | −0.4220 | −0.4220 | −0.4220 |
| 1.00 | 10.2227 | 4.2949 | 2.9748 | 0.7383 | −0.1231 | −0.3398 | −0.4188 | −0.4447 | −0.4466 | −0.4472 | −0.4472 | −0.4472 |
| 2.00 | 8.6475 | 4.1535 | 3.0588 | 0.9962 | −0.0055 | −0.3473 | −0.5233 | −0.6238 | −0.6409 | −0.6522 | −0.6535 | −0.6544 |
| 4.00 | 7.1806 | 3.8570 | 2.9806 | 1.1743 | 0.1277 | −0.3106 | −0.5893 | −0.8083 | −0.8645 | −0.9212 | −0.9319 | −0.9447 |
| 6.00 | 6.4445 | 3.6576 | 2.8918 | 1.2401 | 0.1994 | −0.2775 | −0.6084 | −0.9033 | −0.9909 | −1.0934 | −1.1167 | −1.1487 |
| 8.00 | 5.9875 | 3.5177 | 2.8205 | 1.2724 | 0.2448 | −0.2520 | −0.6148 | −0.9620 | −1.0735 | −1.2150 | −1.2501 | −1.3022 |
| 10.00 | 5.6704 | 3.4134 | 2.7637 | 1.2907 | 0.2765 | −0.2322 | −0.6165 | −1.0023 | −1.1326 | −1.3066 | −1.3522 | −1.4232 |
| 20.00 | 4.8774 | 3.1261 | 2.5947 | 1.3204 | 0.3559 | −0.1746 | −0.6103 | −1.0998 | −1.2855 | −1.5637 | −1.6459 | −1.7877 |
| 40.00 | 4.3242 | 2.9025 | 2.4527 | 1.3260 | 0.4104 | −0.1275 | −0.5957 | −1.1635 | −1.3957 | −1.7697 | −1.8885 | −2.1063 |
| 60.00 | 4.0846 | 2.7994 | 2.3846 | 1.3240 | 0.4336 | −0.1052 | −0.5865 | −1.1895 | −1.4440 | −1.8665 | −2.0047 | −2.2643 |
| 80.00 | 3.9440 | 2.7371 | 2.3427 | 1.3215 | 0.4471 | −0.0916 | −0.5801 | −1.2042 | −1.4724 | −1.9256 | −2.0764 | −2.3635 |
| 100.00 | 3.8491 | 2.6942 | 2.3136 | 1.3192 | 0.4560 | −0.0822 | −0.5754 | −1.2139 | −1.4916 | −1.9666 | −2.1263 | −2.4332 |
| 200.00 | 3.6181 | 2.5870 | 2.2396 | 1.3115 | 0.4776 | −0.0585 | −0.5625 | −1.2365 | −1.5385 | −2.0698 | −2.2532 | −2.6137 |
| 400.00 | 3.4588 | 2.5107 | 2.1861 | 1.3044 | 0.4921 | −0.0415 | −0.5523 | −1.2511 | −1.5708 | −2.1440 | −2.3456 | −2.7474 |
| 600.00 | 3.3894 | 2.4769 | 2.1621 | 1.3008 | 0.4983 | −0.0339 | −0.5476 | −1.2572 | −1.5848 | −2.1772 | −2.3871 | −2.8083 |
| 800.00 | 3.3484 | 2.4567 | 2.1478 | 1.2985 | 0.5020 | −0.0294 | −0.5446 | −1.2608 | −1.5930 | −2.1970 | −2.4120 | −2.8450 |
| 1000.00 | 3.3205 | 2.4429 | 2.1379 | 1.2969 | 0.5044 | −0.0263 | −0.5426 | −1.2631 | −1.5987 | −2.2106 | −2.4291 | −2.8702 |

**Table 1.** *Cont.*

| | b = 2.50 | | | | | | | | | | | |
| --- | --- | --- | --- | --- | --- | --- | --- | --- | --- | --- | --- | --- |
| | **Exceedance Probability** *p* | | | | | | | | | | | |
| $\alpha$ | 0.001 | 0.01 | 0.02 | 0.10 | 0.30 | 0.50 | 0.70 | 0.90 | 0.95 | 0.99 | 0.995 | 0.999 |
| 0.10 | 12.0692 | 1.7751 | 0.6880 | −0.0666 | −0.0880 | −0.0880 | −0.0880 | −0.0880 | −0.0880 | −0.0880 | −0.0880 | −0.0880 |
| 0.20 | 12.7687 | 2.5964 | 1.2456 | −0.0196 | −0.1254 | −0.1273 | −0.1273 | −0.1273 | −0.1273 | −0.1273 | −0.1273 | −0.1273 |
| 0.30 | 12.8544 | 3.0532 | 1.6046 | 0.0500 | −0.1496 | −0.1586 | −0.1590 | −0.1591 | −0.1591 | −0.1591 | −0.1591 | −0.1591 |
| 0.40 | 12.7678 | 3.3500 | 1.8603 | 0.1209 | −0.1643 | −0.1851 | −0.1868 | −0.1869 | −0.1869 | −0.1869 | −0.1869 | −0.1869 |
| 0.50 | 12.6164 | 3.5586 | 2.0536 | 0.1881 | −0.1725 | −0.2076 | −0.2120 | −0.2123 | −0.2123 | −0.2123 | −0.2123 | −0.2123 |
| 0.60 | 12.4387 | 3.7119 | 2.2053 | 0.2503 | −0.1761 | −0.2269 | −0.2350 | −0.2358 | −0.2358 | −0.2358 | −0.2358 | −0.2358 |
| 0.70 | 12.2515 | 3.8278 | 2.3275 | 0.3075 | −0.1764 | −0.2434 | −0.2563 | −0.2580 | −0.2580 | −0.2580 | −0.2580 | −0.2580 |
| 0.80 | 12.0630 | 3.9171 | 2.4277 | 0.3599 | −0.1744 | −0.2575 | −0.2760 | −0.2790 | −0.2790 | −0.2790 | −0.2790 | −0.2790 |
| 0.90 | 11.8772 | 3.9868 | 2.5112 | 0.4081 | −0.1708 | −0.2696 | −0.2942 | −0.2989 | −0.2991 | −0.2991 | −0.2991 | −0.2991 |
| 1.00 | 11.6965 | 4.0417 | 2.5815 | 0.4524 | −0.1660 | −0.2801 | −0.3111 | −0.3180 | −0.3183 | −0.3184 | −0.3184 | −0.3184 |
| 2.00 | 10.2326 | 4.2145 | 2.9186 | 0.7530 | −0.0967 | −0.3301 | −0.4289 | −0.4725 | −0.4779 | −0.4807 | −0.4809 | −0.4810 |
| 4.00 | 8.5558 | 4.0896 | 3.0201 | 1.0138 | 0.0227 | −0.3331 | −0.5308 | −0.6614 | −0.6890 | −0.7126 | −0.7163 | −0.7201 |
| 6.00 | 7.6256 | 3.9254 | 2.9892 | 1.1268 | 0.1006 | −0.3139 | −0.5718 | −0.7717 | −0.8230 | −0.8757 | −0.8860 | −0.8988 |
| 8.00 | 7.0250 | 3.7892 | 2.9406 | 1.1878 | 0.1541 | −0.2938 | −0.5915 | −0.8444 | −0.9163 | −0.9979 | −1.0159 | −1.0403 |
| 10.00 | 6.6000 | 3.6791 | 2.8931 | 1.2250 | 0.1931 | −0.2760 | −0.6019 | −0.8961 | −0.9854 | −1.0937 | −1.1193 | −1.1562 |
| 20.00 | 5.5168 | 3.3486 | 2.7229 | 1.2952 | 0.2960 | −0.2165 | −0.6133 | −1.0273 | −1.1736 | −1.3789 | −1.4358 | −1.5289 |
| 40.00 | 4.7541 | 3.0720 | 2.5591 | 1.3206 | 0.3698 | −0.1617 | −0.6056 | −1.1166 | −1.3157 | −1.6225 | −1.7159 | −1.8812 |
| 60.00 | 4.4248 | 2.9407 | 2.4762 | 1.3243 | 0.4016 | −0.1346 | −0.5974 | −1.1535 | −1.3793 | −1.7409 | −1.8552 | −2.0641 |
| 80.00 | 4.2324 | 2.8606 | 2.4242 | 1.3242 | 0.4200 | −0.1177 | −0.5911 | −1.1745 | −1.4169 | −1.8142 | −1.9426 | −2.1814 |
| 100.00 | 4.1031 | 2.8052 | 2.3877 | 1.3231 | 0.4323 | −0.1058 | −0.5862 | −1.1882 | −1.4424 | −1.8654 | −2.0040 | −2.2651 |
| 200.00 | 3.7900 | 2.6660 | 2.2939 | 1.3169 | 0.4618 | −0.0757 | −0.5717 | −1.2200 | −1.5046 | −1.9958 | −2.1625 | −2.4853 |
| 400.00 | 3.5763 | 2.5666 | 2.2252 | 1.3095 | 0.4815 | −0.0539 | −0.5596 | −1.2405 | −1.5474 | −2.0907 | −2.2794 | −2.6519 |
| 600.00 | 3.4838 | 2.5225 | 2.1943 | 1.3054 | 0.4899 | −0.0441 | −0.5538 | −1.2489 | −1.5660 | −2.1333 | −2.3323 | −2.7285 |
| 800.00 | 3.4293 | 2.4961 | 2.1757 | 1.3028 | 0.4948 | −0.0382 | −0.5502 | −1.2538 | −1.5769 | −2.1589 | −2.3642 | −2.7750 |
| 1000.00 | 3.3925 | 2.4782 | 2.1630 | 1.3009 | 0.4981 | −0.0342 | −0.5476 | −1.2570 | −1.5843 | −2.1764 | −2.3861 | −2.8071 |

### 2.3. Maximum Likelihood Estimation of the Parameters

For the maximum likelihood estimation (MLE), the log-likelihood function for a sample $x = \{x_1, x_2, \cdots, x_n\}$ drawn from the FPEG distribution can be written as

$$ln(L) = n\alpha ln\beta - nlnb - nln\Gamma(\alpha) + \frac{\alpha}{b}\sum_{i=1}^{n}ln(x_i - \delta) - \sum_{i=1}^{n}ln(x_i - \delta) - \beta\sum_{i=1}^{n}(x_i - \delta)^{\frac{1}{b}} \quad (7)$$

where $n$ is the sample size, $\delta \leq x < \infty; \alpha > 0; \beta > 0$.

The MLE parameters can be obtained by taking the derivatives of the log likelihood with respect to parameters, setting them equal to zero, and solving for the parameters. Differentiating Equation (7) partially with respect to each parameter and equating each partial derivative to zero yield

$$-\frac{\alpha}{b}\sum_{i=1}^{n}\frac{1}{x_i - \delta} + \sum_{i=1}^{n}\frac{1}{x_i - \delta} + \frac{\beta}{b}\sum_{i=1}^{n}(x_i - \delta)^{\frac{1}{b}-1} = 0 \quad (8)$$

$$nln\beta - n\Psi(\alpha) + \frac{1}{b}\sum_{i=1}^{n}ln(x_i - \delta) = 0 \quad (9)$$

where $\Psi(\alpha)$ is the digamma function.

$$\frac{n\alpha}{\beta} - \sum_{i=1}^{n}(x_i - \delta)^{\frac{1}{b}} = 0 \quad (10)$$

$$-\frac{n}{b} - \frac{\alpha}{b^2}\sum_{i=1}^{n}ln(x_i - \delta) + \frac{\beta}{b^2}\sum_{i=1}^{n}\left[(x_i - \delta)^{\frac{1}{b}}ln(x_i - \delta)\right] = 0 \quad (11)$$

Equations (8)–(11) can be solved numerically to obtain parameters $\hat{\alpha}, \hat{\beta}, \hat{\delta},$ and $\hat{b}$.

### 2.4. Confidence Intervals of Quantiles

For a given $p$, the design value estimate $x_p$ is a random variable. The $1 - q$ confidence intervals for the population quantiles $x_p$ may be determined by [2,3]

$$x_p^L = x_p - u_{1-q/2}S(x_p), \ x_p^U = x_p + u_{1-q/2}S(x_p) \tag{12}$$

where $u_{1-q/2}$ is the $1 - q/2$ quantile of the standard normal distribution, $x_p$ is the quantile estimator corresponding to the probability of exceedance $p$ which can be determined from Equation (4) or Equation (5), and $S(x_p)$ is the standard deviation or standard error of $x_p$. Such standard error $S(x_p)$ determined by the MLE is given in what follows.

For the FPEG distribution, when parameters $\alpha$, $\beta$, $\delta$, and $b$ are estimated by the MLE, $x_p$ is a function of $\alpha$, $\beta$, $\delta$, and $b$:

$$x_p = x_p(\alpha, \beta, \delta, b, p) \tag{13}$$

The variance in this case is given by [21]

$$Var(x_p) = \left(\frac{\partial x_p}{\partial \delta}\right)^2 Var(\delta) + \left(\frac{\partial x_p}{\partial \alpha}\right)^2 Var(\alpha) + \left(\frac{\partial x_p}{\partial \beta}\right)^2 Var(\beta) + \left(\frac{\partial x_p}{\partial b}\right)^2 Var(b)$$

$$+ 2\frac{\partial x_p}{\partial \delta}\frac{\partial x_p}{\partial \alpha}Cov(\delta, \alpha) + 2\frac{\partial x_p}{\partial \delta}\frac{\partial x_p}{\partial \beta}Cov(\delta, \beta) + 2\frac{\partial x_p}{\partial \delta}\frac{\partial x_p}{\partial b}Cov(\delta, b)$$

$$+ 2\frac{\partial x_p}{\partial \alpha}\frac{\partial x_p}{\partial \beta}Cov(\alpha, \beta) + 2\frac{\partial x_p}{\partial \alpha}\frac{\partial x_p}{\partial b}Cov(\alpha, b) + 2\frac{\partial x_p}{\partial \beta}\frac{\partial x_p}{\partial b}Cov(\beta, b) \tag{14}$$

The variance and covariance matrix of parameters in Equation (13) is the inverse of Fisher's expected information matrix [3].

$$\begin{bmatrix} Var(\delta) & Cov(\delta, \alpha) & Cov(\delta, \beta) & Cov(\delta, b) \\ Cov(\alpha, \delta) & Var(\alpha) & Cov(\alpha, \beta) & Cov(\alpha, b) \\ Cov(\beta, \delta) & Cov(\beta, \alpha) & Var(\beta) & Cov(\beta, b) \\ Cov(b, \delta) & Cov(b, \alpha) & Cov(b, \beta) & Var(b) \end{bmatrix}$$

$$= \begin{bmatrix} E\left(-\frac{\partial^2 lnL}{\partial \delta^2}\right) & E\left(-\frac{\partial^2 lnL}{\partial \delta \partial \alpha}\right) & E\left(-\frac{\partial^2 lnL}{\partial \delta \partial \beta}\right) & E\left(-\frac{\partial^2 lnL}{\partial \delta \partial b}\right) \\ E\left(-\frac{\partial^2 lnL}{\partial \alpha \partial \delta}\right) & E\left(-\frac{\partial^2 lnL}{\partial \alpha^2}\right) & E\left(-\frac{\partial^2 lnL}{\partial \alpha \partial \beta}\right) & E\left(-\frac{\partial^2 lnL}{\partial \alpha \partial b}\right) \\ E\left(-\frac{\partial^2 lnL}{\partial \beta \partial \delta}\right) & E\left(-\frac{\partial^2 lnL}{\partial \beta \partial \alpha}\right) & E\left(-\frac{\partial^2 lnL}{\partial \beta^2}\right) & E\left(-\frac{\partial^2 lnL}{\partial \beta \partial b}\right) \\ E\left(-\frac{\partial^2 lnL}{\partial b \partial \delta}\right) & E\left(-\frac{\partial^2 lnL}{\partial b \partial \alpha}\right) & E\left(-\frac{\partial^2 lnL}{\partial b \partial \beta}\right) & E\left(-\frac{\partial^2 lnL}{\partial b^2}\right) \end{bmatrix} \tag{15}$$

where E represents the expected value;

$$\begin{bmatrix} -\frac{\partial^2 lnL}{\partial \delta^2} & -\frac{\partial^2 lnL}{\partial \delta \partial \alpha} & -\frac{\partial^2 lnL}{\partial \delta \partial \beta} & -\frac{\partial^2 lnL}{\partial \delta \partial b} \\ -\frac{\partial^2 lnL}{\partial \alpha \partial \delta} & -\frac{\partial^2 lnL}{\partial \alpha^2} & -\frac{\partial^2 lnL}{\partial \alpha \partial \beta} & -\frac{\partial^2 lnL}{\partial \alpha \partial b} \\ -\frac{\partial^2 lnL}{\partial \beta \partial \delta} & -\frac{\partial^2 lnL}{\partial \beta \partial \alpha} & -\frac{\partial^2 lnL}{\partial \beta^2} & -\frac{\partial^2 lnL}{\partial \beta \partial b} \\ -\frac{\partial^2 lnL}{\partial b \partial \delta} & -\frac{\partial^2 lnL}{\partial b \partial \alpha} & -\frac{\partial^2 lnL}{\partial b \partial \beta} & -\frac{\partial^2 lnL}{\partial b^2} \end{bmatrix}$$

is the sample information matrix;

$$\begin{bmatrix} E\left(-\frac{\partial^2 lnL}{\partial \delta^2}\right) & E\left(-\frac{\partial^2 lnL}{\partial \delta \partial \alpha}\right) & E\left(-\frac{\partial^2 lnL}{\partial \delta \partial \beta}\right) & E\left(-\frac{\partial^2 lnL}{\partial \delta \partial b}\right) \\ E\left(-\frac{\partial^2 lnL}{\partial \alpha \partial \delta}\right) & E\left(-\frac{\partial^2 lnL}{\partial \alpha^2}\right) & E\left(-\frac{\partial^2 lnL}{\partial \alpha \partial \beta}\right) & E\left(-\frac{\partial^2 lnL}{\partial \alpha \partial b}\right) \\ E\left(-\frac{\partial^2 lnL}{\partial \beta \partial \delta}\right) & E\left(-\frac{\partial^2 lnL}{\partial \beta \partial \alpha}\right) & E\left(-\frac{\partial^2 lnL}{\partial \beta^2}\right) & E\left(-\frac{\partial^2 lnL}{\partial \beta \partial b}\right) \\ E\left(-\frac{\partial^2 lnL}{\partial b \partial \delta}\right) & E\left(-\frac{\partial^2 lnL}{\partial b \partial \alpha}\right) & E\left(-\frac{\partial^2 lnL}{\partial b \partial \beta}\right) & E\left(-\frac{\partial^2 lnL}{\partial b^2}\right) \end{bmatrix}$$

is Fisher's expected information matrix, the elements of which can be determined by taking the expected value of the sample information matrix. Its elements are derived in Equations (A1)–(A16) and Equations (A17)–(A32) in Appendix A, respectively.

$$E\left(-\frac{\partial^2 lnL}{\partial\delta^2}\right) = \frac{\alpha - 2b + b^2}{b^2}\cdot\frac{n\beta^{2b}\Gamma(\alpha - 2b)}{\Gamma(\alpha)} \tag{16}$$

$$E\left(-\frac{\partial^2 lnL}{\partial\delta\partial\alpha}\right) = \frac{1}{b}\cdot\frac{n\beta^b\Gamma(\alpha - b)}{\Gamma(\alpha)} \tag{17}$$

$$E\left(-\frac{\partial^2 lnL}{\partial\delta\partial\beta}\right) = -\frac{(\alpha - b)}{b}\cdot\frac{n\beta^{b-1}\Gamma(\alpha - b)}{\Gamma(\alpha)} \tag{18}$$

$$E\left(-\frac{\partial^2 lnL}{\partial\delta\partial b}\right) = -\frac{1}{b}\cdot\frac{n\beta^b\Gamma(\alpha - b)}{\Gamma(\alpha)} + \frac{1}{b^2}\frac{n\beta^b}{\Gamma(\alpha)}\left[\frac{\partial\Gamma(\alpha - b + 1)}{\partial(\alpha - b + 1)} - ln\beta\cdot\Gamma(\alpha - b + 1)\right] \tag{19}$$

$$E\left(-\frac{\partial^2 lnL}{\partial\alpha\partial\delta}\right) = E\left(-\frac{\partial^2 lnL}{\partial\delta\partial\alpha}\right) \tag{20}$$

$$E\left(-\frac{\partial^2 lnL}{\partial\alpha^2}\right) = n\Psi'(\alpha) \tag{21}$$

$$E\left(-\frac{\partial^2 lnL}{\partial\alpha\partial\beta}\right) = -\frac{n}{\beta} \tag{22}$$

$$E\left(-\frac{\partial^2 lnL}{\partial\alpha\partial b}\right) = \frac{n[-ln\beta + \Psi(\alpha)]}{b} \tag{23}$$

$$E\left(-\frac{\partial^2 lnL}{\partial\beta\partial\delta}\right) = E\left(-\frac{\partial^2 lnL}{\partial\delta\partial\beta}\right) \tag{24}$$

$$E\left(-\frac{\partial^2 lnL}{\partial\beta\partial\alpha}\right) = E\left(-\frac{\partial^2 lnL}{\partial\alpha\partial\beta}\right) \tag{25}$$

$$E\left(-\frac{\partial^2 lnL}{\partial\beta^2}\right) = \frac{n\alpha}{\beta^2} \tag{26}$$

$$E\left(-\frac{\partial^2 lnL}{\partial\beta\partial b}\right) = -\frac{n}{b\beta}\cdot\frac{\frac{\partial\Gamma(\alpha+1)}{\partial(\alpha+1)} - ln\beta\cdot\Gamma(\alpha + 1)}{\Gamma(\alpha)} \tag{27}$$

$$E\left(-\frac{\partial^2 lnL}{\partial b\partial\delta}\right) = E\left(-\frac{\partial^2 lnL}{\partial\delta\partial b}\right) \tag{28}$$

$$E\left(-\frac{\partial^2 lnL}{\partial b\partial\alpha}\right) = E\left(-\frac{\partial^2 lnL}{\partial\alpha\partial b}\right) \tag{29}$$

$$E\left(-\frac{\partial^2 lnL}{\partial b\partial\beta}\right) = E\left(-\frac{\partial^2 lnL}{\partial\beta\partial b}\right) \tag{30}$$

$$E\left(-\frac{\partial^2 lnL}{\partial b^2}\right) = -\frac{n}{b^2} - \frac{2n\alpha[-ln\beta + \Psi(\alpha)]}{b^2} + \frac{2}{b^2}\frac{n\left[\frac{\partial\Gamma(\alpha+1)}{\partial(\alpha+1)} - ln\beta\cdot\Gamma(\alpha + 1)\right]}{\Gamma(\alpha)}$$

$$+ \frac{1}{b^2}\frac{n}{\Gamma(\alpha)}\left[\frac{\partial^2\Gamma(\alpha + 1)}{\partial(\alpha + 1)^2} - 2ln\frac{\partial\Gamma(\alpha + 1)}{\partial(\alpha + 1)} + (ln\beta)^2\cdot\Gamma(\alpha + 1)\right] \tag{31}$$

The derivatives of $x_p$ with respect to the parameters of the FPEG distribution are obtained from Equation (4) as

$$\frac{\partial x_p}{\partial\delta} = 1 \tag{32}$$

$$\frac{\partial x_p}{\partial \alpha} = \frac{b}{\beta^b} t_p^{b-1} \frac{\partial t_p}{\partial \alpha} \tag{33}$$

$$\frac{\partial x_p}{\partial \beta} = -\frac{b}{\beta^{b+1}} t_p^b \tag{34}$$

$$\frac{\partial x_p}{\partial b} = -\frac{1}{\beta^b} ln\beta \cdot t_p^b + \frac{1}{\beta^b} t_p^b ln t_p \tag{35}$$

For $\frac{\partial x_p}{\partial \alpha}$ in Equation (33), $p$ is a constant and is a function of $t_p$ and $\alpha$ in Equation (3). Thus, we can get $\frac{\partial p}{\partial t_p} \frac{\partial t_p}{\partial \alpha} + \frac{\partial p}{\partial \alpha} = 0$, which reduces to

$$\frac{\partial t_p}{\partial \alpha} = -\frac{\partial p}{\partial \alpha} \Big/ \frac{\partial p}{\partial t_p} \tag{36}$$

$$\frac{\partial p}{\partial t_p} = \frac{1}{\Gamma(\alpha)} t_p^{\alpha-1} e^{-t_p} \tag{37}$$

$$\begin{aligned}
\frac{\partial p}{\partial \alpha} &= -\frac{\Gamma'(\alpha)}{\Gamma^2(\alpha)} \int_0^{t_p} t^{\alpha-1} e^{-t} dt + \frac{1}{\Gamma(\alpha)} \int_0^{t_p} t^{\alpha-1} e^{-t} ln t \, dt \\
&= -\Psi(\alpha)\gamma(t_p, \alpha) + \frac{1}{\Gamma(\alpha)} \int_0^{t_p} t^{\alpha-1} e^{-t} ln t \, dt
\end{aligned} \tag{38}$$

where $\Psi(\alpha) = \frac{\Gamma'(\alpha)}{\Gamma^2(\alpha)}$ is the psi function with a different meaning in Equation (9); $\gamma(t_p, \alpha) = \frac{1}{\Gamma(\alpha)} \int_0^{t_p} t^{\alpha-1} e^{-t} dt$ is the incomplete gamma function; $\int_0^{t_p} t^{\alpha-1} e^{-t} ln t \, dt$ can be numerically calculated by the central difference of $\int_0^{t_p} t^{\alpha-1} e^{-t} dt$. Some $\frac{1}{\Gamma(\alpha)} \int_0^{t_p} t^{\alpha-1} e^{-t} ln t \, dt$ values obtained numerically are listed in Table 2.

**Table 2.** $\frac{1}{\Gamma(\alpha)} \int_0^{t_p} t^{\alpha-1} e^{-t} ln t \cdot dt$ values under different probabilities of exceedance $p$.

| $\alpha$ | Exceedance Probability $p$ | | | | | | | | | | | |
|---|---|---|---|---|---|---|---|---|---|---|---|---|
| | 0.001 | 0.01 | 0.02 | 0.10 | 0.30 | 0.50 | 0.70 | 0.90 | 0.95 | 0.99 | 0.995 | 0.999 |
| 0.10 | −10.4252 | −10.4319 | −10.4346 | −10.3799 | −9.8448 | −8.7151 | −6.7615 | −3.3525 | −2.0228 | −0.5655 | −0.3174 | −0.0796 |
| 0.20 | −5.2906 | −5.2998 | −5.3062 | −5.3012 | −5.0361 | −4.4449 | −3.4340 | −1.6940 | −1.0203 | −0.2845 | −0.1596 | −0.0400 |
| 0.30 | −3.5042 | −3.5148 | −3.5232 | −3.5425 | −3.3894 | −2.9959 | −2.3115 | −1.1369 | −0.6840 | −0.1904 | −0.1068 | −0.0267 |
| 0.40 | −2.5632 | −2.5747 | −2.5845 | −2.6200 | −2.5347 | −2.2519 | −1.7404 | −0.8555 | −0.5144 | −0.1431 | −0.0802 | −0.0201 |
| 0.50 | −1.9653 | −1.9777 | −1.9886 | −2.0361 | −1.9990 | −1.7906 | −1.3900 | −0.6845 | −0.4116 | −0.1145 | −0.0642 | −0.0161 |
| 0.60 | −1.5425 | −1.5556 | −1.5673 | −1.6246 | −1.6247 | −1.4713 | −1.1502 | −0.5688 | −0.3423 | −0.0953 | −0.0534 | −0.0134 |
| 0.70 | −1.2220 | −1.2356 | −1.2481 | −1.3136 | −1.3439 | −1.2340 | −0.9737 | −0.4847 | −0.2921 | −0.0814 | −0.0457 | −0.0114 |
| 0.80 | −0.9670 | −0.9811 | −0.9944 | −1.0668 | −1.1227 | −1.0485 | −0.8371 | −0.4204 | −0.2539 | −0.0709 | −0.0398 | −0.0100 |
| 0.90 | −0.7570 | −0.7716 | −0.7854 | −0.8641 | −0.9420 | −0.8981 | −0.7273 | −0.3694 | −0.2237 | −0.0627 | −0.0352 | −0.0088 |
| 1.00 | −0.5793 | −0.5943 | −0.6087 | −0.6930 | −0.7903 | −0.7726 | −0.6365 | −0.3277 | −0.1992 | −0.0560 | −0.0315 | −0.0079 |
| 2.00 | 0.4205 | 0.4024 | 0.3841 | 0.2622 | 0.0411 | −0.0998 | −0.1628 | −0.1194 | −0.0788 | −0.0242 | −0.0139 | −0.0036 |
| 3.00 | 0.9203 | 0.9002 | 0.8795 | 0.7345 | 0.4422 | 0.2155 | 0.0509 | −0.0314 | −0.0295 | −0.0119 | −0.0072 | −0.0020 |
| 4.00 | 1.2535 | 1.2319 | 1.2093 | 1.0474 | 0.7045 | 0.4186 | 0.1857 | 0.0223 | −0.0001 | −0.0048 | −0.0034 | −0.0011 |
| 5.00 | 1.5033 | 1.4805 | 1.4565 | 1.2811 | 0.8988 | 0.5675 | 0.2834 | 0.0603 | 0.0206 | 0.0001 | −0.0008 | −0.0005 |
| 6.00 | 1.7032 | 1.6794 | 1.6541 | 1.4675 | 1.0528 | 0.6848 | 0.3597 | 0.0894 | 0.0363 | 0.0037 | 0.0011 | −0.0001 |
| 7.00 | 1.8698 | 1.8450 | 1.8187 | 1.6224 | 1.1802 | 0.7813 | 0.4220 | 0.1130 | 0.0489 | 0.0066 | 0.0027 | 0.0002 |
| 8.00 | 2.0126 | 1.9870 | 1.9597 | 1.7550 | 1.2888 | 0.8633 | 0.4747 | 0.1327 | 0.0594 | 0.0090 | 0.0039 | 0.0005 |
| 9.00 | 2.1375 | 2.1112 | 2.0830 | 1.8708 | 1.3834 | 0.9344 | 0.5202 | 0.1497 | 0.0683 | 0.0110 | 0.0050 | 0.0008 |
| 10.00 | 2.2486 | 2.2216 | 2.1926 | 1.9736 | 1.4671 | 0.9972 | 0.5602 | 0.1645 | 0.0762 | 0.0127 | 0.0059 | 0.0010 |
| 20.00 | 2.9669 | 2.9353 | 2.9010 | 2.6357 | 2.0024 | 1.3952 | 0.8111 | 0.2554 | 0.1237 | 0.0231 | 0.0112 | 0.0021 |
| 30.00 | 3.3805 | 3.3460 | 3.3084 | 3.0150 | 2.3062 | 1.6189 | 0.9504 | 0.3049 | 0.1493 | 0.0286 | 0.0140 | 0.0027 |
| 40.00 | 3.6722 | 3.6356 | 3.5955 | 3.2817 | 2.5189 | 1.7748 | 1.0469 | 0.3388 | 0.1667 | 0.0323 | 0.0159 | 0.0031 |
| 50.00 | 3.8976 | 3.8594 | 3.8174 | 3.4876 | 2.6826 | 1.8944 | 1.1206 | 0.3645 | 0.1799 | 0.0350 | 0.0173 | 0.0034 |

**Table 2.** *Cont.*

| $\alpha$ | Exceedance Probability $p$ | | | | | | | | | | | |
|---|---|---|---|---|---|---|---|---|---|---|---|---|
| | 0.001 | 0.01 | 0.02 | 0.10 | 0.30 | 0.50 | 0.70 | 0.90 | 0.95 | 0.99 | 0.995 | 0.999 |
| 60.00 | 4.0815 | 4.0418 | 3.9983 | 3.6553 | 2.8157 | 1.9913 | 1.1802 | 0.3852 | 0.1905 | 0.0372 | 0.0184 | 0.0036 |
| 70.00 | 4.2367 | 4.1959 | 4.1509 | 3.7967 | 2.9277 | 2.0729 | 1.2303 | 0.4025 | 0.1993 | 0.0391 | 0.0194 | 0.0038 |
| 80.00 | 4.3710 | 4.3291 | 4.2830 | 3.9190 | 3.0244 | 2.1432 | 1.2734 | 0.4174 | 0.2069 | 0.0406 | 0.0202 | 0.0040 |
| 90.00 | 4.4894 | 4.4466 | 4.3994 | 4.0267 | 3.1096 | 2.2050 | 1.3112 | 0.4305 | 0.2135 | 0.0420 | 0.0209 | 0.0041 |
| 100.00 | 4.5952 | 4.5516 | 4.5035 | 4.1229 | 3.1856 | 2.2601 | 1.3449 | 0.4420 | 0.2194 | 0.0432 | 0.0215 | 0.0042 |
| 200.00 | 5.2903 | 5.2410 | 5.1866 | 4.7540 | 3.6826 | 2.6197 | 1.5640 | 0.5170 | 0.2573 | 0.0510 | 0.0254 | 0.0050 |
| 300.00 | 5.6962 | 5.6436 | 5.5853 | 5.1219 | 3.9715 | 2.8280 | 1.6904 | 0.5599 | 0.2791 | 0.0554 | 0.0277 | 0.0055 |
| 400.00 | 5.9841 | 5.9290 | 5.8680 | 5.3825 | 4.1758 | 2.9752 | 1.7796 | 0.5901 | 0.2943 | 0.0585 | 0.0292 | 0.0058 |
| 500.00 | 6.2072 | 6.1503 | 6.0872 | 5.5845 | 4.3340 | 3.0890 | 1.8485 | 0.6134 | 0.3060 | 0.0609 | 0.0304 | 0.0061 |
| 600.00 | 6.3896 | 6.3311 | 6.2662 | 5.7494 | 4.4631 | 3.1818 | 1.9046 | 0.6324 | 0.3155 | 0.0629 | 0.0314 | 0.0063 |
| 700.00 | 6.5437 | 6.4839 | 6.4176 | 5.8887 | 4.5722 | 3.2601 | 1.9519 | 0.6483 | 0.3236 | 0.0645 | 0.0322 | 0.0064 |
| 800.00 | 6.6772 | 6.6162 | 6.5486 | 6.0094 | 4.6665 | 3.3279 | 1.9929 | 0.6621 | 0.3305 | 0.0659 | 0.0329 | 0.0066 |
| 900.00 | 6.7949 | 6.7329 | 6.6642 | 6.1158 | 4.7497 | 3.3876 | 2.0289 | 0.6743 | 0.3366 | 0.0671 | 0.0335 | 0.0067 |
| 1000.00 | 6.9002 | 6.8374 | 6.7676 | 6.2110 | 4.8241 | 3.4410 | 2.0611 | 0.6851 | 0.3421 | 0.0682 | 0.0341 | 0.0068 |

Finally, the expression of $\frac{\partial t_p}{\partial \alpha}$ in Equation (33) can be obtained by substituting Equations (37) and (38) into Equation (36). Table 3 shows some $\frac{\partial p}{\partial \alpha}$ values obtained numerically.

**Table 3.** Values under different probabilities of exceedance $p$ .

| $\alpha$ | Exceedance Probability $p$ | | | | | | | | | | | |
|---|---|---|---|---|---|---|---|---|---|---|---|---|
| | 0.001 | 0.01 | 0.02 | 0.10 | 0.30 | 0.50 | 0.70 | 0.90 | 0.95 | 0.99 | 0.995 | 0.999 |
| 0.10 | 9.7084 | 7.9363 | 7.0694 | 3.7664 | 0.6446 | 0.0416 | 0.0004 | 0.0000 | 0.0000 | 0.0000 | 0.0000 | 0.0000 |
| 0.20 | 5.9082 | 4.9696 | 4.5486 | 3.0424 | 1.2760 | 0.3800 | 0.0489 | 0.0004 | 0.0000 | 0.0000 | 0.0000 | 0.0000 |
| 0.30 | 4.6026 | 3.9077 | 3.6104 | 2.5960 | 1.3989 | 0.6421 | 0.1800 | 0.0085 | 0.0011 | 0.0000 | 0.0000 | 0.0000 |
| 0.40 | 3.9326 | 3.3560 | 3.1165 | 2.3249 | 1.4068 | 0.7822 | 0.3124 | 0.0353 | 0.0080 | 0.0002 | 0.0000 | 0.0000 |
| 0.50 | 3.5200 | 3.0148 | 2.8093 | 2.1448 | 1.3882 | 0.8584 | 0.4166 | 0.0775 | 0.0247 | 0.0015 | 0.0004 | 0.0000 |
| 0.60 | 3.2374 | 2.7812 | 2.5983 | 2.0165 | 1.3647 | 0.9026 | 0.4946 | 0.1263 | 0.0501 | 0.0051 | 0.0018 | 0.0002 |
| 0.70 | 3.0301 | 2.6099 | 2.4434 | 1.9203 | 1.3420 | 0.9298 | 0.5533 | 0.1751 | 0.0809 | 0.0120 | 0.0051 | 0.0007 |
| 0.80 | 2.8704 | 2.4782 | 2.3242 | 1.8452 | 1.3215 | 0.9475 | 0.5983 | 0.2210 | 0.1140 | 0.0222 | 0.0107 | 0.0018 |
| 0.90 | 2.7428 | 2.3733 | 2.2293 | 1.7849 | 1.3034 | 0.9596 | 0.6336 | 0.2626 | 0.1472 | 0.0354 | 0.0186 | 0.0040 |
| 1.00 | 2.6380 | 2.2874 | 2.1515 | 1.7351 | 1.2876 | 0.9680 | 0.6619 | 0.3000 | 0.1793 | 0.0508 | 0.0287 | 0.0073 |
| 2.00 | 2.1140 | 1.8622 | 1.7677 | 1.4875 | 1.1980 | 0.9933 | 0.7908 | 0.5174 | 0.4011 | 0.2223 | 0.1716 | 0.0927 |
| 3.00 | 1.9000 | 1.6917 | 1.6146 | 1.3887 | 1.1589 | 0.9973 | 0.8365 | 0.6130 | 0.5127 | 0.3440 | 0.2902 | 0.1950 |
| 4.00 | 1.7760 | 1.5941 | 1.5271 | 1.3326 | 1.1362 | 0.9985 | 0.8613 | 0.6683 | 0.5797 | 0.4254 | 0.3738 | 0.2775 |
| 5.00 | 1.6926 | 1.5288 | 1.4688 | 1.2953 | 1.1210 | 0.9991 | 0.8775 | 0.7052 | 0.6253 | 0.4834 | 0.4347 | 0.3415 |
| 6.00 | 1.6314 | 1.4812 | 1.4264 | 1.2682 | 1.1099 | 0.9994 | 0.8890 | 0.7321 | 0.6588 | 0.5271 | 0.4813 | 0.3921 |
| 7.00 | 1.5842 | 1.4445 | 1.3937 | 1.2474 | 1.1014 | 0.9996 | 0.8978 | 0.7528 | 0.6847 | 0.5615 | 0.5182 | 0.4330 |
| 8.00 | 1.5462 | 1.4152 | 1.3676 | 1.2308 | 1.0946 | 0.9997 | 0.9048 | 0.7693 | 0.7056 | 0.5894 | 0.5484 | 0.4669 |
| 9.00 | 1.5148 | 1.3909 | 1.3460 | 1.2172 | 1.0890 | 0.9997 | 0.9105 | 0.7830 | 0.7227 | 0.6126 | 0.5735 | 0.4955 |
| 10.00 | 1.4882 | 1.3705 | 1.3279 | 1.2057 | 1.0843 | 0.9998 | 0.9153 | 0.7944 | 0.7372 | 0.6323 | 0.5950 | 0.5201 |
| 20.00 | 1.3450 | 1.2609 | 1.2307 | 1.1444 | 1.0591 | 0.9999 | 0.9408 | 0.8557 | 0.8151 | 0.7397 | 0.7125 | 0.6569 |
| 30.00 | 1.2817 | 1.2128 | 1.1880 | 1.1176 | 1.0482 | 1.0000 | 0.9518 | 0.8824 | 0.8493 | 0.7875 | 0.7650 | 0.7191 |
| 40.00 | 1.2439 | 1.1842 | 1.1627 | 1.1017 | 1.0416 | 1.0000 | 0.9583 | 0.8983 | 0.8696 | 0.8160 | 0.7965 | 0.7564 |
| 50.00 | 1.2181 | 1.1647 | 1.1455 | 1.0909 | 1.0372 | 1.0000 | 0.9628 | 0.9091 | 0.8834 | 0.8354 | 0.8179 | 0.7820 |
| 60.00 | 1.1990 | 1.1503 | 1.1327 | 1.0829 | 1.0339 | 1.0000 | 0.9660 | 0.9171 | 0.8936 | 0.8498 | 0.8338 | 0.8009 |
| 70.00 | 1.1842 | 1.1391 | 1.1229 | 1.0767 | 1.0314 | 1.0000 | 0.9686 | 0.9233 | 0.9015 | 0.8609 | 0.8461 | 0.8156 |
| 80.00 | 1.1722 | 1.1301 | 1.1149 | 1.0718 | 1.0294 | 1.0000 | 0.9706 | 0.9282 | 0.9079 | 0.8699 | 0.8560 | 0.8275 |
| 90.00 | 1.1623 | 1.1226 | 1.1083 | 1.0676 | 1.0277 | 1.0000 | 0.9723 | 0.9323 | 0.9132 | 0.8773 | 0.8643 | 0.8373 |
| 100.00 | 1.1539 | 1.1163 | 1.1027 | 1.0642 | 1.0263 | 1.0000 | 0.9737 | 0.9358 | 0.9177 | 0.8836 | 0.8712 | 0.8456 |
| 200.00 | 1.1080 | 1.0821 | 1.0726 | 1.0453 | 1.0186 | 1.0000 | 0.9814 | 0.9547 | 0.9418 | 0.9177 | 0.9089 | 0.8908 |
| 300.00 | 1.0874 | 1.0670 | 1.0592 | 1.0370 | 1.0151 | 1.0000 | 0.9848 | 0.9630 | 0.9525 | 0.9328 | 0.9256 | 0.9108 |
| 400.00 | 1.0749 | 1.0579 | 1.0512 | 1.0320 | 1.0131 | 1.0000 | 0.9869 | 0.9679 | 0.9589 | 0.9418 | 0.9356 | 0.9228 |

Table 3. Cont.

| $\alpha$ | Exceedance Probability $p$ | | | | | | | | | | | |
|---|---|---|---|---|---|---|---|---|---|---|---|---|
| | 0.001 | 0.01 | 0.02 | 0.10 | 0.30 | 0.50 | 0.70 | 0.90 | 0.95 | 0.99 | 0.995 | 0.999 |
| 500.00 | 1.0662 | 1.0517 | 1.0457 | 1.0286 | 1.0117 | 1.0000 | 0.9883 | 0.9713 | 0.9632 | 0.9480 | 0.9424 | 0.9309 |
| 600.00 | 1.0597 | 1.0471 | 1.0417 | 1.0261 | 1.0107 | 1.0000 | 0.9893 | 0.9738 | 0.9664 | 0.9525 | 0.9474 | 0.9369 |
| 700.00 | 1.0544 | 1.0435 | 1.0386 | 1.0242 | 1.0099 | 1.0000 | 0.9901 | 0.9758 | 0.9689 | 0.9560 | 0.9513 | 0.9416 |
| 800.00 | 1.0502 | 1.0406 | 1.0360 | 1.0226 | 1.0092 | 1.0000 | 0.9907 | 0.9773 | 0.9709 | 0.9589 | 0.9545 | 0.9454 |
| 900.00 | 1.0465 | 1.0382 | 1.0339 | 1.0213 | 1.0087 | 1.0000 | 0.9912 | 0.9786 | 0.9726 | 0.9612 | 0.9571 | 0.9485 |
| 1000.00 | 1.0434 | 1.0361 | 1.0321 | 1.0202 | 1.0083 | 1.0000 | 0.9917 | 0.9797 | 0.9740 | 0.9632 | 0.9593 | 0.9511 |

## 3. Data and Case Study

Annual precipitation data from eight sites in the Weihe watershed, China, (1959–2007) were applied to compute the parameters, quantiles, and confidence intervals for the FPEG distribution. All data were obtained from the National Climate of China Meteorological Administration (http://data.cma.cn (accessed on 29 July 2021)) and are complete. The sites and some statistical characteristics of data are summarized in Table 4. It was seen that for annual precipitation from these eight sites, the values of skewness were lower than 1 and the values of kurtosis were higher than 3. All the annual precipitation records also had very low first-order serial correlation coefficients. Using Anderson's test of independence, results showed that these gauge data are independent at the 90% confidence level. Hence, they are considered suitable for precipitation frequency analysis. One of the advantages of the FPEG distribution is that it accommodates a wide range of skewness and kurtosis values, which is one reason it was applied to these sites.

Table 4. Characteristics of data used for parameter estimation of case study sites.

| Site Name | Average | Standard Deviation | Variation Coefficient | Skewness Coefficient | Kurtosis Coefficient | Autocorrelation Coefficient |
|---|---|---|---|---|---|---|
| Binxian | 539.37959 | 126.87725 | 0.23523 | 0.67198 | 3.75030 | 0.01027 |
| Changwu | 578.74082 | 131.48669 | 0.22719 | 0.51381 | 3.19707 | −0.20924 |
| Chunhua | 576.70408 | 140.49133 | 0.24362 | 0.52984 | 3.84797 | 0.03607 |
| Liquan | 530.83673 | 136.19730 | 0.25657 | 0.61866 | 3.45238 | −0.02273 |
| Qianxian | 528.00612 | 126.60962 | 0.23979 | 0.44268 | 3.24396 | 0.04246 |
| Xianyang | 516.14898 | 125.58901 | 0.24332 | 0.86722 | 3.64090 | −0.05607 |
| Xingping | 560.22041 | 140.74120 | 0.25122 | 0.45382 | 3.12778 | −0.03360 |
| Yongshou | 579.65306 | 125.07565 | 0.21578 | 0.28827 | 3.09300 | −0.19330 |

### 3.1. Parameters Estimation

Since there is no explicit solution for the parameters in Equations (8)–(11), the three dimensional Levenberg-Marquardt algorithm was used to obtain a numerical solution for the MLE estimates of $\alpha$, $\beta$, $\delta$, and $b$. The procedure is summarized as follows [22].

(1) From Equation (10), one gets $\beta = \dfrac{n\alpha}{\sum_{i=1}^{n} ln(x_i - \delta)^{\frac{1}{b}}}$. Substituting this quantity into Equations (8), (9), and (11), respectively, the result is the system of nonlinear Equations as

$$F_1(\delta, \alpha, b) = -\frac{\alpha}{b} \sum_{i=1}^{n} \frac{1}{x_i - \delta} + \sum_{i=1}^{n} \frac{1}{x_i - \delta} + \frac{1}{b} \frac{n\alpha}{\sum_{i=1}^{n} ln(x_i - \delta)^{\frac{1}{b}}} \sum_{i=1}^{n} (x_i - \delta)^{\frac{1}{b}-1} = 0 \tag{39}$$

$$F_2(\delta, \alpha, b) = nln\left[\frac{n\alpha}{\sum_{i=1}^{n} ln(x_i - \delta)^{\frac{1}{b}}}\right] - n\Psi(\alpha) + \frac{1}{b} \sum_{i=1}^{n} (x_i - \delta) = 0 \tag{40}$$

$$F_3(\delta, \alpha, b) = -\frac{n}{b} - \frac{\alpha}{b^2} \sum_{i=1}^{n} ln(x_i - \delta) + \frac{1}{b^2} \frac{n\alpha}{\sum_{i=1}^{n} ln(x_i - \delta)^{\frac{1}{b}}} \sum_{i=1}^{n} \left[(x_i - \delta)^{\frac{1}{b}} ln(x_i - \delta)\right] = 0 \tag{41}$$

(2)  Employ the three dimensional Levenberg-Marquardt algorithm to solve for parameters $\delta$, $\alpha$ and $b$:

$$\mathbf{y}_{i+1} = \mathbf{y}_i - \left[\mathbf{J}^T(\mathbf{y}_i)\mathbf{J}(\mathbf{y}_i) + c_i\mathbf{I}\right]^{-1}\mathbf{J}^T(\mathbf{y}_i)\mathbf{F}(\mathbf{y}_i) \tag{42}$$

where $c_i \geq 0$ is the scaling factor; $\mathbf{y}_i$ and $\mathbf{y}_{i+1}$ are the parameter matrix estimates $\mathbf{y}_i = \begin{bmatrix} \delta_i \\ \alpha_i \\ b_i \end{bmatrix}$ and $\mathbf{y}_{i+1} = \begin{bmatrix} \delta_{i+1} \\ \alpha_{i+1} \\ b_{i+1} \end{bmatrix}$ at iteration $i$ and $i+1$, respectively; $\mathbf{I}$ is the three dimensional identity matrix; $(\mathbf{y}_i) = \begin{bmatrix} F_1 \\ F_2 \\ F_3 \end{bmatrix}_{\mathbf{y}=\mathbf{y}_i} = \begin{bmatrix} F_1(\delta_i, \alpha_i, b_i) \\ F_2(\delta_i, \alpha_i, b_i) \\ F_3(\delta_i, \alpha_i, b_i) \end{bmatrix}$;

$\mathbf{J}(\mathbf{y}_i) = \begin{bmatrix} \frac{\partial F_1}{\partial \delta_i} & \frac{\partial F_1}{\partial \alpha_i} & \frac{\partial F_1}{\partial b_i} \\ \frac{\partial F_2}{\partial \delta_i} & \frac{\partial F_2}{\partial \alpha_i} & \frac{\partial F_2}{\partial b_i} \\ \frac{\partial F_3}{\partial \delta_i} & \frac{\partial F_3}{\partial \alpha_i} & \frac{\partial F_3}{\partial b_i} \end{bmatrix}$ is the Jacobian matrix at iteration $\mathbf{y}_i$, the elements of the Jacobian matrix can be numerically calculated by central difference or their first derivatives derived in Equations (A73)–(A85); $\mathbf{J}^T(\mathbf{y}_i)$ is the transpose matrix of $\mathbf{J}(\mathbf{y}_i)$. Throughout this paper the iterative procedure was repeated until the relative change in all parameters was less than 0.01%, that is, $\max\left|\frac{\mathbf{y}_{i+1}-\mathbf{y}_i}{\mathbf{y}_{i+1}}\right| < 0.01\%$.

(3)  After obtaining parameters $\hat{\delta}$, $\hat{\alpha}$ and $\hat{b}$, and substituting these quantities in $\hat{\beta} = \frac{n\hat{\alpha}}{\sum_{i=1}^{n} ln\left(x_i-\hat{\delta}\right)^{\frac{1}{\hat{b}}}}$, one obtains parameter $\hat{\beta}$.

The values of the distribution parameters are given in Table 5. For eight sites, the values of $\alpha$ fell in the range (72, 92), the values of $\beta$ were higher than 4, and the values of $\delta$ were lower than 0.1, with one of them being even as high as 1.75. The sixth, seventh and eighth columns show the computed quantities of the left side functions in Equations (8), (9), and (11). It is seen that these computed quantities were close to zero, indicating a satisfactory performance of the three dimensional Levenberg-Marquardt algorithm.

**Table 5.** Parameter values estimated for case study sites.

| Site Name | $\delta$ | $\alpha$ | $\beta$ | $b$ | $G_1$ | $G_2$ | $G_3$ |
|---|---|---|---|---|---|---|---|
| Binxian | 0.31712 | 86.09054 | 4.57213 | 2.13793 | 0.00114 | 0.00046 | 0.00057 |
| Changwu | 0.01000 | 88.50553 | 4.38302 | 2.11213 | −0.00091 | −0.00124 | 0.08451 |
| Chunhua | 0.01000 | 72.51987 | 4.59741 | 2.29802 | −0.00322 | −0.00126 | 0.00931 |
| Liquan | 0.01000 | 77.27239 | 4.64334 | 2.22515 | −0.00120 | −0.00163 | 0.09101 |
| Qianxian | 0.01000 | 82.18295 | 4.64635 | 2.17684 | −0.00335 | −0.00219 | 0.07195 |
| Xianyang | 1.75373 | 84.11609 | 4.70719 | 2.16017 | 0.00156 | 0.00101 | 0.00134 |
| Xingping | 0.01000 | 78.07368 | 4.54844 | 2.22002 | −0.00266 | −0.00238 | 0.08988 |
| Yongshou | 0.01000 | 91.62207 | 4.34586 | 2.08314 | −0.00403 | −0.00260 | 0.09089 |

*3.2. Goodness-of-Fit Tests and Confidence Interval Calculation*

Goodness-of-fit tests are designed to measure the agreement between a theoretical probability distribution and an empirical distribution for a random sample. Here, we used the Kolmogorov–Smirnov (K–S) test $D_n$ for the goodness-of-fit test of the FPEG distribution. The K–S test $D_n$ is also called empirical distribution function test statistic, because it measures the distance between a continuous distribution function and the empirical distribution function.

Let $x(1) < x(2) < \cdots < x(n)$ be order statistics for a sample size $n$ whose population is defined by a continuous cumulative distribution function $F(x)$ and $F_0(x_i)$ be a specified distribution that contains a set of parameters $\theta$ ($\hat{\theta}$ is the value estimated from a sample size $n$). For an annual precipitation series, the null hypothesis $H_0$ that the true distribution was $F_0$ with parameters $\theta$ was tested. The K–S test $D_n$ can be expressed as [2]:

$$D_n = \max_{0 \le i \le n} \hat{\delta}_i \tag{43}$$

$$\hat{\delta}_i = \max\left[\frac{i}{n} - F_0\left(x_i; \hat{\theta}\right), F_0\left(x_i; \hat{\theta}\right) - \frac{i-1}{n}\right] \tag{44}$$

The sample values of the K–S test statistic $D_n$ are shown in Table 6. The critical value $D_n^*$ of the FPEG distribution (at the significance level $a = 0.05$, for sample size $n$) was 0.1940. It is seen that the statistics of observed annual precipitation were all less than their corresponding critical value, respectively, so that annual precipitation series were all accepted by the K–S test.

**Table 6.** Sample values of K–S test statistic $D_n$ for case study sites.

| Site Name | $D_n$ | Site Name | $D_n$ |
|---|---|---|---|
| Binxian | 0.05100 | Qianxian | 0.10557 |
| Changwu | 0.06521 | Xianyang | 0.12884 |
| Chunhua | 0.10638 | Xingping | 0.07415 |
| Liquan | 0.09806 | Yongshou | 0.08772 |

For the FPEG distribution and the values of standard errors of the quantile estimates using the above methods, the 95 percent confidence intervals may be set at $\mp 1.96$ standard errors around the $x_p$ values. Table 7 shows the quantiles and confidence interval widths estimated by the above methods for different probabilities of exceedance. For example, $p = 70\%$ annual precipitation at Binxian site was 465.56 mm. Using Equation (12), a 95% confidence interval for the $p = 70\%$ annual precipitation was 430.66 mm to 500.47 mm, its width was 69.80 mm. From Table 7, It is seen that for $p = 30\text{–}95\%$ the confidence interval widths estimated were much less than those for $p < 30\%$ and $p > 95\%$.

**Table 7.** Quantiles and confidence intervals for case study sites.

| Site Name | $p$ (%) | $x_p$ | $S(x_p)$ | $x_p^L$ | $x_p^U$ | Confidence Interval Width |
|---|---|---|---|---|---|---|
| | 0.10 | 1045.17 | 146.32 | 758.39 | 1331.95 | 573.57 |
| | 1.00 | 888.98 | 75.00 | 741.98 | 1035.98 | 293.99 |
| | 2.00 | 838.16 | 57.31 | 725.84 | 950.49 | 224.65 |
| | 5.00 | 766.47 | 39.90 | 688.26 | 844.68 | 156.42 |
| | 10.00 | 707.12 | 31.20 | 645.97 | 768.26 | 122.29 |
| | 30.00 | 595.60 | 22.11 | 552.27 | 638.93 | 86.66 |
| | 50.00 | 527.27 | 18.80 | 490.42 | 564.11 | 73.69 |
| Binxian | 70.00 | 465.56 | 17.81 | 430.66 | 500.47 | 69.80 |
| | 90.00 | 387.13 | 19.30 | 349.30 | 424.96 | 75.65 |
| | 95.00 | 353.58 | 21.17 | 312.10 | 395.06 | 82.97 |
| | 98.00 | 318.74 | 25.62 | 268.52 | 368.96 | 100.44 |
| | 99.00 | 297.12 | 30.87 | 236.62 | 357.63 | 121.00 |
| | 99.50 | 278.42 | 37.63 | 204.67 | 352.17 | 147.50 |
| | 99.90 | 242.90 | 57.79 | 129.63 | 356.17 | 226.54 |

**Table 7.** *Cont.*

| Site Name | $p$ (%) | $x_p$ | $S(x_p)$ | $x_p^L$ | $x_p^U$ | Confidence Interval Width |
|---|---|---|---|---|---|---|
| Changwu | 0.10 | 1099.36 | 149.27 | 806.79 | 1391.93 | 585.14 |
| | 1.00 | 939.48 | 77.49 | 787.60 | 1091.36 | 303.76 |
| | 2.00 | 887.33 | 59.34 | 771.03 | 1003.63 | 232.60 |
| | 5.00 | 813.64 | 41.44 | 732.43 | 894.85 | 162.42 |
| | 10.00 | 752.51 | 32.47 | 688.87 | 816.15 | 127.28 |
| | 30.00 | 637.34 | 23.09 | 592.08 | 682.60 | 90.52 |
| | 50.00 | 566.52 | 19.69 | 527.94 | 605.11 | 77.17 |
| | 70.00 | 502.40 | 18.71 | 465.74 | 539.06 | 73.33 |
| | 90.00 | 420.60 | 20.35 | 380.70 | 460.49 | 79.79 |
| | 95.00 | 385.49 | 22.37 | 341.64 | 429.34 | 87.70 |
| | 98.00 | 348.95 | 27.16 | 295.71 | 402.18 | 106.47 |
| | 99.00 | 326.23 | 32.78 | 261.99 | 390.48 | 128.49 |
| | 99.50 | 306.54 | 40.01 | 228.13 | 384.95 | 156.83 |
| | 99.90 | 269.06 | 61.57 | 148.39 | 389.73 | 241.35 |
| Chunhua | 0.10 | 1162.49 | 211.46 | 748.04 | 1576.94 | 828.90 |
| | 1.00 | 974.96 | 101.77 | 775.50 | 1174.41 | 398.92 |
| | 2.00 | 914.94 | 76.86 | 764.31 | 1065.58 | 301.27 |
| | 5.00 | 831.20 | 52.59 | 728.12 | 934.28 | 206.17 |
| | 10.00 | 762.75 | 40.51 | 683.35 | 842.15 | 158.80 |
| | 30.00 | 636.53 | 28.02 | 581.61 | 691.45 | 109.83 |
| | 50.00 | 560.91 | 23.41 | 515.04 | 606.79 | 91.75 |
| | 70.00 | 493.93 | 21.72 | 451.36 | 536.50 | 85.14 |
| | 90.00 | 410.79 | 22.92 | 365.86 | 455.72 | 89.86 |
| | 95.00 | 376.01 | 24.77 | 327.46 | 424.56 | 97.10 |
| | 98.00 | 340.44 | 29.41 | 282.79 | 398.09 | 115.30 |
| | 99.00 | 318.69 | 35.02 | 250.06 | 387.32 | 137.27 |
| | 99.50 | 300.07 | 42.31 | 217.14 | 382.99 | 165.85 |
| | 99.90 | 265.27 | 64.10 | 139.64 | 390.91 | 251.28 |
| Liquan | 0.10 | 1087.94 | 172.63 | 749.60 | 1426.28 | 676.68 |
| | 1.00 | 912.30 | 84.92 | 745.86 | 1078.74 | 332.88 |
| | 2.00 | 855.69 | 64.42 | 729.42 | 981.96 | 252.54 |
| | 5.00 | 776.33 | 44.40 | 689.31 | 863.34 | 174.03 |
| | 10.00 | 711.09 | 34.41 | 643.65 | 778.54 | 134.89 |
| | 30.00 | 589.85 | 24.05 | 542.71 | 636.99 | 94.28 |
| | 50.00 | 516.51 | 20.24 | 476.83 | 556.19 | 79.36 |
| | 70.00 | 451.00 | 18.95 | 413.86 | 488.15 | 74.29 |
| | 90.00 | 368.87 | 20.23 | 329.21 | 408.53 | 79.31 |
| | 95.00 | 334.18 | 22.00 | 291.06 | 377.30 | 86.25 |
| | 98.00 | 298.47 | 26.34 | 246.84 | 350.10 | 103.25 |
| | 99.00 | 276.50 | 31.52 | 214.72 | 338.28 | 123.55 |
| | 99.50 | 257.60 | 38.23 | 182.67 | 332.53 | 149.85 |
| | 99.90 | 222.05 | 58.26 | 107.86 | 336.24 | 228.39 |
| Qianxian | 0.10 | 1038.34 | 155.31 | 733.94 | 1342.74 | 608.80 |
| | 1.00 | 879.33 | 78.10 | 726.25 | 1032.41 | 306.16 |
| | 2.00 | 827.80 | 59.49 | 711.21 | 944.40 | 233.19 |
| | 5.00 | 755.31 | 41.23 | 674.50 | 836.13 | 161.63 |
| | 10.00 | 695.48 | 32.12 | 632.53 | 758.43 | 125.90 |
| | 30.00 | 583.58 | 22.63 | 539.24 | 627.93 | 88.69 |
| | 50.00 | 515.39 | 19.16 | 477.84 | 552.93 | 75.09 |
| | 70.00 | 454.10 | 18.06 | 418.72 | 489.49 | 70.77 |
| | 90.00 | 376.65 | 19.44 | 338.55 | 414.76 | 76.21 |
| | 95.00 | 343.71 | 21.24 | 302.07 | 385.35 | 83.28 |
| | 98.00 | 309.62 | 25.60 | 259.45 | 359.78 | 100.34 |
| | 99.00 | 288.54 | 30.75 | 228.27 | 348.81 | 120.53 |
| | 99.50 | 270.35 | 37.40 | 197.04 | 343.66 | 146.62 |
| | 99.90 | 235.95 | 57.26 | 123.73 | 348.17 | 224.44 |

**Table 7.** *Cont.*

| Site Name | $p$ (%) | $x_p$ | $S(x_p)$ | $x_p^L$ | $x_p^U$ | Confidence Interval Width |
|---|---|---|---|---|---|---|
| Xianyang | 0.10 | 1019.75 | 145.65 | 734.28 | 1305.22 | 570.95 |
| | 1.00 | 863.49 | 73.92 | 718.61 | 1008.37 | 289.76 |
| | 2.00 | 812.76 | 56.39 | 702.25 | 923.28 | 221.03 |
| | 5.00 | 741.30 | 39.17 | 664.53 | 818.07 | 153.53 |
| | 10.00 | 682.23 | 30.56 | 622.33 | 742.13 | 119.80 |
| | 30.00 | 571.52 | 21.59 | 529.21 | 613.84 | 84.63 |
| | 50.00 | 503.88 | 18.32 | 467.98 | 539.78 | 71.80 |
| | 70.00 | 442.96 | 17.30 | 409.04 | 476.87 | 67.83 |
| | 90.00 | 365.75 | 18.69 | 329.11 | 402.38 | 73.27 |
| | 95.00 | 332.82 | 20.46 | 292.72 | 372.92 | 80.19 |
| | 98.00 | 298.69 | 24.70 | 250.27 | 347.11 | 96.84 |
| | 99.00 | 277.55 | 29.72 | 219.31 | 335.80 | 116.49 |
| | 99.50 | 259.29 | 36.18 | 188.37 | 330.21 | 141.84 |
| | 99.90 | 224.68 | 55.48 | 115.95 | 333.41 | 217.46 |
| Xingping | 0.10 | 1134.72 | 179.57 | 782.76 | 1486.67 | 703.91 |
| | 1.00 | 953.89 | 88.64 | 780.15 | 1127.62 | 347.47 |
| | 2.00 | 895.57 | 67.29 | 763.68 | 1027.45 | 263.76 |
| | 5.00 | 813.76 | 46.41 | 722.80 | 904.71 | 181.91 |
| | 10.00 | 746.48 | 36.00 | 675.93 | 817.03 | 141.10 |
| | 30.00 | 621.32 | 25.19 | 571.96 | 670.68 | 98.73 |
| | 50.00 | 545.52 | 21.22 | 503.94 | 587.11 | 83.17 |
| | 70.00 | 477.77 | 19.88 | 438.80 | 516.74 | 77.94 |
| | 90.00 | 392.72 | 21.25 | 351.07 | 434.38 | 83.31 |
| | 95.00 | 356.77 | 23.13 | 311.44 | 402.09 | 90.65 |
| | 98.00 | 319.73 | 27.71 | 265.41 | 374.04 | 108.63 |
| | 99.00 | 296.92 | 33.18 | 231.89 | 361.95 | 130.06 |
| | 99.50 | 277.29 | 40.26 | 198.38 | 356.20 | 157.81 |
| | 99.90 | 240.34 | 61.40 | 120.01 | 360.68 | 240.67 |
| Yongshou | 0.10 | 1071.01 | 140.57 | 795.50 | 1346.52 | 551.02 |
| | 1.00 | 921.08 | 74.15 | 775.74 | 1066.41 | 290.67 |
| | 2.00 | 872.03 | 56.93 | 760.46 | 983.61 | 223.16 |
| | 5.00 | 802.60 | 39.89 | 724.41 | 880.79 | 156.37 |
| | 10.00 | 744.87 | 31.35 | 683.44 | 806.31 | 122.87 |
| | 30.00 | 635.76 | 22.38 | 591.89 | 679.63 | 87.75 |
| | 50.00 | 568.40 | 19.14 | 530.89 | 605.92 | 75.03 |
| | 70.00 | 507.21 | 18.25 | 471.44 | 542.99 | 71.55 |
| | 90.00 | 428.82 | 19.95 | 389.72 | 467.93 | 78.21 |
| | 95.00 | 395.06 | 21.98 | 351.97 | 438.15 | 86.18 |
| | 98.00 | 359.81 | 26.78 | 307.32 | 412.30 | 104.98 |
| | 99.00 | 337.85 | 32.38 | 274.38 | 401.32 | 126.94 |
| | 99.50 | 318.77 | 39.59 | 241.19 | 396.36 | 155.17 |
| | 99.90 | 282.37 | 61.06 | 162.70 | 402.04 | 239.35 |

## 4. Conclusions

The use of the FPEG distribution has received only limited attention from the hydrologic community, but some investigations in China suggest that this distribution performs well in modeling hydrological data. The MLE is proposed for determining the parameters and confidence intervals of the FPEG distribution. It involves parameter estimation and asymptotic variances of quantile estimators. The parameter estimation formulas constitute a system of nonlinear equations that have tedious forms. However, this should not be an insurmountable difficultly with the Levenberg–Marquardt algorithm, given the available numerical tools and computer power. An analytical expression of sample information matrix and Fisher's expected information matrix, and derivatives of design value with respect to the parameters were then derived. The asymptotic variances of the MLE quantile estimators for the FPEG distribution were expressed as a function of the probability (return period), parameters and sample size. Such variances can be employed for estimating the confidence intervals of the FPEG distribution quantiles. The FPEG distribution is applied to precipitation data of the Weihe watershed in China. The observed annual precipitation data were all accepted by the Kolmogorov–Smirnov test. These results showed that the FPEG distribution is a good candidate for modelling annual precipitation data. We expect that our results will provide guidance for estimating design values of random variables in other parts of world. In addition, Bayesian inference is a very good method for inferring the estimation of parameters from quantile parameters of the FPEG distribution, and will be studied further.

**Author Contributions:** Conceptualization, methodology, writing—original draft, S.S.; Data curation, writing—original draft, Y.K. and X.S.; Writing-review & editing, V.P.S. All authors have read and agreed to the published version of the manuscript.

**Funding:** This research was funded by National Natural Science Foundation of China, grant number 52079110.

**Institutional Review Board Statement:** Not applicable.

**Informed Consent Statement:** Not applicable.

**Data Availability Statement:** The data that support the findings of this study are available from the National Meteorological Information Center of China Meteorological Administration (CMA) and can be accessed via fpt upon requesting the corresponding author.

**Acknowledgments:** The observed hydrologic data are from the National Meteorological Information Center of China Meteorological Administration (CMA). The authors acknowledge the financial support of National Natural Science Foundation of China (Grant Nos. 52079110). The authors also wish to express their cordial gratitude to the editors, and anonymous reviewers for their illuminating comments which have greatly helped to improve the quality of this manuscript.

**Conflicts of Interest:** The authors declare no conflict of interest.

## Appendix A

*Appendix A.1. Sample Information Matrix*

The partial derivatives of the log likelihood model in Equation (15) with respect to parameters $\alpha$, $\beta$, $\delta$, and $b$ can be expressed as

$$\frac{\partial^2 L}{\partial \delta^2} = \frac{\partial}{\partial \delta}\left[-\frac{\alpha}{b}\sum_{i=1}^{n}\frac{1}{x_i - \delta} + \sum_{i=1}^{n}\frac{1}{x_i - \delta} + \frac{\beta}{b}\sum_{i=1}^{n}(x_i - \delta)^{\frac{1}{b}-1}\right]$$

$$= -\frac{\alpha}{b}\sum_{i=1}^{n}\frac{1}{(x_i - \delta)^2} + \sum_{i=1}^{n}\frac{1}{(x_i - \delta)^2} - \frac{\beta}{b}\left(\frac{1}{b}-1\right)\sum_{i=1}^{n}(x_i - \delta)^{\frac{1}{b}-2} \tag{A1}$$

$$\frac{\partial^2 L}{\partial \delta \partial \alpha} = \frac{\partial}{\partial \alpha}\left[-\frac{\alpha}{b}\sum_{i=1}^{n}\frac{1}{x_i - \delta} + \sum_{i=1}^{n}\frac{1}{x_i - \delta} + \frac{\beta}{b}\sum_{i=1}^{n}(x_i - \delta)^{\frac{1}{b}-1}\right] = -\frac{1}{b}\sum_{i=1}^{n}\frac{1}{x_i - \delta} \tag{A2}$$

$$\frac{\partial^2 L}{\partial \delta \partial \beta} = \frac{\partial}{\partial \beta}\left[-\frac{\alpha}{b}\sum_{i=1}^{n}\frac{1}{x_i - \delta} + \sum_{i=1}^{n}\frac{1}{x_i - \delta} + \frac{\beta}{b}\sum_{i=1}^{n}(x_i - \delta)^{\frac{1}{b}-1}\right] = \frac{1}{b}\sum_{i=1}^{n}(x_i - \delta)^{\frac{1}{b}-1} \tag{A3}$$

$$\frac{\partial^2 L}{\partial \delta \partial b} = \frac{\partial}{\partial b}\left[-\frac{\alpha}{b}\sum_{i=1}^{n}\frac{1}{x_i - \delta} + \sum_{i=1}^{n}\frac{1}{x_i - \delta} + \frac{\beta}{b}\sum_{i=1}^{n}(x_i - \delta)^{\frac{1}{b}-1}\right]$$

$$= \frac{\alpha}{b^2}\sum_{i=1}^{n}\frac{1}{x_i - \delta} - \frac{\beta}{b^2}\sum_{i=1}^{n}(x_i - \delta)^{\frac{1}{b}-1} - \frac{\beta}{b^3}\sum_{i=1}^{n}(x_i - \delta)^{\frac{1}{b}-1}ln(x_i - \delta) \tag{A4}$$

$$\frac{\partial^2 L}{\partial \alpha \partial \delta} = \frac{\partial}{\partial \delta}\left[nln\beta - n\psi(\alpha) + \frac{1}{b}\sum_{i=1}^{n}ln(x_i - \delta)\right] = \frac{1}{b}\sum_{i=1}^{n}\frac{-1}{x_i - \delta} = -\frac{1}{b}\sum_{i=1}^{n}\frac{1}{x_i - \delta} \tag{A5}$$

$$\frac{\partial^2 L}{\partial \alpha^2} = \frac{\partial}{\partial \alpha}\left[nln\beta - n\psi(\alpha) + \frac{1}{b}\sum_{i=1}^{n}ln(x_i - \delta)\right] = -n\psi'(\alpha) \tag{A6}$$

$$\frac{\partial^2 L}{\partial \alpha \partial \beta} = \frac{\partial}{\partial \beta}\left[ nln\beta - n\psi(\alpha) + \frac{1}{b}\sum_{i=1}^{n} ln(x_i - \delta) \right] = \frac{n}{\beta} \tag{A7}$$

$$\frac{\partial^2 L}{\partial \alpha \partial b} = \frac{\partial}{\partial b}\left[ nln\beta - n\psi(\alpha) + \frac{1}{b}\sum_{i=1}^{n} ln(x_i - \delta) \right] = -\frac{1}{b^2}\sum_{i=1}^{n} ln(x_i - \delta) \tag{A8}$$

$$\frac{\partial^2 L}{\partial \beta \partial \delta} = \frac{\partial}{\partial \delta}\left[ \frac{n\alpha}{\beta} - \sum_{i=1}^{n}(x_i - \delta)^{\frac{1}{b}} \right] = -\sum_{i=1}^{n}\frac{1}{b}(x_i - \delta)^{\frac{1}{b}-1}(-1) = \frac{1}{b}\sum_{i=1}^{n}(x_i - \delta)^{\frac{1}{b}-1} \tag{A9}$$

$$\frac{\partial^2 L}{\partial \beta \partial \alpha} = \frac{\partial}{\partial \alpha}\left[ \frac{n\alpha}{\beta} - \sum_{i=1}^{n}(x_i - \delta)^{\frac{1}{b}} \right] = \frac{n}{\beta} \tag{A10}$$

$$\frac{\partial^2 L}{\partial \beta^2} = \frac{\partial}{\partial \beta}\left[ \frac{n\alpha}{\beta} - \sum_{i=1}^{n}(x_i - \delta)^{\frac{1}{b}} \right] = -\frac{n\alpha}{\beta^2} \tag{A11}$$

$$\frac{\partial^2 L}{\partial \beta \partial b} = \frac{\partial}{\partial b}\left[ \frac{n\alpha}{\beta} - \sum_{i=1}^{n}(x_i - \delta)^{\frac{1}{b}} \right] = -\sum_{i=1}^{n}(x_i - \delta)^{\frac{1}{b}}ln(x_i - \delta)\frac{-1}{b^2}$$

$$= \frac{1}{b^2}\sum_{i=1}^{n}(x_i - \delta)^{\frac{1}{b}}ln(x_i - \delta) \tag{A12}$$

$$\frac{\partial^2 L}{\partial b \partial \delta} = \frac{\partial}{\partial \delta}\left\{ -\frac{n}{b} - \frac{\alpha}{b^2}\sum_{i=1}^{n} ln(x_i - \delta) + \frac{\beta}{b^2}\sum_{i=1}^{n}\left[ (x_i - \delta)^{\frac{1}{b}}ln(x_i - \delta) \right] \right\}$$

$$= \frac{\alpha}{b^2}\sum_{i=1}^{n}\frac{1}{x_i - \delta} + \frac{\beta}{b^2}\sum_{i=1}^{n}\left[ -\frac{1}{b}(x_i - \delta)^{\frac{1}{b}-1}ln(x_i - \delta) - (x_i - \delta)^{\frac{1}{b}-1} \right]$$

$$= \frac{\alpha}{b^2}\sum_{i=1}^{n}\frac{1}{(x_i - \delta)} - \frac{\beta}{b^2}\sum_{i=1}^{n}(x_i - \delta)^{\frac{1}{b}-1} - \frac{\beta}{b^3}\sum_{i=1}^{n}\left[ (x_i - \delta)^{\frac{1}{b}-1}ln(x_i - \delta) \right] \tag{A13}$$

$$\frac{\partial^2 L}{\partial b \partial \alpha} = \frac{\partial}{\partial \alpha}\left\{ -\frac{n}{b} - \frac{\alpha}{b^2}\sum_{i=1}^{n} ln(x_i - \delta) + \frac{\beta}{b^2}\sum_{i=1}^{n}\left[ (x_i - \delta)^{\frac{1}{b}}ln(x_i - \delta) \right] \right\} = -\frac{1}{b^2}\sum_{i=1}^{n} ln(x_i - \delta) \tag{A14}$$

$$\frac{\partial^2 L}{\partial b \partial \beta} = \frac{\partial}{\partial \beta}\left\{ -\frac{n}{b} - \frac{\alpha}{b^2}\sum_{i=1}^{n} ln(x_i - \delta) + \frac{\beta}{b^2}\sum_{i=1}^{n}\left[ (x_i - \delta)^{\frac{1}{b}}ln(x_i - \delta) \right] \right\}$$

$$= \frac{1}{b^2}\sum_{i=1}^{n}\left[ (x_i - \delta)^{\frac{1}{b}}ln(x_i - \delta) \right] \tag{A15}$$

$$\frac{\partial^2 L}{\partial b^2} = \frac{\partial}{\partial b}\left\{ -\frac{n}{b} - \frac{\alpha}{b^2}\sum_{i=1}^{n} ln(x_i - \delta) + \frac{\beta}{b^2}\sum_{i=1}^{n}\left[ (x_i - \delta)^{\frac{1}{b}}ln(x_i - \delta) \right] \right\}$$

$$= \frac{n}{b^2} + \frac{2\alpha}{b^3}\sum_{i=1}^{n} ln(x_i - \delta) - \frac{2\beta}{b^3}\sum_{i=1}^{n}\left[ (x_i - \delta)^{\frac{1}{b}}ln(x_i - \delta) \right]$$

$$- \frac{\beta}{b^4}\sum_{i=1}^{n}\left[ (x_i - \delta)^{\frac{1}{b}}ln^2(x_i - \delta) \right] \tag{A16}$$

*Appendix A.2. Fisher's Expected Information Matrix*

Multiplying Equations (A1)–(A16) and taking mathematical expectation, the elements of Fisher's expected information matrix can be obtained as:

$$E\left( -\frac{\partial^2 L}{\partial \delta^2} \right) = \frac{\alpha}{b}\sum_{i=1}^{n} E\left[ \frac{1}{(x_i - \delta)^2} \right] - \sum_{i=1}^{n} E\left[ \frac{1}{(x_i - \delta)^2} \right] + \frac{\beta}{b}\left( \frac{1}{b} - 1 \right)\sum_{i=1}^{n} E\left[ (x_i - \delta)^{\frac{1}{b}-2} \right] \tag{A17}$$

$$E\left( -\frac{\partial^2 L}{\partial \delta \partial \alpha} \right) = \frac{1}{b}\sum_{i=1}^{n} E\left( \frac{1}{x_i - \delta} \right) \tag{A18}$$

$$E\left( -\frac{\partial^2 L}{\partial \delta \partial \beta} \right) = -\frac{1}{b}\sum_{i=1}^{n} E\left[ (x_i - \delta)^{\frac{1}{b}-1} \right] \tag{A19}$$

$$E\left( -\frac{\partial^2 L}{\partial \delta \partial b} \right) = -\frac{\alpha}{b^2}\sum_{i=1}^{n} E\left( \frac{1}{x_i - \delta} \right) + \frac{\beta}{b^2}\sum_{i=1}^{n} E\left[ (x_i - \delta)^{\frac{1}{b}-1} \right] + \frac{\beta}{b^3}\sum_{i=1}^{n}\left[ (x_i - \delta)^{\frac{1}{b}-1}ln(x_i - \delta) \right] \tag{A20}$$

$$E\left(-\frac{\partial^2 L}{\partial\alpha\partial\delta}\right) = \frac{1}{b}\sum_{i=1}^{n} E\left(\frac{1}{x_i - \delta}\right) \tag{A21}$$

$$E\left(-\frac{\partial^2 L}{\partial\alpha^2}\right) = n\psi'(\alpha) \tag{A22}$$

$$E\left(-\frac{\partial^2 L}{\partial\alpha\partial\beta}\right) = -\frac{n}{\beta} \tag{A23}$$

$$E\left(-\frac{\partial^2 L}{\partial\alpha\partial b}\right) = \frac{1}{b^2}\sum_{i=1}^{n} E[ln(x_i - \delta)] \tag{A24}$$

$$E\left(-\frac{\partial^2 L}{\partial\beta\partial\delta}\right) = -\frac{1}{b}\sum_{i=1}^{n} E(x_i - \delta)^{\frac{1}{b}-1} \tag{A25}$$

$$E\left(-\frac{\partial^2 L}{\partial\beta\partial\alpha}\right) = -\frac{n}{\beta} \tag{A26}$$

$$E\left(-\frac{\partial^2 L}{\partial\beta^2}\right) = \frac{n\alpha}{\beta^2} \tag{A27}$$

$$E\left(-\frac{\partial^2 L}{\partial\beta\partial b}\right) = -\frac{1}{b^2}\sum_{i=1}^{n} E\left[(x_i - \delta)^{\frac{1}{b}}ln(x_i - \delta)\right] \tag{A28}$$

$$E\left(-\frac{\partial^2 L}{\partial b\partial\delta}\right) = -\frac{\alpha}{b^2}\sum_{i=1}^{n} E\left(\frac{1}{X - \delta}\right) + \frac{\beta}{b^2}\sum_{i=1}^{n} E\left[(x_i - \delta)^{\frac{1}{b}-1}\right]$$
$$+ \frac{\beta}{b^3}\sum_{i=1}^{n} E\left[(x_i - \delta)^{\frac{1}{b}-1}ln(x_i - \delta)\right] \tag{A29}$$

$$E\left(-\frac{\partial^2 L}{\partial b\partial\alpha}\right) = \frac{1}{b^2}\sum_{i=1}^{n} E[ln(x_i - \delta)] \tag{A30}$$

$$E\left(-\frac{\partial^2 L}{\partial b\partial\beta}\right) = -\frac{1}{b^2}\sum_{i=1}^{n} E\left[(x_i - \delta)^{\frac{1}{b}}ln(x_i - \delta)\right] \tag{A31}$$

$$E\left(-\frac{\partial^2 L}{\partial b^2}\right) = -\frac{n}{b^2} - \frac{2\alpha}{b^3}\sum_{i=1}^{n} E[ln(x_i - \delta)] + \frac{2\beta}{b^3}\sum_{i=1}^{n} E\left[(x_i - \delta)^{\frac{1}{b}}ln(x_i - \delta)\right]$$
$$+ \frac{\beta}{b^4}\sum_{i=1}^{n} E\left[(x_i - \delta)^{\frac{1}{b}}ln^2(x_i - \delta)\right] \tag{A32}$$

Because the above equations have some unknown mathematical expectations, these expectations need to be derived first.

$$E\left[\frac{1}{(x_i - \delta)^r}\right], \ r = 1, 2, \cdots$$

$$E\left[\frac{1}{(x_i - \delta)^r}\right] = \int_\delta^\infty \frac{1}{(x - \delta)^r}\frac{\beta^\alpha}{b\Gamma(\alpha)}(x - \delta)^{\frac{\alpha}{b}-1}e^{-\beta(x-\delta)^{\frac{1}{b}}}dx = \int_\delta^\infty \frac{\beta^\alpha}{b\Gamma(\alpha)}(x - \delta)^{\frac{\alpha}{b}-r-1}e^{-\beta(x-\delta)^{\frac{1}{b}}}dx \tag{A33}$$

Let $t = \beta(x - \delta)^{\frac{1}{b}}$, then $dx = \frac{b}{\beta^b}t^{b-1}dt$. Therefore, Equation (A33) becomes

$$E\left[\frac{1}{(x_i - \delta)^r}\right] = \int_0^\infty \frac{\beta^\alpha}{b\Gamma(\alpha)}\frac{1}{\beta^{\alpha-b(r+1)}}t^{\alpha-b(r+1)}e^{-t}\frac{b}{\beta^b}t^{b-1}dt = \frac{\beta^{br}\Gamma(\alpha - br)}{\Gamma(\alpha)} \tag{A34}$$

In particular,

$$E\left[\frac{1}{(x - \delta)^2}\right] = \frac{\beta^{2b}\Gamma(\alpha - 2b)}{\Gamma(\alpha)} \tag{A35}$$

$$E\left[\frac{1}{x - \delta}\right] = \frac{\beta^b\Gamma(\alpha - b)}{\Gamma(\alpha)} \tag{A36}$$

$$E\left[(x_i - \delta)^r\right], \ r = 1, 2, \cdots$$

$$E\left[(x_i - \delta)^r\right] = \int_\delta^\infty (x - \delta)^r\frac{\beta^\alpha}{b\Gamma(\alpha)}(x - \delta)^{\frac{\alpha}{b}-1}e^{-\beta(x-\delta)^{\frac{1}{b}}}dx = \int_\delta^\infty \frac{\beta^\alpha}{b\Gamma(\alpha)}(x - \delta)^{\frac{\alpha}{b}+r-1}e^{-\beta(x-\delta)^{\frac{1}{b}}}dx \tag{A37}$$

Let $t = \beta(x - \delta)^{\frac{1}{b}}$, then $dx = \frac{b}{\beta^b} t^{b-1} dt$, Equation (A37) yields

$$E[(x_i - \delta)^r] = \int_0^\infty \frac{\beta^\alpha}{b\Gamma(\alpha)} \left(\frac{1}{\beta^b} t^b\right)^{\frac{\alpha}{b} + r - 1} e^{-t} \frac{b}{\beta^b} t^{b-1} dt$$

$$= \int_0^\infty \frac{\beta^\alpha}{b\Gamma(\alpha)} \frac{1}{\beta^{\alpha + b(r-1)}} t^{\alpha + b(r-1)} e^{-t} \frac{b}{\beta^b} t^{b-1} dt = \int_0^\infty \frac{\beta^{-br}}{\Gamma(\alpha)} t^{\alpha + br - 1} e^{-t} dt$$

$$= \frac{\beta^{-br}\Gamma(\alpha + br)}{\Gamma(\alpha)} \tag{A38}$$

In particular,

$$E\left[(x_i - \delta)^{\frac{1}{b} - 1}\right] = \frac{\beta^{-b(\frac{1}{b} - 1)}\Gamma\left[\alpha + b\left(\frac{1}{b} - 1\right)\right]}{\Gamma(\alpha)} = \frac{\beta^{b-1}\Gamma(\alpha - b + 1)}{\Gamma(\alpha)} \tag{A39}$$

$$E\left[(x_i - \delta)^{\frac{1}{b} - 2}\right] = \frac{\beta^{-b(\frac{1}{b} - 2)}\Gamma\left[\alpha + b\left(\frac{1}{b} - 2\right)\right]}{\Gamma(\alpha)} = \frac{\beta^{2b-1}\Gamma(\alpha - 2b + 1)}{\Gamma(\alpha)} \tag{A40}$$

$E\{[ln(x_i - \delta)]^r\}, r = 1, 2, \cdots$

$$E\{[ln(x_i - \delta)]^r\} = \int_\delta^\infty [ln(x - \delta)]^r \frac{\beta^\alpha}{b\Gamma(\alpha)} (x - \delta)^{\frac{\alpha}{b} - 1} e^{-\beta(x-\delta)^{\frac{1}{b}}} dx \tag{A41}$$

Let $= \beta(x - \delta)^{\frac{1}{b}}$, then $dx = \frac{b}{\beta^b} t^{b-1} dt$, Equation (A41) produces

$$E\{[ln(x_i - \delta)]^r\} = \int_0^\infty \left[ln\frac{t^b}{\beta^b}\right]^r \frac{\beta^\alpha}{b\Gamma(\alpha)} \left(\frac{t^b}{\beta^b}\right)^{\frac{\alpha}{b} - 1} e^{-t} \frac{b}{\beta^b} t^{b-1} dt$$

$$= \int_0^\infty \left[ln\frac{t^b}{\beta^b}\right]^r \frac{\beta^\alpha}{b\Gamma(\alpha)} \frac{1}{\beta^{\alpha - b}} t^{\alpha - b} e^{-t} \frac{b}{\beta^b} t^{b-1} dt = \frac{1}{\Gamma(\alpha)} \int_0^\infty (blnt - bln\beta)^r t^{\alpha - 1} e^{-t} dt$$

$$= \frac{b^r}{\Gamma(\alpha)} \int_0^\infty (lnt - ln\beta)^r t^{\alpha - 1} e^{-t} dt \tag{A42}$$

Following $(x + a)^n = \sum_{k=0}^n \binom{n}{k} x^k a^{n-k}$, Equation (A42) can be written as

$$E\{[ln(x_i - \delta)]^r\} = \frac{b^r}{\Gamma(\alpha)} \int_0^\infty \sum_{k=0}^r \binom{r}{k} (lnt)^k (-ln\beta)^{r-k} t^{\alpha - 1} e^{-t} dt$$

$$= \frac{b^r}{\Gamma(\alpha)} (-ln\beta)^r \int_0^\infty t^{\alpha - 1} e^{-t} dt + \frac{b^r}{\Gamma(\alpha)} \sum_{k=1}^r \binom{r}{k} (-ln\beta)^{r-k} \int_0^\infty (lnt)^k t^{\alpha - 1} e^{-t} dt$$

$$= \frac{b^r}{\Gamma(\alpha)} (-ln\beta)^r \Gamma(\alpha) + \frac{b^r}{\Gamma(\alpha)} \sum_{k=1}^r \binom{r}{k} (-ln\beta)^{r-k} \int_0^\infty (lnt)^k t^{\alpha - 1} e^{-t} dt$$

$$= b^r(-ln\beta)^r + \frac{b^r}{\Gamma(\alpha)} \sum_{k=1}^r \binom{r}{k} (-ln\beta)^{r-k} \int_0^\infty (lnt)^k t^{\alpha - 1} e^{-t} dt$$

$$= b^r(-ln\beta)^r + \frac{b^r}{\Gamma(\alpha)} \sum_{k=1}^r \binom{r}{k} (-ln\beta)^{r-k} \frac{\partial^k \Gamma(\alpha)}{\partial \alpha^k} \tag{A43}$$

In particular,

$$E[ln(x_i - \delta)] = b(-ln\beta) + \frac{b}{\Gamma(\alpha)} \frac{\partial \Gamma(\alpha)}{\partial \alpha} = -bln\beta + b\psi(\alpha) \tag{A44}$$

$E[(x_i - \delta)^r ln(x_i - \delta)], r = 1, 2,$

$$E[(x_i - \delta)^r ln(x_i - \delta)] = \int_\delta^\infty ln(x - \delta) \frac{\beta^\alpha}{b\Gamma(\alpha)} (x - \delta)^{\frac{\alpha}{b} + r - 1} e^{-\beta(x-\delta)^{\frac{1}{b}}} dx \tag{A45}$$

Let $t = \beta(x-\delta)^{\frac{1}{b}}$, then $dx = \frac{b}{\beta^b} t^{b-1} dt$, Equation (A45) becomes

$$E\left[(x_i - \delta)^r ln(x_i - \delta)\right] = \int_0^\infty ln \frac{t^b}{\beta^b} \frac{\beta^\alpha}{b\Gamma(\alpha)} \left(\frac{t^b}{\beta^b}\right)^{\frac{\alpha}{b}+r-1} e^{-t} \frac{b}{\beta^b} t^{b-1} dt$$

$$= \int_0^\infty ln \frac{t^b}{\beta^b} \frac{\beta^{-br}}{\Gamma(\alpha)} t^{\alpha+br-1} e^{-t} dt = \frac{\beta^{-br}}{\Gamma(\alpha)} \int_0^\infty ln \frac{t^b}{\beta^b} t^{\alpha+br-1} e^{-t} dt$$

$$= \frac{b\beta^{-br}}{\Gamma(\alpha)} \left(\int_0^\infty lnt \cdot t^{\alpha+br-1} e^{-t} dt - ln\beta \int_0^\infty t^{\alpha+br-1} e^{-t} dt\right)$$

$$= \frac{b\beta^{-br}}{\Gamma(\alpha)} \left[\frac{\partial \Gamma(\alpha + br)}{\partial(\alpha + br)} - ln\beta \Gamma(\alpha + br)\right] \tag{A46}$$

In particular,

$$E\left[(x-\delta)^{\frac{1}{b}-1} ln(x-\delta)\right] = \frac{b\beta^{b-1}}{\Gamma(\alpha)} \left[\frac{\partial \Gamma(\alpha - b + 1)}{\partial(\alpha - b + 1)} - ln\beta \Gamma(\alpha - b + 1)\right] \tag{A47}$$

$$E\left[(x-\delta)^{\frac{1}{b}} ln(x-\delta)\right] = \frac{b\beta^{-1}}{\Gamma(\alpha)} \left[\frac{\partial \Gamma(\alpha + 1)}{\partial(\alpha + 1)} - ln\beta \Gamma(\alpha + 1)\right] \tag{A48}$$

$$E\left[(x_i - \delta)^{\frac{1}{b}} ln^2(x_i - \delta)\right]$$

$$E\left[(x_i - \delta)^{\frac{1}{b}} ln^2(x_i - \delta)\right] = \int_\delta^\infty ln^2(x-\delta) \frac{\beta^\alpha}{b\Gamma(\alpha)} (x-\delta)^{\frac{\alpha+1}{b}-1} e^{-\beta(x-\delta)^{\frac{1}{b}}} dx \tag{A49}$$

Let $t = \beta(x-\delta)^{\frac{1}{b}}$, then $dx = \frac{b}{\beta^b} t^{b-1} dt$, Equation (A49) becomes

$$E\left[(x_i - \delta)^{\frac{1}{b}} ln^2(x_i - \delta)\right] = \int_0^\infty \left(ln \frac{t^b}{\beta^b}\right)^2 \frac{\beta^\alpha}{b\Gamma(\alpha)} \left(\frac{t^b}{\beta^b}\right)^{\frac{\alpha+1}{b}-1} e^{-t} \frac{b}{\beta^b} t^{b-1} dt$$

$$= \int_0^\infty (blnt - bln\beta)^2 \frac{\beta^{-1}}{b\Gamma(\alpha)} t^\alpha e^{-t} dt = \frac{b^2\beta^{-1}}{\Gamma(\alpha)} \int_0^\infty (lnt - ln\beta)^2 t^\alpha e^{-t} dt$$

$$= \frac{b^2\beta^{-1}}{\Gamma(\alpha)} \int_0^\infty \left[(lnt)^2 - 2lnt \cdot ln\beta + (ln\beta)^2\right] t^\alpha e^{-t} dt$$

$$= \frac{b^2\beta^{-1}}{\Gamma(\alpha)} \left[\int_0^\infty (lnt)^2 t^\alpha e^{-t} dt - 2ln\beta \int_0^\infty lnt \cdot t^\alpha e^{-t} dt + (ln\beta)^2 \int_0^\infty t^\alpha e^{-t} dt\right]$$

$$= \frac{b^2\beta^{-1}}{\Gamma(\alpha)} \left[\frac{\partial^2 \Gamma(\alpha + 1)}{\partial(\alpha + 1)^2} - 2ln\beta \cdot \frac{\partial \Gamma(\alpha + 1)}{\partial(\alpha + 1)} + (ln\beta)^2 \Gamma(\alpha + 1)\right] \tag{A50}$$

$$E\left[\beta(x_i - \delta)^{\frac{1}{b}}\right]$$

$$E\left[\beta(x_i - \delta)^{\frac{1}{b}}\right] = \int_\delta^\infty \beta(x-\delta)^{\frac{1}{b}} \frac{\beta^\alpha}{b\Gamma(\alpha)} (x-\delta)^{\frac{\alpha}{b}-1} e^{-\beta(x-\delta)^{\frac{1}{b}}} dx$$

$$= \int_\delta^\infty \frac{\beta^{\alpha+1}}{b\Gamma(\alpha)} (x-\delta)^{\frac{\alpha+1}{b}-1} e^{-\beta(x-\delta)^{\frac{1}{b}}} dx \tag{A51}$$

Let $t = \beta(x-\delta)^{\frac{1}{b}}$, then $dx = \frac{b}{\beta^b} t^{b-1} dt$, Equation (A51) yields

$$E\left[\beta(x_i - \delta)^{\frac{1}{b}}\right] = \int_0^\infty \frac{\beta^{\alpha+1}}{b\Gamma(\alpha)} \frac{1}{\beta^{\alpha+1-b}} t^{\alpha+1-b} e^{-t} \frac{b}{\beta^b} t^{b-1} dt$$

$$= \frac{1}{\Gamma(\alpha)} \int_0^\infty t^\alpha e^{-t} dt = \frac{\Gamma(\alpha + 1)}{\Gamma(\alpha)} = \frac{\alpha\Gamma(\alpha)}{\Gamma(\alpha)} = \alpha \tag{A52}$$

In particular,

$$E\left[\beta(X - \delta)^{\frac{1}{b}}\right] = \alpha \tag{A53}$$

$$E\left[\frac{1}{\beta(x_i-\delta)^{\frac{1}{b}}}\right]$$

$$E\left[\frac{1}{\beta(x_i-\delta)^{\frac{1}{b}}}\right]=\int_{\delta}^{\infty}\frac{1}{\beta(x-\delta)^{\frac{1}{b}}}\frac{\beta^{\alpha}}{b\Gamma(\alpha)}(x-\delta)^{\frac{\alpha}{b}-1}e^{-\beta(x-\delta)^{\frac{1}{b}}}dx$$

$$=\int_{\delta}^{\infty}\frac{\beta^{\alpha-1}}{b\Gamma(\alpha)}(x-\delta)^{\frac{\alpha-1}{b}-1}e^{-\beta(x-\delta)^{\frac{1}{b}}}dx \tag{A54}$$

Let $t=\beta(x-\delta)^{\frac{1}{b}}$, then $dx=\frac{b}{\beta^b}t^{b-1}dt$, Equation (A54) yields

$$E\left[\frac{1}{\beta(x_i-\delta)^{\frac{1}{b}}}\right]=\int_0^{\infty}\frac{\beta^{\alpha-1}}{b\Gamma(\alpha)}\frac{1}{\beta^{\alpha-1-b}}t^{\alpha-1-b}e^{-t}\frac{b}{\beta^b}t^{b-1}dt=\int_0^{\infty}\frac{1}{\Gamma(\alpha)}t^{\alpha-2}e^{-t}dt$$

$$=\frac{\Gamma(\alpha-1)}{\Gamma(\alpha)}=\frac{\Gamma(\alpha-1)}{\Gamma(\alpha-1+1)}=\frac{\Gamma(\alpha-1)}{(\alpha-1)\Gamma(\alpha-1)}=\frac{1}{(\alpha-1)} \tag{A55}$$

In particular,

$$E\left[\frac{1}{\beta(X-\delta)^{\frac{1}{b}}}\right]=\frac{1}{(\alpha-1)} \tag{A56}$$

Substitution of equations of Equations (A33)–(A56) into Equations (A17)–(A32), we can get the elements of the expected information matrix.

$$E\left(-\frac{\partial^2 L}{\partial\delta^2}\right)=\frac{\alpha}{b}\frac{n\beta^{2b}\Gamma(\alpha-2b)}{\Gamma(\alpha)}-\frac{n\beta^{2b}\Gamma(\alpha-2b)}{\Gamma(\alpha)}+\frac{n\beta}{b}\left(\frac{1}{b}-1\right)\frac{\beta^{2b-1}\Gamma(\alpha-2b+1)}{\Gamma(\alpha)}$$

$$=\frac{\alpha}{b}\frac{n\beta^{2b}\Gamma(\alpha-2b)}{\Gamma(\alpha)}-\frac{n\beta^{2b}\Gamma(\alpha-2b)}{\Gamma(\alpha)}+\frac{(1-b)(\alpha-2b)}{b^2}\frac{n\beta^{2b}\Gamma(\alpha-2b)}{\Gamma(\alpha)}$$

$$=\frac{\alpha b-b^2+\alpha-2b-\alpha b+2b^2}{b^2}\frac{n\beta^{2b}\Gamma(\alpha-2b)}{\Gamma(\alpha)}=\frac{\alpha-2b+b^2}{b^2}\frac{n\beta^{2b}\Gamma(\alpha-2b)}{\Gamma(\alpha)} \tag{A57}$$

$$E\left(-\frac{\partial^2 L}{\partial\delta\partial\alpha}\right)=\frac{1}{b}\sum_{i=1}^{n}E\left(\frac{1}{x_i-\delta}\right)=\frac{1}{b}\frac{n\beta^b\Gamma(\alpha-b)}{\Gamma(\alpha)} \tag{A58}$$

$$E\left(-\frac{\partial^2 L}{\partial\delta\partial\beta}\right)=-\frac{1}{b}\sum_{i=1}^{n}E\left[(x_i-\delta)^{\frac{1}{b}-1}\right]=-\frac{1}{b}\frac{n\beta^{b-1}\Gamma(\alpha-b+1)}{\Gamma(\alpha)}$$

$$=-\frac{(\alpha-b)}{b}\frac{n\beta^{b-1}\Gamma(\alpha-b)}{\Gamma(\alpha)} \tag{A59}$$

$$E\left(-\frac{\partial^2 L}{\partial\delta\partial b}\right)=-\frac{\alpha}{b^2}\frac{n\beta^b\Gamma(\alpha-b)}{\Gamma(\alpha)}+\frac{\beta}{b^2}\frac{n\beta^{b-1}\Gamma(\alpha-b+1)}{\Gamma(\alpha)}$$

$$+\frac{\beta}{b^3}\frac{nb\beta^{b-1}}{\Gamma(\alpha)}\left[\frac{\partial\Gamma(\alpha-b+1)}{\partial(\alpha-b+1)}-\ln\beta\cdot\Gamma(\alpha-b+1)\right]$$

$$=-\frac{\alpha}{b^2}\frac{n\beta^b\Gamma(\alpha-b)}{\Gamma(\alpha)}+\frac{\alpha-b}{b^2}\frac{n\beta^b\Gamma(\alpha-b)}{\Gamma(\alpha)}$$

$$+\frac{1}{b^2}\frac{n\beta^b}{\Gamma(\alpha)}\left[\frac{\partial\Gamma(\alpha-b+1)}{\partial(\alpha-b+1)}-\ln\beta\cdot\Gamma(\alpha-b+1)\right]$$

$$=-\frac{1}{b}\frac{n\beta^b\Gamma(\alpha-b)}{\Gamma(\alpha)}+\frac{1}{b^2}\frac{n\beta^b}{\Gamma(\alpha)}\left[\frac{\partial\Gamma(\alpha-b+1)}{\partial(\alpha-b+1)}-\ln\beta\cdot\Gamma(\alpha-b+1)\right] \tag{A60}$$

$$E\left(-\frac{\partial^2 L}{\partial\alpha\partial\delta}\right)=\frac{1}{b}\frac{n\beta^b\Gamma(\alpha-b)}{\Gamma(\alpha)} \tag{A61}$$

$$E\left(-\frac{\partial^2 L}{\partial\alpha^2}\right)=n\psi'(\alpha) \tag{A62}$$

$$E\left(-\frac{\partial^2 L}{\partial\alpha\partial\beta}\right)=-\frac{n}{\beta} \tag{A63}$$

$$E\left(-\frac{\partial^2 L}{\partial\alpha\partial b}\right) = \frac{n[-ln\beta + \psi(\alpha)]}{b} \tag{A64}$$

$$E\left(-\frac{\partial^2 L}{\partial\beta\partial\delta}\right) = -\frac{\alpha - b}{b}\frac{n\beta^{b-1}\Gamma(\alpha - b)}{\Gamma(\alpha)} \tag{A65}$$

$$E\left(-\frac{\partial^2 L}{\partial\beta\partial\alpha}\right) = -\frac{n}{\beta} \tag{A66}$$

$$E\left(-\frac{\partial^2 L}{\partial\beta^2}\right) = \frac{n\alpha}{\beta^2} \tag{A67}$$

$$E\left(-\frac{\partial^2 L}{\partial\beta\partial b}\right) = -\frac{n}{b\beta}\frac{\frac{\partial\Gamma(\alpha+1)}{\partial(\alpha+1)} - ln\beta\cdot\Gamma(\alpha + 1)}{\Gamma(\alpha)} \tag{A68}$$

$$E\left(-\frac{\partial^2 L}{\partial b\partial\delta}\right) = -\frac{\alpha}{b^2}\frac{n\beta^b\Gamma(\alpha - b)}{\Gamma(\alpha)} + \frac{\beta}{b^2}\frac{n\beta^{b-1}\Gamma(\alpha - b + 1)}{\Gamma(\alpha)}$$

$$+ \frac{\beta}{b^3}\frac{n\beta^{b-1}}{\Gamma(\alpha)}\left[\frac{\partial\Gamma(\alpha - b + 1)}{\partial(\alpha - b + 1)} - ln\beta\cdot\Gamma(\alpha - b + 1)\right]$$

$$= -\frac{\alpha}{b^2}\frac{n\beta^b\Gamma(\alpha - b)}{\Gamma(\alpha)} + \frac{\alpha - b}{b^2}\frac{n\beta^b\Gamma(\alpha - b)}{\Gamma(\alpha)} + \frac{1}{b^2}\frac{n\beta^b}{\Gamma(\alpha)}\left[\frac{\partial\Gamma(\alpha - b + 1)}{\partial(\alpha - b + 1)} - ln\beta\cdot\Gamma(\alpha - b + 1)\right]$$

$$= -\frac{1}{b}\frac{n\beta^b\Gamma(\alpha - b)}{\Gamma(\alpha)} + \frac{1}{b^2}\frac{n\beta^b}{\Gamma(\alpha)}\left[\frac{\partial\Gamma(\alpha - b + 1)}{\partial(\alpha - b + 1)} - ln\beta\cdot\Gamma(\alpha - b + 1)\right] \tag{A69}$$

$$E\left(-\frac{\partial^2 L}{\partial b\partial\alpha}\right) = \frac{n[-ln\beta + \psi(\alpha)]}{b} \tag{A70}$$

$$E\left(-\frac{\partial^2 L}{\partial b\partial\beta}\right) = -\frac{1}{b^2}\frac{bn\beta^{-1}}{\Gamma(\alpha)}\left[\frac{\partial\Gamma(\alpha + 1)}{\partial(\alpha + 1)} - ln\beta\cdot\Gamma(\alpha + 1)\right] = -\frac{1}{b\beta}\frac{n\left[\frac{\partial\Gamma(\alpha+1)}{\partial(\alpha+1)} - ln\beta\cdot\Gamma(\alpha + 1)\right]}{\Gamma(\alpha)} \tag{A71}$$

$$E\left(-\frac{\partial^2 L}{\partial b^2}\right) = -\frac{n}{b^2} - \frac{2\alpha}{b^3}n[-bln\beta + b\psi(\alpha)] + \frac{2\beta}{b^3}\frac{bn\beta^{-1}}{\Gamma(\alpha)}\left[\frac{\partial\Gamma(\alpha + 1)}{\partial(\alpha + 1)} - ln\beta\cdot\Gamma(\alpha + 1)\right]$$

$$+ \frac{\beta}{b^4}\frac{b^2 n\beta^{-1}}{\Gamma(\alpha)}\left[\frac{\partial^2\Gamma(\alpha + 1)}{\partial(\alpha + 1)^2} - 2ln\beta\frac{\partial\Gamma(\alpha + 1)}{\partial(\alpha + 1)} + (ln\beta)^2\cdot\Gamma(\alpha + 1)\right]$$

$$= -\frac{n}{b^2} - \frac{2n\alpha[-ln\beta + \psi(\alpha)]}{b^2} + \frac{2}{b^2}\frac{n\left[\frac{\partial\Gamma(\alpha+1)}{\partial(\alpha+1)} - ln\beta\cdot\Gamma(\alpha + 1)\right]}{\Gamma(\alpha)}$$

$$+ \frac{1}{b^2}\frac{n}{\Gamma(\alpha)}\left[\frac{\partial^2\Gamma(\alpha + 1)}{\partial(\alpha + 1)^2} - 2ln\beta\frac{\partial\Gamma(\alpha + 1)}{\partial(\alpha + 1)} + (ln\beta)^2\cdot\Gamma(\alpha + 1)\right] \tag{A72}$$

*Appendix A.3. Jacobian Matrix*

$$\frac{\partial F_1(\delta, \alpha, b)}{\partial\delta} = \frac{\partial}{\partial\delta}\left[-\frac{\alpha}{b}\sum_{i=1}^{n}\frac{1}{x_i - \delta} + \sum_{i=1}^{n}\frac{1}{x_i - \delta} + \frac{1}{b}\frac{n\alpha}{\sum_{i=1}^{n}ln(x_i - \delta)^{\frac{1}{b}}}\sum_{i=1}^{n}(x_i - \delta)^{\frac{1}{b} - 1}\right]$$

$$= -\frac{\alpha}{b}\sum_{i=1}^{n}\frac{1}{(x_i - \delta)^2} + \sum_{i=1}^{n}\frac{1}{(x_i - \delta)^2}$$

$$+ \frac{n\alpha}{b}\frac{-\left(\frac{1}{b} - 1\right)\sum_{i=1}^{n}(x_i - \delta)^{\frac{1}{b} - 2}\sum_{i=1}^{n}ln(x_i - \delta)^{\frac{1}{b}} + \frac{1}{b}\sum_{i=1}^{n}(x_i - \delta)^{\frac{1}{b} - 1}\sum_{i=1}^{n}\frac{1}{(x_i - \delta)^{\frac{1}{b}}}(x_i - \delta)^{\frac{1}{b} - 1}}{\left[\sum_{i=1}^{n}ln(x_i - \delta)^{\frac{1}{b}}\right]^2},$$

$$= -\frac{\alpha}{b}\sum_{i=1}^{n}\frac{1}{(x_i - \delta)^2} + \sum_{i=1}^{n}\frac{1}{(x_i - \delta)^2}$$

$$+ \frac{n\alpha}{b}\frac{\left(1 - \frac{1}{b}\right)\sum_{i=1}^{n}(x_i - \delta)^{\frac{1}{b} - 2}\sum_{i=1}^{n}ln(x_i - \delta)^{\frac{1}{b}} + \frac{1}{b}\sum_{i=1}^{n}(x_i - \delta)^{\frac{1}{b} - 1}\sum_{i=1}^{n}\frac{1}{x_i - \delta}}{\left[\sum_{i=1}^{n}ln(x_i - \delta)^{\frac{1}{b}}\right]^2} \tag{A73}$$

$$\frac{\partial F_1(\delta,\alpha,b)}{\partial \alpha} = \frac{\partial}{\partial \alpha}\left[ -\frac{\alpha}{b}\sum_{i=1}^{n}\frac{1}{x_i-\delta} + \sum_{i=1}^{n}\frac{1}{x_i-\delta} + \frac{1}{b}\frac{n\alpha}{\sum_{i=1}^{n}ln(x_i-\delta)^{\frac{1}{b}}}\sum_{i=1}^{n}(x_i-\delta)^{\frac{1}{b}-1} \right]$$

$$= -\frac{1}{b}\sum_{i=1}^{n}\frac{1}{x_i-\delta} + \frac{1}{b}\frac{n}{\sum_{i=1}^{n}ln(x_i-\delta)^{\frac{1}{b}}}\sum_{i=1}^{n}(x_i-\delta)^{\frac{1}{b}-1} \tag{A74}$$

$$\frac{\partial F_1(\delta,\alpha,b)}{\partial b} = \frac{\partial}{\partial b}\left[ -\frac{\alpha}{b}\sum_{i=1}^{n}\frac{1}{x_i-\delta} + \sum_{i=1}^{n}\frac{1}{x_i-\delta} + \frac{1}{b}\frac{n\alpha}{\sum_{i=1}^{n}ln(x_i-\delta)^{\frac{1}{b}}}\sum_{i=1}^{n}(x_i-\delta)^{\frac{1}{b}-1} \right]$$

$$= \frac{\partial}{\partial b}\left[ -\frac{\alpha}{b}\sum_{i=1}^{n}\frac{1}{x_i-\delta} + \sum_{i=1}^{n}\frac{1}{x_i-\delta} + \frac{n\alpha}{b}\frac{\sum_{i=1}^{n}(x_i-\delta)^{\frac{1}{b}-1}}{\sum_{i=1}^{n}ln(x_i-\delta)^{\frac{1}{b}}} \right]$$

$$= \frac{\alpha}{b^2}\sum_{i=1}^{n}\frac{1}{x_i-\delta} - \frac{n\alpha}{b^2}\frac{\sum_{i=1}^{n}(x_i-\delta)^{\frac{1}{b}-1}}{\sum_{i=1}^{n}ln(x_i-\delta)^{\frac{1}{b}}}$$

$$+\frac{n\alpha}{b}\frac{\frac{-1}{b^2}\sum_{i=1}^{n}\left[(x_i-\delta)^{\frac{1}{b}-1}ln(x_i-\delta)\right]\cdot\sum_{i=1}^{n}ln(x_i-\delta)^{\frac{1}{b}} + \frac{1}{b^2}\sum_{i=1}^{n}(x_i-\delta)^{\frac{1}{b}-1}\cdot\sum_{i=1}^{n}\frac{1}{(x_i-\delta)^{\frac{1}{b}}}(x_i-\delta)^{\frac{1}{b}}ln(x_i-\delta)}{\left[\sum_{i=1}^{n}ln(x_i-\delta)^{\frac{1}{b}}\right]^2}$$

$$= \frac{\alpha}{b^2}\sum_{i=1}^{n}\frac{1}{x_i-\delta} - \frac{n\alpha}{b^2}\frac{\sum_{i=1}^{n}(x_i-\delta)^{\frac{1}{b}-1}}{\sum_{i=1}^{n}ln(x_i-\delta)^{\frac{1}{b}}}$$

$$+\frac{n\alpha}{b^3}\frac{-\sum_{i=1}^{n}\left[(x_i-\delta)^{\frac{1}{b}-1}ln(x_i-\delta)\right]\cdot\sum_{i=1}^{n}ln(x_i-\delta)^{\frac{1}{b}} + \sum_{i=1}^{n}(x_i-\delta)^{\frac{1}{b}-1}\cdot\sum_{i=1}^{n}ln(x_i-\delta)}{\left[\sum_{i=1}^{n}ln(x_i-\delta)^{\frac{1}{b}}\right]^2} \tag{A75}$$

$$\frac{\partial F_2(\delta,\alpha,b)}{\partial \delta} = \frac{\partial}{\partial \delta}\left[ nln\frac{n\alpha}{\sum_{i=1}^{n}ln(x_i-\delta)^{\frac{1}{b}}} - n\psi(\alpha) + \frac{1}{b}\sum_{i=1}^{n}ln(x_i-\delta) \right]$$

$$= \frac{\partial}{\partial \delta}\left[ nln(n\alpha) - nln\left(\sum_{i=1}^{n}ln(x_i-\delta)^{\frac{1}{b}}\right) - n\psi(\alpha) + \frac{1}{b}\sum_{i=1}^{n}ln(x_i-\delta) \right]$$

$$= -n\frac{1}{\sum_{i=1}^{n}ln(x_i-\delta)^{\frac{1}{b}}}\sum_{i=1}^{n}\left(\frac{1}{(x_i-\delta)^{\frac{1}{b}}}(x_i-\delta)^{\frac{1}{b}-1}\frac{-1}{b}\right) - \frac{1}{b}\sum_{i=1}^{n}\frac{1}{x_i-\delta}$$

$$= \frac{n}{b}\frac{1}{\sum_{i=1}^{n}ln(x_i-\delta)^{\frac{1}{b}}}\sum_{i=1}^{n}\left(\frac{1}{x_i-\delta}\right) - \frac{1}{b}\sum_{i=1}^{n}\frac{1}{x_i-\delta} \tag{A76}$$

$$\frac{\partial F_2(\delta,\alpha,b)}{\partial \alpha} = \frac{\partial}{\partial \alpha}\left[ nln\frac{n\alpha}{\sum_{i=1}^{n}ln(x_i-\delta)^{\frac{1}{b}}} - n\psi(\alpha) + \frac{1}{b}\sum_{i=1}^{n}ln(x_i-\delta) \right]$$

$$= \frac{\partial}{\partial \alpha}\left[ nln(n\alpha) - nln\left(\sum_{i=1}^{n}ln(x_i-\delta)^{\frac{1}{b}}\right) - n\psi(\alpha) + \frac{1}{b}\sum_{i=1}^{n}ln(x_i-\delta) \right]$$

$$= n\frac{n}{n\alpha} - n\psi'(\alpha) = \frac{n}{\alpha} - n\psi'(\alpha) \tag{A77}$$

$$\frac{\partial F_2(\delta,\alpha,b)}{\partial b} = \frac{\partial}{\partial b}\left[ nln\frac{n\alpha}{\sum_{i=1}^{n}ln(x_i-\delta)^{\frac{1}{b}}} - n\psi(\alpha) + \frac{1}{b}\sum_{i=1}^{n}ln(x_i-\delta) \right]$$

$$= \frac{\partial}{\partial b}\left[ nln(n\alpha) - nln\left(\sum_{i=1}^{n}ln(x_i-\delta)^{\frac{1}{b}}\right) - n\psi(\alpha) + \frac{1}{b}\sum_{i=1}^{n}ln(x_i-\delta) \right]$$

$$= -n \frac{1}{\sum_{i=1}^{n} ln(x_i - \delta)^{\frac{1}{b}}} \sum_{i=1}^{n} \left( \frac{1}{(x_i - \delta)^{\frac{1}{b}}} (x_i - \delta)^{\frac{1}{b}} ln(x_i - \delta) \frac{-1}{b^2} \right) - \frac{1}{b^2} \sum_{i=1}^{n} ln(x_i - \delta)$$

$$= \frac{n}{b^2} \frac{1}{\sum_{i=1}^{n} ln(x_i - \delta)^{\frac{1}{b}}} \sum_{i=1}^{n} ln(x_i - \delta) - \frac{1}{b^2} \sum_{i=1}^{n} ln(x_i - \delta) \tag{A78}$$

$$\frac{\partial F_3(\delta, \alpha, b)}{\partial \delta} = \frac{\partial}{\partial \delta} \left[ -\frac{n}{b} - \frac{\alpha}{b^2} \sum_{i=1}^{n} ln(x_i - \delta) + \frac{1}{b^2} ln \frac{n\alpha}{\sum_{i=1}^{n} ln(x_i - \delta)^{\frac{1}{b}}} \sum_{i=1}^{n} \left[ (x_i - \delta)^{\frac{1}{b}} ln(x_i - \delta) \right] \right] \tag{A79}$$

Let $G_1 = ln \frac{n\alpha}{\sum_{i=1}^{n} ln(x_i - \delta)^{\frac{1}{b}}}$ and $G_2 = \sum_{i=1}^{n} \left[ (x_i - \delta)^{\frac{1}{b}} ln(x_i - \delta) \right]$, Equation (A79) becomes to

$$\frac{\partial F_3(\delta, \alpha, b)}{\partial \delta} = \frac{\partial}{\partial \delta} \left\{ -\frac{n}{b} - \frac{\alpha}{b^2} \sum_{i=1}^{n} ln(x_i - \delta) + \frac{1}{b^2} G_1 G_2 \right\}$$

$$= \frac{\alpha}{b^2} \sum_{i=1}^{n} \frac{1}{x_i - \delta} + \frac{1}{b^2} \left( \frac{\partial G_1}{\partial \delta} G_2 + G_1 \frac{\partial G_2}{\partial \delta} \right) \tag{A80}$$

where

$$\frac{\partial G_1}{\partial \delta} = \frac{\partial}{\partial \delta} ln \frac{n\alpha}{\sum_{i=1}^{n} ln(x_i - \delta)^{\frac{1}{b}}} = \frac{\partial}{\partial \delta} \left[ ln(n\alpha) - ln \sum_{i=1}^{n} ln(x_i - \delta)^{\frac{1}{b}} \right]$$

$$= -\frac{1}{\sum_{i=1}^{n} ln(x_i - \delta)^{\frac{1}{b}}} \sum_{i=1}^{n} \frac{1}{(x_i - \delta)^{\frac{1}{b}}} (x_i - \delta)^{\frac{1}{b} - 1} \frac{-1}{b}$$

$$= \frac{1}{b} \frac{1}{\sum_{i=1}^{n} ln(x_i - \delta)^{\frac{1}{b}}} \sum_{i=1}^{n} \frac{1}{(x_i - \delta)};$$

$$\frac{\partial G_2}{\partial \delta} = \frac{\partial}{\partial \delta} \sum_{i=1}^{n} \left[ (x_i - \delta)^{\frac{1}{b}} ln(x_i - \delta) \right]$$

$$= (x_i - \delta)^{\frac{1}{b} - 1} \frac{-1}{b} ln(x_i - \delta) - (x_i - \delta)^{\frac{1}{b}} \frac{1}{x_i - \delta}$$

$$= -\frac{1}{b} (x_i - \delta)^{\frac{1}{b} - 1} ln(x_i - \delta) - (x_i - \delta)^{\frac{1}{b} - 1} \tag{A81}$$

$$\frac{\partial F_3(\delta, \alpha, b)}{\partial \alpha} = \frac{\partial}{\partial \alpha} \left[ -\frac{n}{b} - \frac{\alpha}{b^2} \sum_{i=1}^{n} ln(x_i - \delta) + \frac{1}{b^2} ln \frac{n\alpha}{\sum_{i=1}^{n} ln(x_i - \delta)^{\frac{1}{b}}} \sum_{i=1}^{n} \left[ (x_i - \delta)^{\frac{1}{b}} ln(x_i - \delta) \right] \right]$$

$$= \frac{\partial}{\partial \alpha} \left\{ -\frac{n}{b} - \frac{\alpha}{b^2} \sum_{i=1}^{n} ln(x_i - \delta) + \frac{1}{b^2} ln(n\alpha) \sum_{i=1}^{n} \left[ (x_i - \delta)^{\frac{1}{b}} ln(x_i - \delta) \right] \right.$$

$$\left. - \frac{1}{b^2} ln \sum_{i=1}^{n} ln(x_i - \delta)^{\frac{1}{b}} \sum_{i=1}^{n} \left[ (x_i - \delta)^{\frac{1}{b}} ln(x_i - \delta) \right] \right\}$$

$$= -\frac{1}{b^2} \sum_{i=1}^{n} ln(x_i - \delta) + \frac{1}{b^2} \frac{n}{n\alpha} \sum_{i=1}^{n} \left[ (x_i - \delta)^{\frac{1}{b}} ln(x_i - \delta) \right]$$

$$= -\frac{1}{b^2} \sum_{i=1}^{n} ln(x_i - \delta) + \frac{1}{\alpha b^2} \sum_{i=1}^{n} \left[ (x_i - \delta)^{\frac{1}{b}} ln(x_i - \delta) \right] \tag{A82}$$

$$\frac{\partial F_3(\delta, \alpha, b)}{\partial b} = \frac{\partial}{\partial b} \left[ -\frac{n}{b} - \frac{\alpha}{b^2} \sum_{i=1}^{n} ln(x_i - \delta) + \frac{1}{b^2} ln \frac{n\alpha}{\sum_{i=1}^{n} ln(x_i - \delta)^{\frac{1}{b}}} \sum_{i=1}^{n} \left[ (x_i - \delta)^{\frac{1}{b}} ln(x_i - \delta) \right] \right] \tag{A83}$$

Let $G_1 = ln\dfrac{n\alpha}{\sum_{i=1}^{n} ln(x_i-\delta)^{\frac{1}{b}}}$ and $G_2 = \sum_{i=1}^{n}\left[(x_i-\delta)^{\frac{1}{b}}ln(x_i-\delta)\right]$, Equation (A83) becomes to

$$\frac{\partial F_3(\delta,\alpha,b)}{\partial b} = \frac{\partial}{\partial b}\left[-\frac{n}{b}-\frac{\alpha}{b^2}\sum_{i=1}^{n}ln(x_i-\delta)+\frac{1}{b^2}G_1G_2\right] \tag{A84}$$

$$\frac{\partial F_3(\delta,\alpha,b)}{\partial b} = \frac{\partial}{\partial b}\left[-\frac{n}{b}-\frac{\alpha}{b^2}\sum_{i=1}^{n}ln(x_i-\delta)+\frac{1}{b^2}G_1G_2\right]$$

$$= \frac{n}{b^2}+\frac{2\alpha}{b^3}\sum_{i=1}^{n}ln(x_i-\delta)-\frac{2}{b^3}G_1G_2+\frac{1}{b^2}\frac{\partial(G_1G_2)}{\partial b}$$

$$= \frac{n}{b^2}+\frac{2\alpha}{b^3}\sum_{i=1}^{n}ln(x_i-\delta)-\frac{2}{b^3}G_1G_2+\frac{1}{b^2}\left(\frac{\partial G_1}{\partial b}G_2+G_1\frac{\partial G_2}{\partial b}\right) \tag{A85}$$

where

$$\frac{\partial G_1}{\partial b} = \frac{\partial}{\partial b}ln\frac{n\alpha}{\sum_{i=1}^{n}ln(x_i-\delta)^{\frac{1}{b}}} = \frac{\partial}{\partial b}\left[ln(n\alpha)-\sum_{i=1}^{n}ln(x_i-\delta)^{\frac{1}{b}}\right]$$

$$= -\sum_{i=1}^{n}\frac{1}{(x_i-\delta)^{\frac{1}{b}}}(x_i-\delta)^{\frac{1}{b}}ln(x_i-\delta)\frac{-1}{b^2} = \frac{1}{b^2}\sum_{i=1}^{n}ln(x_i-\delta).$$

$$\frac{\partial G_2}{\partial b} = \frac{\partial}{\partial b}\sum_{i=1}^{n}\left[(x_i-\delta)^{\frac{1}{b}}ln(x_i-\delta)\right]$$

$$= \sum_{i=1}^{n}\left[(x_i-\delta)^{\frac{1}{b}}ln(x_i-\delta)\frac{-1}{b^2}ln(x_i-\delta)\right]$$

$$= -\frac{1}{b^2}\sum_{i=1}^{n}\left[(x_i-\delta)^{\frac{1}{b}}ln^2(x_i-\delta)\right].$$

## References


1. Chowdhury, J.U.; Stedinger, J.R. Confidence interval for design floods with estimated skew coefficient. *J. Hydrol. Eng.* **1991**, *117*, 811–831. [CrossRef]
2. Meylan, P.; Favre, A.C.; Musy, A. *Predictive Hydrology: A Frequency Analysis Approach*; Science Publishers: Jersey, UK, 2012.
3. Rao, A.R.; Hamed, K.H. *Flood Frequency Analysis*; CRC Press LLC: Washington, DC, USA, 2000.
4. Tung, Y.K.; Yen, B.C.; Melching, C. *Hydrosystems Engineering Reliability Assessment and Risk Analysis*; McGraw-Hill: New York, NY, USA, 2006.
5. Shin, H. *Uncertainty Assessment of Quantile Estimators Based on the Generalized Logistic Distribution*; Yonsei University: Seoul, Korea, 2009.
6. Shin, H.; Nam, W.; Heo, J.H. Confidence Intervals of Quantiles for the Generalized Logistic Distribution; AHR CONGRESS. 2005. Available online: http://hydro-eng.yonsei.ac.kr/data_files/Confidenceintervalsofquantilesforthegeneralizedlogisticdistribution. pdf (accessed on 16 September 2005).
7. Whitley, R.J.; Hromadka, T.V. Chowdhury and Stedinger's approximate confidence intervals for design floods for a single site. *Stoch. Hydrol. Hydraul.* **1997**, *11*, 51–63. [CrossRef]
8. Ashkar, F.; Quarda, T.B.M.J. Approximate confidence intervals for quantiles of gamma and generalized gamma distribution. *J. Hydrol. Eng.* **1998**, *3*, 43–51. [CrossRef]
9. Heo, J.H.; Salas, J.D.; Kim, K.D. Estimation of confidence intervals of quantiles for the Weibull distribution. *Stochastic Environ. Res. Risk Assess.* **2001**, *15*, 284–309. [CrossRef]
10. Khalidou, M.B.; Carlos, D.D.; Alin, C. Confidence intervals of quantiles in hydrology computed by an analytical method. *Nat. Hazard.* **2001**, *24*, 1–12. [CrossRef]
11. Serinaldi, F. Analytical confidence intervals for index flow flow duration curves. *Water Resour. Res.* **2011**, *47*, W02542. [CrossRef]
12. Cohn, T.A.; Lane, W.L.; Stedinger, J.R. Confidence intervals for expected moments algorithm flood quantile estimates. *Water Resour. Res.* **2001**, *37*, 1695–1706. [CrossRef]
13. Burn, D.H. The use of resampling for estimating confidence intervals for single site and pooled frequency analysis. *Hydrol. Sci. J.* **2003**, *48*, 5–38. [CrossRef]
14. Dunn, P.K. Bootstrap confidence intervals for predicted rainfall quantiles. *Int. J. Climatol.* **2001**, *21*, 89–94. [CrossRef]
15. Rust, H.W.; Kallache, M.; Schellnhuber, H.J.; Kropp, J.P. Confidence intervals for flood return level estimates assuming long-range dependence. In *Extremis: Disruptive Events and Trends in Climate and Hydrology*; Springer: Berlin/Heidelberg, Germany, 2011.

16. Schendela, T.; Thongwichian, R. Flood frequency analysis: Confidence interval estimation by test inversion bootstrapping. *Adv. Water Resour.* **2015**, *83*, 1–9. [CrossRef]
17. Sveinsson, O.G.B.; Salas, J.D.; Boes, D.C. Uncertainty of quantile estimators using the population index flood method. *Water Resour. Res.* **2003**, *39*, 1206. [CrossRef]
18. Sun, J.; Qin, D.; Sun, H. *Generalized Probability Distribution in Hydrometeorology*; China Water and Power Press: Beijing, China, 2001.
19. Wang, H.; Qin, D.; Sun, J.; Wang, J. Study on the general model of hydrological frequency analysis. *Sci. China Ser. E Eng. Mater. Sci.* **2001**, *44*, 52–61. [CrossRef]
20. Kendall, M.G.; Stuart, A. *The Advanced Theory of Statistics. Vol. 2 Inference and Relationship*; Charles Griffin: London, UK, 1979.
21. Kite, G.W. *Frequency and Risk Analysis in Hydrology*; Water Resources Publications: Littleton, CO, USA, 1977.
22. Gavin, H.P. The Levenberg-Marquardt method for nonlinear least squares curve-fitting problems. *Comput. Sci.* **2013**, 1–17. Available online: http://people.duke.edu/~{}hpgavin/ce281/lm.pdf (accessed on 29 July 2021).

MDPI

*Article*

# Nonstationary Bayesian Modeling of Extreme Flood Risk and Return Period Affected by Climate Variables for Xiangjiang River Basin, in South-Central China

Hang Zeng [1,2,*], Jiaqi Huang [1,3], Zhengzui Li [4], Weihou Yu [4] and Hui Zhou [4]

1 School of Hydraulic and Environmental Engineering, Changsha University of Science & Technology, Changsha 410114, China; huangjiaqi@stu.csust.edu.cn
2 Key Laboratory of Dongting Lake Aquatic Eco-Environmental Control and Restoration of Hunan Province, Changsha 410114, China
3 Key Laboratory of Water-Sediment Sciences and Water Disaster Prevention of Hunan Province, Changsha 410114, China
4 Hydrology and Water Resources Survey Center of Hunan Province, Changsha 410008, China; hnlzz@139.com (Z.L.); yweihou@163.com (W.Y.); 2527092@163.com (H.Z.)
* Correspondence: hzeng1989@csust.edu.cn; Tel.: +86-18890028903

**Abstract:** The accurate design flood of hydraulic engineering is an important precondition to ensure the safety of residents, and the high precision estimation of flood frequency is a vital perquisite. The Xiangjiang River basin, which is the largest river in Hunan Province of China, is highly inclined to floods. This paper aims to investigate the annual maximum flood peak (AMFP) risk of Xiangjiang River basin under the climate context employing the Bayesian nonstationary time-varying moment models. Two climate covariates, i.e., the average June-July-August Artic Oscillation and sea level pressure in the Northwest Pacific Ocean, are selected and found to exhibit significant positive correlation with AMFP through a rigorous statistical analysis. The proposed models are tested with three cases, namely, stationary, linear-temporal and climate-based conditions. The results both indicate that the climate-informed model demonstrates the best performance as well as sufficiently explain the variability of extreme flood risk. The nonstationary return periods estimated by the expected number of exceedances method are larger than traditional ones built on the stationary assumption. In addition, the design flood could vary with the climate drivers which has great implication when applied in the context of climate change. This study suggests that nonstationary Bayesian modelling with climatic covariates could provide useful information for flood risk management.

**Keywords:** extreme flood risk; climatic factors; nonstationary frequency analysis; Bayesian modeling; nonstationary return period; Xiangjiang River basin

**Citation:** Zeng, H.; Huang, J.; Li, Z.; Yu, W.; Zhou, H. Nonstationary Bayesian Modeling of Extreme Flood Risk and Return Period Affected by Climate Variables for Xiangjiang River Basin, in South-Central China. *Water* **2022**, *14*, 66. https://doi.org/10.3390/w14010066

Academic Editors: Momcilo Markus and Sajjad Ahmad

Received: 16 October 2021
Accepted: 24 December 2021
Published: 31 December 2021

**Publisher's Note:** MDPI stays neutral with regard to jurisdictional claims in published maps and institutional affiliations.

## 1. Introduction

Extreme flood events with immediate and widespread devastation on natural systems and society have been extensively studied over recent decades [1]. The flood frequency analysis is a critical and effective tool to estimate flood risk and is commonly accepted by hydrologists. In recent decades, with climate change and anthropogenic activities, the stationary flood frequency analysis method is frequently questionable relative to the heterogeneous flood population; meanwhile, the nonstationary extreme flood risk analysis with various methodological frameworks has been extensively explored over recent decades. Among these approaches, the extreme flood risk analysis considering environmental factors could assist decision makers and engineers to better understand the nonstationary behaviors of extreme flood events and to take flood control measures quickly.

In general, extreme flood events have occurred all around the world, with a wide variety of regional characteristics. Among the several nonstationary flood frequency approaches, the time-varying moment method is widely utilized. Through incorporating factors into the covariates of model parameters, the time-varying moment method could quantify and even forecast the variation of extreme flood risks under changing environments [2–6]. For instance, Lopez and Frances (2013) applied the generalized additive models for location, scale and shape (GAMLSS) on flood risk impacted by multiple climate indices such as Arctic Oscillation (AO), North Atlantic Oscillation (NAO), Mediterranean Oscillation and so on [7]. Ficchi et al. (2021) reported that the Indian Ocean Dipole (IOD) and Tropical South Atlantic climate mode are equally as important as El Nino-Southern Oscillation (ENSO) for driving changes in the frequency of impactful floods across Africa [8]. Zhou (2020) quantified the contributions of changing climate and reservoir storage on the nonstationarity in flood risks worldwide [9]. Zhang et al. (2015) estimated the annual maximum stream-flow quantiles linearly conditional on Pacific Decadal Oscillation (PDO) in South China [10]. Kundzewicz et al. (2019b) reviewed the linkages between climate variability and water-abundance like high river discharge, flood magnitude and flood loss across the world [11] and Kundzewicz et al. (2020) analyzed the non-stationarity of flood frequency over China [12]. Not only the standard climate indices with limited spatial climate information as climate forcing on extreme flood risk, but also the large-scale ocean-atmosphere fields play a critically important role in the interpretation of the nonstationary behavior on flood. Zeng et al. (2017) used sea surface temperature anomaly (SSTa) in Northern Indian Ocean and Western Pacific Ocean incorporated into the location and scale parameters of lognormal distribution and general extreme value distribution (GEV) to forecast the flood risk in North China using a Bayesian framework [13]. Renard and Lall (2014) employed the large-scale oceanic fields to effectively predict the flood risk in regional nonstationary flood frequency [14]. In this paper, the large-scale climate patterns including the standard climate variability indices and oceanic or atmospheric fields are considered to access the climatic drivers of nonstationary extreme flood frequency and design flood in Xiangjiang River basin in South-Central China.

Indeed, the time-varying moment method is widely proposed to implement the nonstationary flood frequency analysis under the changing environment. Nevertheless, the flood quantiles or return periods varying with covariates yearly is not convincing and convenient in real-world applications. For the sake of accommodating nonstationary conditions, several hydrologic design flood methods [15] have been proposed in recent years, such as the expected waiting time (EWT) [16,17], the expected number of exceedances (ENE) [18,19], design life level (DLL) [20] and equivalent reliability (ER) [21]. Yan et al. (2017) compared and applied the above four design flood methods on annual maximum flood series in several basins [15]. The results indicated that ENE, DLL and ER yielded very similar design flood values for both increasing and decreasing trends. Yan et al. (2020) focused on applying the EWT method to estimate the design flood quantiles and proposed the extrapolation time to guarantee the convergence of the EWT [22]. Given the advantages and disadvantages of these methods [23,24], we choose the ENE method to calculate design floods under a changing climate and the estimated stationary design floods serve as a benchmark for comparison.

The Xiangjiang River is the largestof Hunan Province in South-Central China and its basin is the dominating inflow basin of Dongting Lake, which is the second largest freshwater lake in China, located in the Yangtze River basin. Characterized by the subtropical monsoon climate, the Xiangjiang River basin is highly prone to floods during the summer season from June to August. Mao et al. (2000) reported that the middle and lower reaches of the Xiangjiang River basin have experienced 29 flooding disasters during the years from 1949 to 1998, which is about one flood per 1.7 years on average [25]. Extreme pluvial floods happened in the years of 1954, 1976, 1994, 1996 and 1998. In July 2006, the Southern Hunan Province suffered a 500-year rainstorm caused by the Severe Tropical Storm Bilis and the triggered torrential flood brought casualties and economic loss exceeding 78 hundred million China Yuan [26]. In the summer of 2017, the main stream of the Xiangjiang River basin reached a historic recorded water level due to the influence of a heavy rainfall combined with the flood control measures that were largely relying on the embankment on the riverside. Therefore, extreme flood risk analysis under changing climate is a vital and urgent issue for the Xiangjiang River basin. Although a great amount of research on the extreme flood occurrence probability and design flood under changing climate in the Yangtze River basin has been undertaken, very few studies have been conducted on extreme flood/rainfall frequency analysis triggered by large-scale climate patterns in the Xiangjiang River Basin.

In the Xiangjiang River basin, as floods are primarily driven by abundant precipitation in the wet seasons, we focus on identifying the significant climate covariates for extreme flood peaks and quantitatively estimating the climate contribution to the probability of extreme flood events and design flood using the nonstationary modeling framework. Moreover, in at-site extreme flood frequency analysis, there generally exists relatively large uncertainty. This is especially the case for the probability distribution with more parameters. Hence, the Bayesian statistical model is developed to access the uncertainty of flood risks and the updated return periods in stationary and nonstationary contexts. The results will provide a protocol for the selection of covariates in terms of nonstationary conditions and also show implications for controlling floods and reinforcing the projects of engineers and scientists.

The paper is organized as follows. In Section 2, we describe the study area, the data set and the screening process for selecting climatic factors. Section 3 presents the methodologies and theoretical analysis of the nonstationary model construction, along with a brief description of Bayesian modeling framework and nonstationary hydrologic design method. The results and discussions are demonstrated in Section 4. Finally, the article concludes in Section 5.

## 2. Study Area and Data

The Xiangjiang River basin (Figure 1) is bound by $24°$–$29°$ N and $110°30'$–$114°$ E in South-Central China with a drainage area of 94,660 km$^2$ and a total length of 856 km. Due to the southeast monsoon humid climate in summer, the basin is frequently exposed to extraordinary rainstorm with average annual precipitation ranging from 1200 to 1700 mm and large floods in the rainy season (April to September) leading to high water levels. Because of the low-lying terrain in the middle and lower reaches of the Xiangjiang River, the metropolitan area of Chang-Zhu-Tan, which is the most important economic belt in Hunan Province, does not have appropriate conditions for building a large dam and can, therefore, only rely on the construction of riverside dikes for flood control. In recent years, rainstorm flooding of the Xiangjiang River has continuously reached historic records, having a strong negative impact on the residents. Therefore, analysis of flood risk and uncertainty is extremely important for basin-wide flood risk reduction and management.

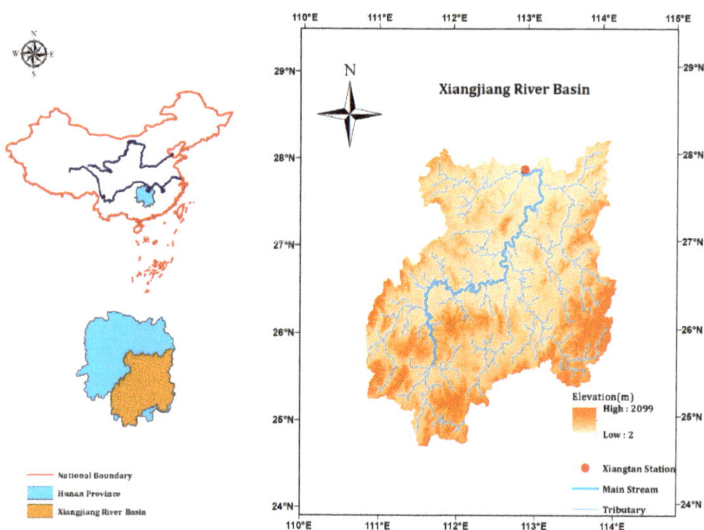

**Figure 1.** The location of Xiangjiang River basin and Xiangtan hydrological station.

*2.1. Extreme Flood*

The annual maximum flood peak series (AMFP), spanning the period of 1959–2017 and measured by the Xiangtan hydrological station (Figure 1) as well as the outflow control station of the Xiangjiang River basin, were collected from the Hydrology and Water Resources Survey Bureau of Hunan Province.

In general, the time-varying characteristic of AMFP is investigated using the Mann–Kendall test and the results indicate that there is no significant trend. Nevertheless, the Xiangjiang River Basin has abundant flow as rainfall is the main source. Annual extreme flood events, taking place during the flood season (June-July-August-September, JJAS), are generally due to heavy rainstorms corresponding to various induced climatic factors. Hence, it is interesting and important to conduct an impact assessment of the climate variables on the flood peak risks in the Xiangjiang River basin.

*2.2. Identification of Significant Climate Factors*

Identifying appropriate factors is a crucial step to better explain the variation in extreme flood frequency and to significantly improve the performance of nonstationary modeling. Slowly varying climate conditions associated with oceanic temperature may influence the development of the monsoon systems that affect the Xiangjiang River basin in the JJAS season. Thus, in this section, we explore the potential climate impact factors, in volving various large-scale low-frequency climate indices: sea surface temperature anomaly (SSTa) conditions extracted from the Hadley Center SST dataset on a $1° \times 1°$ grid [27]; and sea level pressure anomalies (SLPa) extracted from the Hadley Centre's mean sea level pressure data set on a $5° \times 5°$ grid [28]. Simultaneously, as suggested by Zeng et al. (2017) [13], we propose a stepwise selection method to screen the potential climate drivers employing rank correlation and a generalized linear model (GLM) to identify the orthogonal factors which have the best significant correlation with AMFP and the best flood risk modelling performance.

It is extensively recognized that the large-scale atmospheric circulation pattern climate indices, such as ENSO, including Southern Oscillation Index (SOI), Niño 3.4, Niño 12, Niño 3 and Niño 4, PDO, NAO and AO, can significantly influence hydrological variables (flood and precipitation) in China [29–31]. Then, the above climate indices, SLPa and SSTa, are selected as the possible influencing drivers. A three-month moving average from January to February of the next year of all the candidate factors are computed to access the rank correlation with AMFP using the Spearman rank correlation test. The results in Table 1 illustrate that the Spearman rank correlations between flood peak series and the ENSO, PDO, NAO are typically low and statistically non-significant. In addition, it demonstrates that only AO during JJA has significant positive rank correlation with AMFP. Strong teleconnection between precipitation/flood anomalies in the Yangtze River Basin and AO has been found in previous studies [29,32,33]. In addition, the findings of Yang (2011) and Gong and Wang (2003) suggested that a statistically significant positive correlation was detected between the AO and the precipitation over South China [34,35]. The AO, a primary mode of internal atmospheric dynamics over the extratropical northern hemisphere with a quasi-barotropic structure from the surface to the lower stratosphere, exerts great influences on global climate processes [36,37]. Consequently, the average AO from June to August is used as the first candidate factor.

**Table 1.** Spearman rank correlation results between climate indices and AMFP ($\alpha = 0.05$).

| Climate Indices | Coefficient | JFM | FMA | MAM | AMJ | MJJ | JJA | JAS | ASO | SON | OND |
|---|---|---|---|---|---|---|---|---|---|---|---|
| Niño3 | Rho | 0.16 | 0.10 | 0.04 | −0.05 | −0.09 | −0.12 | −0.12 | −0.11 | −0.08 | −0.06 |
| | *p*-value | 0.23 | 0.45 | 0.74 | 0.74 | 0.51 | 0.38 | 0.38 | 0.42 | 0.55 | 0.67 |
| Niño4 | Rho | 0.09 | 0.09 | 0.05 | 0.02 | −0.06 | −0.11 | −0.15 | −0.15 | −0.16 | −0.15 |
| | *p*-value | 0.50 | 0.49 | 0.71 | 0.86 | 0.66 | 0.43 | 0.26 | 0.26 | 0.22 | 0.26 |
| Niño3.4 | Rho | 0.14 | 0.11 | 0.07 | 0.04 | −0.06 | −0.11 | −0.15 | −0.15 | −0.13 | −0.11 |
| | *p*-value | 0.28 | 0.41 | 0.59 | 0.76 | 0.67 | 0.39 | 0.26 | 0.26 | 0.31 | 0.40 |
| Niño12 | Rho | 0.10 | 0.04 | −0.01 | −0.03 | −0.06 | −0.05 | −0.02 | 0.05 | 0.05 | 0.04 |
| | *p*-value | 0.47 | 0.74 | 0.94 | 0.82 | 0.64 | 0.73 | 0.86 | 0.72 | 0.71 | 0.77 |
| SOI | Rho | −0.16 | −0.18 | −0.13 | −0.02 | 0.08 | 0.04 | 0.03 | 0.05 | 0.10 | 0.17 |
| | *p*-value | 0.24 | 0.17 | 0.31 | 0.89 | 0.56 | 0.77 | 0.83 | 0.74 | 0.47 | 0.19 |
| NAO | Rho | 0.13 | −0.02 | −0.22 | −0.05 | 0.04 | 0.10 | −0.08 | −0.24 | −0.24 | −0.17 |
| | *p*-value | 0.32 | 0.91 | 0.09 | 0.73 | 0.78 | 0.45 | 0.55 | 0.06 | 0.06 | 0.20 |
| PDO | Rho | 0.24 | 0.19 | 0.11 | 0.10 | 0.04 | 0.02 | −0.07 | −0.08 | −0.06 | −0.03 |
| | *p*-value | 0.07 | 0.16 | 0.41 | 0.47 | 0.74 | 0.87 | 0.61 | 0.55 | 0.64 | 0.82 |
| AO | Rho | 0.19 | 0.11 | −0.02 | 0.04 | 0.19 | 0.32 | 0.15 | −0.14 | −0.23 | −0.14 |
| | *p*-value | 0.16 | 0.41 | 0.87 | 0.78 | 0.16 | <u>0.01</u> | 0.26 | 0.28 | 0.08 | 0.29 |

Note: JFM, FMA, MAM, AMJ and so on, are the three-month moving average values, respectively. For example, JJA, means the average value of the climate variable during June, July and August. The value with underline "_" represents the *p* value less than 0.05.

With respect to the SLPa and SSTa, we utilize the correlation map [14] to determine the climatic factors. Figure 2a,b demonstrates the Spearman rank correlation between AMFP and the SSTa fields during JFM, and SLPa fields during JJA, respectively. We find that the JFM SSTa blue region (35° N–45° N, 150° W–170° W) reveals significant negative rank correlation with AMFP. In contrast, the two JJA SLPa orange regions (one is the continent region for 35° N–50° N, 110° E–135° E called SLPa1, another is the ocean for 25° N–40° N, 165° E–180° E named SLPa2) show significant positive rank correlation with AMFP. A principal component analysis (PCA) is employed to acquire a set of orthogonal factors on the above JFM SSTa, JJA SLPa1 and JJA SLPa2, respectively, since the large-scale climate variables with fields may be mutually correlated.

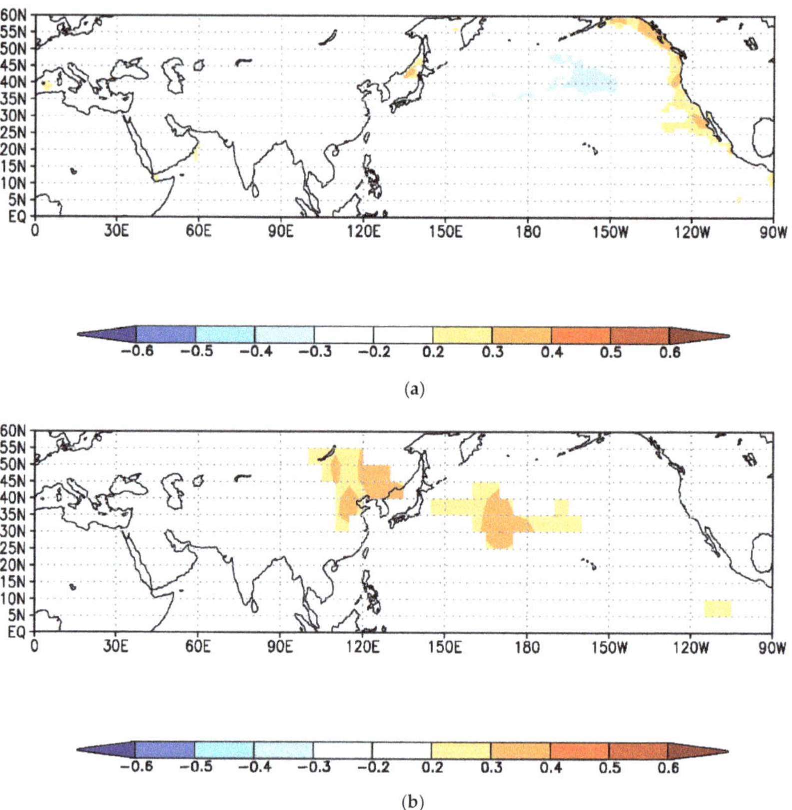

**Figure 2.** (**a**) Rank correlation analysis between AMFP and JFM SSTa (**a**), JJA SLPa (**b**). The blue SSTa region and orange SLPa regions have significant negative and positive correlation with AMFP, respectively, at 90% confidence interval (i.e., $p < 0.1$).

Therefore, three sub-sets of the ten leading principal components named by SSTa-PCs, SLPa1-PCs and SLPa2-PCs, respectively, and the average seasonal (JJA) AO are then considered as potential factors. These factors are further investigated to search the best combination by a generalized linear model (GLM) [13] with an extended leave-one-out cross validation (LOOCV) method implemented by the R package 'bestglm' [38] to choose the candidate climatic factors. For the sake of obtaining the best combination, we propose four GLMs: (1) the first GLM with the SSTa-PCs and AO; (2) the second GLM with SLPa1-PCs and AO; (3) the third GLM with SLPa2-PCs and AO; and (4) the last GLM with AO and the five leading PCs of SLPa1 and SLPa2. The optimal combination, composited by JJA AO and the first principal component accounting for 73.52% of the total variance of JJA SLPa2 region (denoted by SLPa2-PC1), is selected as it has the minimum $R$ square value (the method theory can be referred to [38]). The scatter plots between JJA AO, JJA SLPa2-PC1 and AMFP are displayed in Figure 3 and their Spearman correlations are exhibited in Table 2.

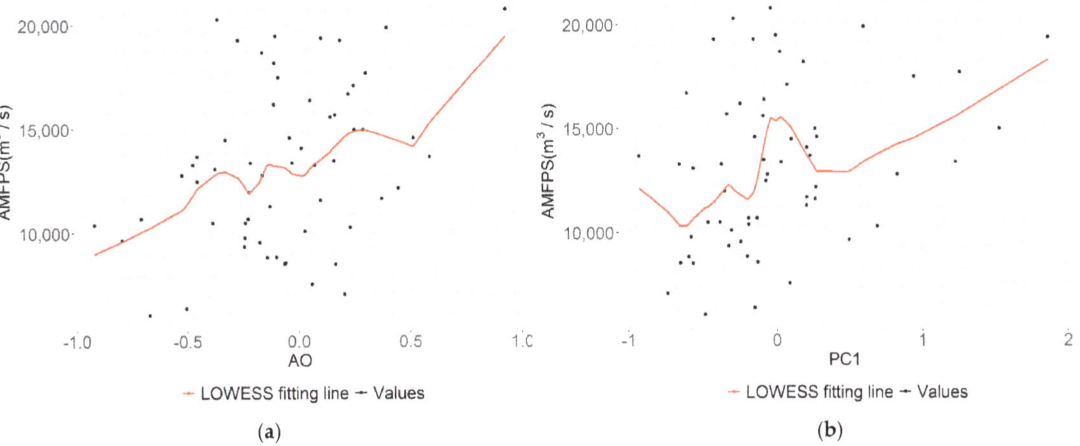



**Figure 3.** The scatter plots of AMFP versus JJA AO (**a**) and JJA SLPa2-PC1 (**b**) respectively.

**Table 2.** Spearman correlation coefficients between climatic factors and AMFP.

| Climate Factors | AO | SLPa2-PC1 |
|---|---|---|
| Coefficients | 0.32 | 0.35 |
| *p*-value | 0.01 | 0.006 |

Flooding of the Xiangjiang river basin in Central and Southern China is greatly impacted by the East Asian summer monsoon, which is driven by the SLP contrast between the Asian low and the Pacific high [39]. The positive correlation teleconnection between JJA SLPa2 and the summer flood in the Xiangjiang river basin suggests that the high pressure in the Pacific Ocean would lead to more water vapor transporting to the Xiangjiang river basin [40]. In consequence, the abovementioned average seasonal JJA AO and SLPa2-PC1 are capable of describing the observed variability of extreme flood events in the Xiangjiang River basin and are selected to reflect the impact of climate change on flood risk.

## 3. Methodology

### 3.1. Choice of Distribution

After ascertaining the climate drivers for the flood peaks generation, the next step is to quantify the flood risk changes with climate change. According to the review of a suitable probability distribution function on flood frequency analysis by [41,42] and all the pdfs of the Interagency Advisory Committee on Water Data (1982) [43], the two-parameter log normal LN2 and log Pearson type III are the only two distributions which do not exhibit significant bias in observed flood frequencies. Since LN2 in nonstationary condition is more parsimonious and is one of the most widely used distributions in hydrology, we choose the LN2 to fit the observed AMFP. The probability density function (PDF) is given by:

$$f(x|\mu,\sigma) = \frac{1}{x\sigma\sqrt{2\pi}}\exp\left[-\frac{(\ln x - \mu)^2}{2\sigma^2}\right], x > 0; \sigma > 0 \tag{1}$$

where $\mu$ and $\sigma$ are the location and scale parameters, respectively. To capture the climate impact, we develop climate variables incorporated into the estimates of the location and scale parameters.

### 3.2. Nonstationary Models Construction

In the nonstationary condition, the parameters $\mu$ (mean, location) and $\sigma$ (standard deviation, scale) of LN2 are made time-dependent by incorporating time or climatic factors as covariates [44]. In this study, in consideration of the significant linear relationship shown in scatter plots in Figure 3 and the parsimonious nonstationary modelling requirement, the parameters are selected to be the linear functions of covariates. In addition, given the AMFP doesn't exhibit a significant trend, the stationary condition (all parameters in the LN2 are constant) and the nonstationary model using time as the covariate are taken as the benchmarks to conduct the comparative analysis. Consequently, the stationary and nonstationary combinations of parameters are given as follows.

Case 1. Time-invariant model. All the parameters are kept constant, which is represented by

$$\text{LN}_\text{S}: \ \mu = \mu_0, \ \ln(\sigma) = \sigma_0 \tag{2}$$

Case 2. Linear temporal model. Only the parameter location is modeling as a function of time, or both the location and scale linearly vary with time. The two models are given by

$$\text{LNt-1}: \ \mu = \mu_0 + \mu_1 t, \ \ln(\sigma) = \sigma_0 \tag{3}$$

$$\text{LNt-2}: \ \mu = \mu_0 + \mu_1 t, \ \ln(\sigma) = \sigma_0 + \sigma_1 t \tag{4}$$

Case 3. Climate-informed model. Nonstationarity is incorporated by allowing the location, or both the location and scale parameters, to vary as a function of climatic factors (JJA AO and SLPa2-PC1). The models are illustrated as follows,

$$\text{LN}_\text{C}\text{-1}: \ \mu = \mu_0 + \mu_1 * \text{AO} + \mu_2 * \text{PC1}, \ \ln(\sigma) = \sigma_0 \tag{5}$$

$$\text{LN}_\text{C}\text{-2}: \mu = \mu_0 + \mu_1 \times \text{AO} + \mu_2 \times \text{PC1}, \ \ln(\sigma) = \sigma_0 + \sigma_1 \times \text{AO} + \sigma_2 \times \text{PC1} \tag{6}$$

where in the Case 1, Case 2 and Case 3, the letters s, t and c represent the stationary, time-varying and climate-based models respectively. The $\mu_i(i = 0, 1, 2)$ and $\sigma_i(i = 0, 1, 2)$ are the estimated regression parameters. For the sake of obtaining the positive values of $\sigma$, we use the natural logarithm $\ln \sigma$.

### 3.3. Bayesian Inference

In this study, we use a Bayesian inference framework to fit observed AMFP to the LN2 distribution. Bayesian-based Markov Chain Monte Carlo (MCMC) sampling with the No-U-Turn sampler variant of Hamiltonian Monte Carlo algorithm [45,46] in the RStan [47] is implemented to estimate the regression parameters associated with the covariates. The Bayesian-based MCMC approach provides full posterior distribution estimates of the parameters, which incorporate constructions from noninformative prior distributions and four chains of 10,000 iteration lengths in a statistically consistent way. The Bayesian theory rule is shown below,

$$f(\theta|X) \propto f(X|\theta)f(\theta) \tag{7}$$

where $f(\theta|X)$ is the posterior distribution of all parameters $\theta$ conditional on the flood peaks X. The $f(X|\theta)$ is the likelihood function and $f(\theta)$ is the prior distribution. In the RStan implement, the R-hat suggested by [48] is used to diagnose the convergence of chains, with the value for less than 1.05.

### 3.4. Models Selection Criteria

As suggested by [13], the Akaike Information Criteria (AIC) [49] and Bayesian Information Criteria (BIC) [50] are extensively used to access model performance and overfitting based on point-estimates of the parameters. In light of the full posterior distribution in Bayesian framework with more powerful and precise information, we adopt the Deviance Information Criterion (DIC) [51] to evaluate the Bayesian model goodness-of-fit and complexity, which won't discard adequate information by AIC and BIC. The DIC [52] is defined below:

$$\text{DIC} = \overline{D} + p_D \tag{8}$$

$$\overline{D} = E_\theta[D(\theta)] = E_\theta[-2\log p(x|\theta)] \tag{9}$$

$$p_D = \overline{D} - D(\overline{\theta}) \tag{10}$$

where the first term $\overline{D}$ in Equation (8), interpreted as a Bayesian measure of model fit, is defined as the posterior expectation of the deviance in Equation (9) while the $p(x|\theta)$ is the likelihood function of the observations $x$. The second term $p_D$ in Equation (8), used to measure the model complexity, is defined as the difference between the posterior mean of the deviance and the deviance evaluated at the posterior mean $(\overline{\theta})$ of the parameters shown by Equation (10).

In summary, the better model fits the observations, which require the larger logarithmic likelihood value $p(x|\theta)$ corresponding to the smaller the value for $\overline{D}$ and the smaller penalty represented by $p_D$. Hence, the models with smaller DIC values are preferred. In order to comprehensively compare the models, the AIC test is also employed.

### 3.5. Nonstationary Return Period

Under the stationary condition, the return period $T = 1/p$ is corresponding to a specific exceedance probability $p$. Nevertheless, the estimated parameters $\mu$ and $\sigma$ are time-varying vectors in the nonstationary flood frequency models. In recent years, four nonstationary design methods (namely ENE, EWT, DLL and ER), have been proposed to address the challenge of nonstationary return period (i.e., hydrologic design) in changing environments [15,23]. The last three methods have different limitations during application. The EWT method requires the time extended infinitely leading to failed convergence sometimes. In addition, both the DLL and ER method need to take into consideration the design life of the project. Consequently, for the nonstationary frequency analysis of the hydrological station in this study, we utilize the ENE method to estimate the nonstationary return period and corresponding design flood.

In the ENE method suggested by [53], assuming that $X$ represents the AMFP, $X_T$ is the design flood in nonstationary condition, and $F_i(x)$ is the cumulative distribution function for the $i$ year. Then, during the return period $T$ years, the $N$ is defined as the number of exceedances of the flood variable $X_i$ over the design flood $X_T$, which is given by

$$N = \sum_{i=1}^{T} I(X_i > X_T), I(\bullet) = \begin{cases} 1, X_i > X_T \\ 0, X_i \leq X_T \end{cases} \tag{11}$$

where $I(\bullet)$ is an indicator function. Since in the ENE method theory the expected number of exceedances in the $T$-year equals to one, then, the expected value of $N$ is calculated by

$$E(N) = \sum_{i=1}^{T} E[I(X_i > X_T)] = \sum_{i=1}^{T} P(X_i > X_T) = \sum_{i=1}^{T} (1 - F_i(X_T)) = 1 \tag{12}$$

Hence, the nonstationary $T$-year return period design flood values can be solved by Equation (12) and vice versa.

## 4. Results and Discussion

### 4.1. Stationary, Time-Covariate and Physical-Based Nonstationary Comparison

As for the aforementioned nonstationary models construction in Section 3.2, the stationary condition (Case 1) and the time-varying nonstationary model (Case 2) are used to carry out the comparative analysis with the climate-based nonstationary model (Case 3) for AMFP in the Xiangjiang River basin. The parameters of each model for three cases in Section 3.2 are estimated by maximizing the Bayesian posterior likelihood function and are tabulated in Table 3. The parameters $\mu_1, \sigma_1$ and $\mu_2, \sigma_2$ before the covariates statistically and quantitatively describe the influence of JJA AO and SLPa2-PC1, respectively, on AMFP. Furthermore, the posterior PDF of the above parameters should be significantly larger or lower than zero value (no more than 10% of their mass crossing 0), which indicate that the covariates have a significant effect. Therefore, in LNc-2 model, as the $\sigma_1 = -0.025$ is very close to 0 and the boxplot in Figure 4 shows a non-significant effect, the new model LNc-3 with removing the $\sigma_1$ is constructed.

**Table 3.** Models fitting results for AMFP in different assumptions.

| Assumption | Models | $\mu_0$ | $\mu_1$ | $\mu_2$ | $\sigma_0$ | $\sigma_1$ | $\sigma_2$ | DIC | AIC |
|---|---|---|---|---|---|---|---|---|---|
| Case 1 | LNs * | 9.45 | | | −1.17 | | | 1145.7 | 1148.9 |
| Case 2 | LNt-1 | 9.37 | 0.0026 | | −1.17 | | | 1145.5 | 1147.5 |
| | LNt-2 * | 9.36 | 0.0031 | | −0.94 | −0.0087 | | 1145.1 | 1145.2 |
| Case 3 | LN$_C$-1 | 9.47 | 0.26 | 0.15 | −1.29 | | | 1135.8 | 1135.6 |
| | LN$_C$-2 | 9.47 | 0.26 | 0.15 | −1.31 | −0.025 | −0.34 | 1134.6 | 1135.2 |
| | LN$_C$-3 * | 9.47 | 0.25 | 0.15 | −1.31 | | −0.34 | 1133.3 | 1133.1 |

Note: The values with underline "_" represent the posterior PDF of the parameter, which is not significantly different from 0. The models with '*' are the best models of each case.

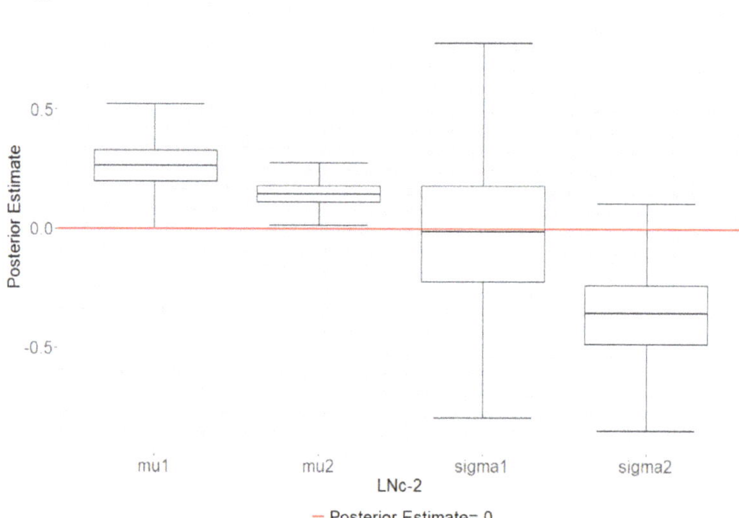


**Figure 4.** The boxplot of parameters in LNc-2 model.

In addition, goodness-of-fit test is performed for all the models by adopting the P-P plots (Probability-probability) and Kolmogorov–Smirnov (K–S) test [54] at the 5% level. The PP plots in Figure 5 demonstrate that the theoretical probabilities of all the optimal models fit well with the empirical probabilities. The statistic D values of the K–S test computed suggest that all the models are accepted at 95% confidence level and perform well in depicting probability distribution behaviors for observations. Meanwhile, such results indicate that the LN distribution fitting to AMFP is reasonable and appropriate. Also, the DIC and AIC are applied to compare the model performance for three cases and are summarized in Table 3. The results suggest that the LN$_C$-3 with climate variables covariates is the model with best fit.

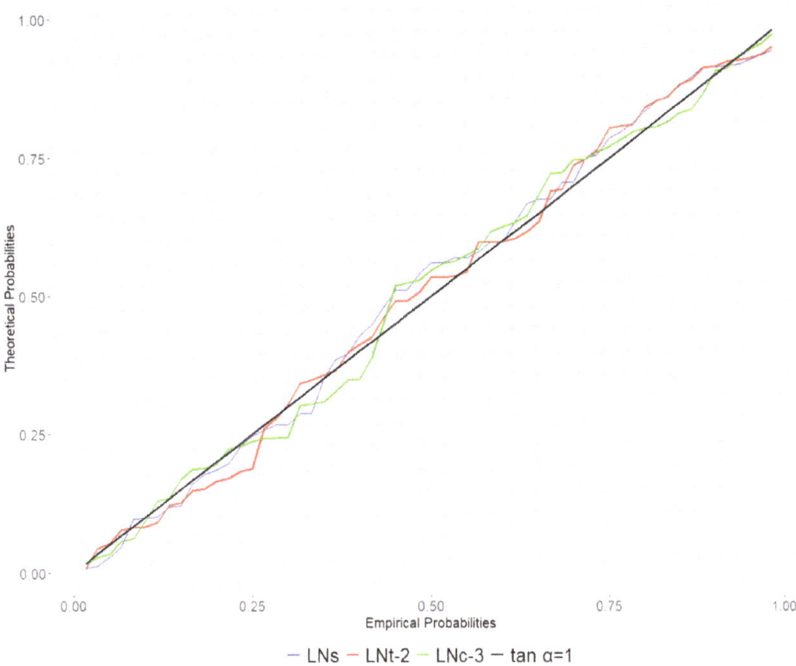



**Figure 5.** The PP-plot of the optimal models for each case.

To obtain the preferred model for each case, the DIC and AIC are compared in Table 3. The above results showed that the model LNt-2 and the model $LN_C$-3 performed the best in Case 2 and Case 3 which represent the time-varying and physical-based nonstationary conditions, respectively. Then the 0.99 quantiles and the corresponding uncertainty comparison of flood extremes based on the hypotheses of stationarity, temporal nonstationarity and climate-informed risk, denoted by the model of $LN_S$, LNt-2 and $LN_C$-3, respectively, are provided in Figure 6. The 100-year return period design flood under stationary assumption keeps invariant and 0.99th quantiles of the time-varying model LNt-2 with non-significant time-covariate exhibit only a slight increase, which is in accordance with the result that suggests no temporal trend in AMFP. The $LN_C$-3 model with the evidently smallest DIC and AIC values, whose 0.99th quantiles display obvious fluctuations, follow the change pattern of observations. Moreover, the 0.99 quantiles uncertainty results indicate that the stationary model kept unchanged doesn't capture the variation of true flood risk, especially for large extreme flood events. The uncertainty boundary of the time-covariate model is the largest, which brings relatively complicated and indistinct flood risk range. Nevertheless, the climate-covariates nonstationary model, which is more robust than both the stationary model and time-varying model, demonstrates better uncertainty performance particularly for large extreme flood events with larger uncertainty range (e.g., year 1998 and 2017). The capability of portraying the variation of the observed flood risk is critically important for policymakers and engineers to avoid overestimating and underestimating the potential risk for flood control.

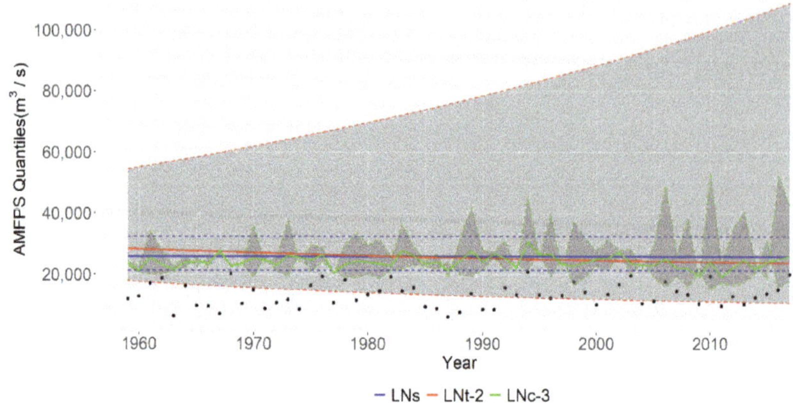



**Figure 6.** The 0.99 quantiles (solid lines) and uncertainty with 90% credibility intervals (areas inside the dashed line) of each model. Black dots are the observed AMFP in each year.

### 4.2. Variability and Uncertainty of Flood Risk

To distinguish nonstationary patterns in annual extreme flood events of the Xiangjiang River basin, we investigate the detailed linkage between climate variables and flood variability analyzed in Section 2.2. On the basis of the above analysis, the $LN_C$-3 nonstationary model, with the mean as linear function of JJA AO and SLPa2-PC1 and the log variance as linear function of JJA SLPa2-PC1, is selected as the optimal model among all candidate models. Furthermore, the centile curve is chosen to confirm whether the optimal model explains the physical variation in extreme flood events. The percentages of observation points below the 5th, 25th, 50th, 75th and 95th centile curves are 5.1%, 30.5%, 44.1%, 71.2% and 96.6% in $LN_C$-3 (Figure 7). The decent coverage rate illustrates that the climate-informed model performs satisfactorily in modeling the variability of the flood observations and their dependence structure over climate variables.

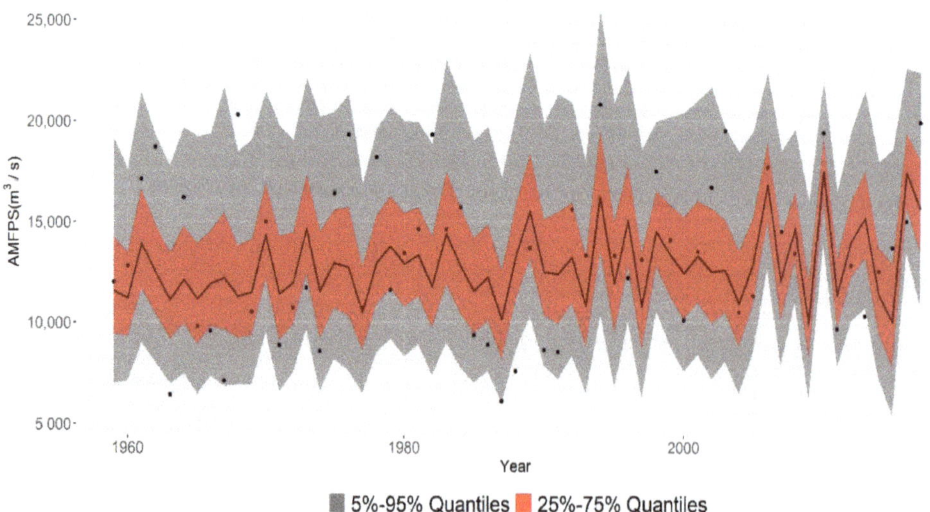



**Figure 7.** The centiles curve plots of the optimal model $LN_C$-3.

Figure 8 displays the median, 0.95 and 0.99 flood quantiles (i.e., 2-year, 20-year and 100-year events) in the physical-based condition with the $LN_C$-3 model against the large-scale climate variables JJA AO and SLPa2-PC1, respectively. As seen, the extreme flood risk increases with the climate covariates and the increasing degree with AO is greater than with SLPa2-PC1, which is suggested by the positive variation in mean with bigger coefficient of AO but relatively slight negative in variance on SLPa2-PC1 in $LN_C$-3 model. It is interesting to note that the flood quantiles evolve with the two important climatic factors with a similar pattern as that between AMFP and two drivers. Consequently, it indicates that the surface atmospheric pressure difference between the northern middle latitude and North Pole compared to the sea level pressure in the Northwest Pacific Ocean have strong teleconnection and influence on extreme flooding of the Xiangjiang River basin. To sum up, if the AO and sea level pressure in the Northwest Pacific Ocean are high, in the same season, the more extreme flooding of the Xiangjiang River will occur during summer. The climate-based optimal nonstationary modeling of results improves the quality of flood quantiles estimation compared to traditional frequency analysis and non-significant time-varying models. The significant inter-relations between the flood risk system and large-scale climate patterns will aid decision makers and scientists to better understand variability in extreme flood risk.

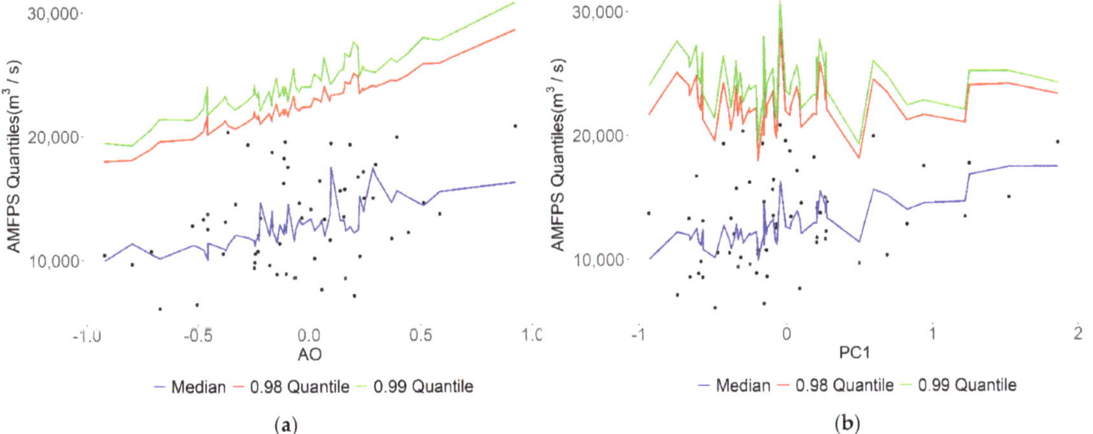



**Figure 8.** The median, 0.95 and 0.99, quantiles of AMFP plotted against JJA AO (**a**) and JJA SLPa2-PC1 (**b**).

*4.3. Return Period and Associated Uncertainty Analysis*

The occurrence of extreme flood events is fully estimated by climate-covariates nonstationary frequency analysis. Then, the corresponding return periods are calculated by the appropriate nonstationary design method ENE for at-site flood (details refer to Section 3.5) based on the optimal model $LN_C$-3. Considering the limited sample size of climatic factors and flood, we investigate the 5-, 10-, 20- and 50-year return periods, which correspond to flood control standards cover most of the major cities in the Xiangjiang River basin and most regions in Changsha, the provincial capital city of Hunan Province. Figure 9 shows the return level differences between conventional stationary model $LN_S$ and time-based nonstationary model LNt-2 compared to climate-based nonstationary model $LN_C$-3, respectively. Specifically, for LNt-2 and $LN_S$ models, the differences are discernable but within 20% during the 10-year and 50-year return levels, and are minor in recurrence periods shorter than 10-year or longer than 50-year. However, there is a gradually increasing gap from 5-year to 59-year return periods between traditional and nonstationary climate-informed models.

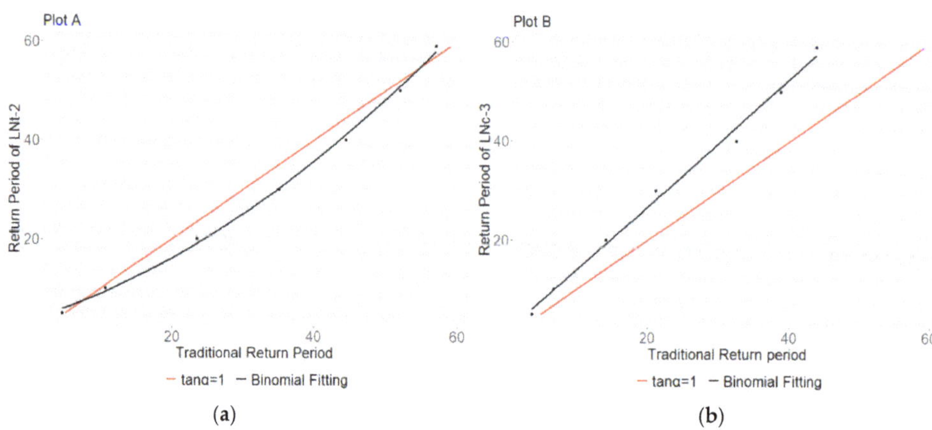



**Figure 9.** The return periods comparison between stationary model $LN_S$ and time-based model LNt-2 (**a**) and climate-based model $LN_C$-3 (**b**).

Furthermore, we also explore the change rate (%) in the return periods of AMFP between climate-covariates nonstationary and stationary models, which is given by the following,

$$v = \frac{nrp - srp}{srp} \tag{13}$$

where $nrp$ and $srp$ are the return periods of the nonstationary and stationary model, respectively. The change rates of 5-, 10-, 20- and 50-year return levels between $LN_C$-3 and $LN_S$ are 25.2%, 32.2%, 29.1% and 22.6%, respectively, which coincide with the variation trends (i.e., $-6.34\%$, $-6.94\%$, $-5.16\%$ and $-3.26\%$ respectively) of design flood values in Table 4. Combing with the plot in Figure 9b, it is demonstrated that the occurrence probability of the extreme flood event with nonstationary climate model is obviously smaller than the stationary benchmark. This has great implication for flood control operation and design.

**Table 4.** The design flood values comparison between stationary and climate-based nonstationary conditions Unit: m$^3$/s.

| Models | 5-Year | 10-Year | 20-Year | 50-Year |
|---|---|---|---|---|
| $LN_S$ | 16,431 | 18,806 | 21,023 | 23,833 |
| $LN_C$-3 | 15,356 | 17,501 | 19,938 | 23,056 |
| Variation (%) | $-6.34$ | $-6.94$ | $-5.16$ | $-3.26$ |

We also compare the statistical characteristics between the models $LN_S$ and $LN_C$-3. Firstly, based on the fitting results of models $LN_S$ and $LN_C$-3, the exceedance probabilities of annual extreme flood events during 1959–2017 are investigated. In the comparison of two models, the average exceedance probabilities value of 50.03% for model $LN_C$-3 is slightly larger than the value of 49.16% for model $LN_S$, which reveals that the occurrence probabilities of most annual extreme flood events during 1959–2017 have a small increasing trend influenced by climate factors. Secondly, however, on the basis of the expected number of exceedances (ENE) method, the return periods of model $LN_C$-3 are bigger than the ones for stationary conditions. Thus, based on the design flood values under 5-, 10-, 20-, and 50-year return periods of model $LN_C$-3 (Table 4), we investigate the exceedance probabilities of each design flood for the corresponding first 5-, 10-, 20-, and 50-years respectively during 1959–2017 (for instance, for the design flood value of 15,356 m$^3$/s over a 5-year return period, we calculate the exceedance probabilities of the '15,356 m$^3$/s' flood event in first 5 years) using models $LN_S$ and $LN_C$-3. The annual exceedance probabilities differences between the two models are illustrated in Figure 10. As can be seen, for each design flood, all the multi-year average exceedance probabilities of model $LN_C$-3 are smaller than the stationary model $LN_S$ benchmark.

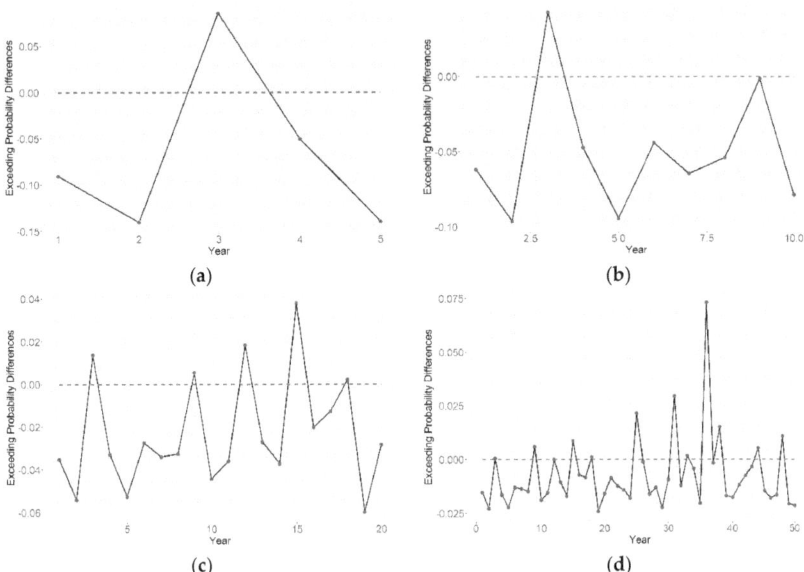

**Figure 10.** The annual exceedance probabilities differences of model $LN_C$-3 compared to the model $LN_S$ corresponding to each design flood. (**a**) 5-year design flood; (**b**) 10-year design flood; (**c**) 20-year design flood; (**d**) 50-year design flood.

The above two conclusions are opposite, prompting us to explore the reasons why the annual maximum flood peak series in the Xiangjiang River Basin exhibit different results. Thus, as the parameters of model $LN_C$-3 are distinct every year from 1959 to 2017, the location and scale parameters for $LN_C$-3 are averaged, respectively, to investigate the climate-based model's statistical characteristic in an average sense. In contrast to the model $LN_S$ location parameter of 9.45 and scale parameter of $-1.17$, the model $LN_C$-3 average location parameter of 9.453 is slightly larger and the scale parameter of $-1.29$ is visibly smaller. Furthermore, the probability density distribution curves and exceedance probability distribution curves for two models are demonstrated in Figure 11. The figures clearly introduce that with the extreme flood event values larger than approximately 12,500 m$^3$/s, the average exceedance probabilities of model $LN_C$-3 are less than the ones in $LN_S$.

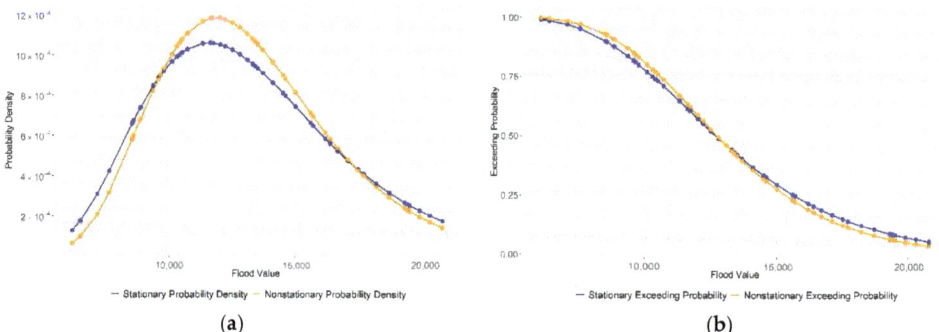

**Figure 11.** The probability density distribution curves and exceedance probability distribution curves for models LN$_S$ (the blue curves) and average LN$_C$-3 (the orange curves). (**a**) Probability density distribution curve; (**b**) Exceedance probability distribution curve.

Consequently, from the mathematical point of view, since medium and small extreme flood events are prone to occur, while large flood events are relatively infrequent during the period of 1959–2017, the average exceedance probability for model LN$_C$-3 is slightly larger than the value of model LN$_S$. Owing to the 5-year design flood with 15,356 m$^3$/s over 12,500 m$^3$/s, the return periods of model LN$_C$-3 are generally bigger than the ones under stationary conditions. On the physical aspect being attributed to the two climate factors' effects, the extreme flood events during the period of 1959–2017 display a mild increasing trend in mean but comparatively descend significantly in variance. In summary, the relatively large decrease in variance contributes to the reduction of exceedance probabilities in model LN$_C$-3.

Regarding the uncertainty with 50% confidence intervals of nonstationary climate-covariate models for AMFP compared to traditional return levels illustrated in Figure 12, it is remarkably large and fully exceeds the 1:1 line. The phenomenon is highly generated by the ENE method because of its calculative restriction relying on covariate length. Hence, the computation will definitely bring about higher levels of uncertainties [22]. However, the abovementioned results indicate that the occurrence periods of AMFP estimated by a climate-informed model are longer than return periods under a stationary model, which will be conservative for decision makers in considering the significant climatic factors in current conditions. However, since the design floods or return periods rely on the climate covariates, it should not be overlooked that the flood risk would be abnormal if the climatic factors are unusual in future uncommon conditions. Consequently, it is worth noticing that the great uncertainty is a warning for engineers and planners in flood control.

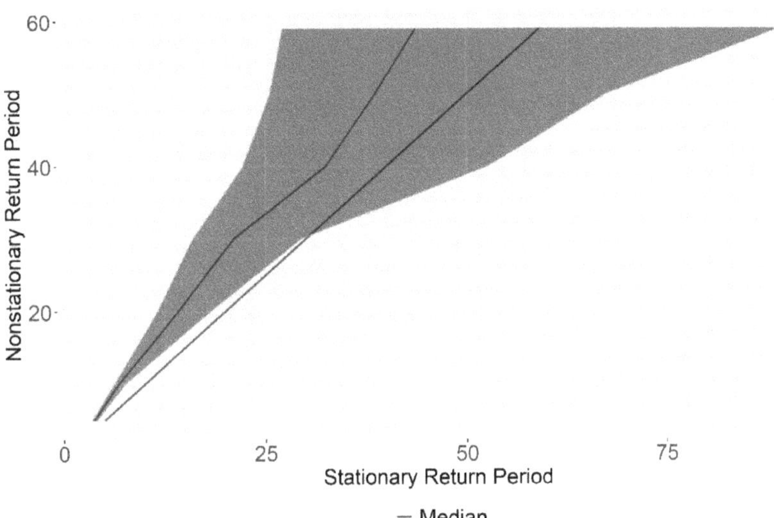



**Figure 12.** The uncertainty with 50% credibility intervals between $LN_C$-3 and LNs models.

## 5. Conclusions

The study contributes to describing the characteristics of extreme flood risks under a changing climate, using nonstationary flood frequency analysis models in the Xiangjiang River basin in South-Central China. The paper aims to determine the most significant correlated climatic factors with annual maximum flood peak series, to identify the best model for extreme flood frequency analysis considering different criteria, and to estimate the uncertainties of extreme flood quantiles and return periods based on a Bayesian modeling inference framework. The following main conclusions can be drawn from this study.

(1) While the stationary frequency analysis is commonly used for hydraulic structure design, recent multiple studies highlighted the increase in the nonstationary pattern of extreme events. In this study, we performed stationary and nonstationary frequency based on the annual maximum flood peak series covering a period of 1959–2017 in the Xiangjiang River basin. Prior to implementing the nonstationary frequency analysis, the selection of physical covariates is an important step. Since most of the extreme floods conventionally occur during the summer season, from June to August, mainly caused by precipitation from the East Asian monsoon, we consider the physical impacts primarily on climatic factors including the eight large-scale low-frequency standard climate indices and two oceanic-atmospheric climate patterns SSTa and SLPa. The identification screening process of potential climate covariates is divided into two steps: the Spearman rank correlation test with AMFP and constructing GLMs to obtain the best combination of climatic impactors. Overall, two distinct climate covariates, which are Arctic Oscillation and the most informative factor, PC1, derived from the SLPa in the Northwest Pacific Ocean during the period of June to August, are identified as the statistically significant positive correlation with AMFP. The abovementioned best processes for screening significant climate drivers can serve as a protocol to apply on other basins.

(2) The stationary model and nonstationary flood frequency models with time or climate covariates are evaluated for AMFP employing the two-parameter lognormal distribution, which is an excellent and parsimonious model for representing the distribution of AMFP, and is consistent with the recommendation of other researchers [15]. The results show that extreme flood events follow the nonstationary climate pattern, namely, the optimal model is the nonstationary climate-covariates model with a linearly positive effect on location parameters of two climatic factors and a linearly negative

coefficient on scale parameter only for SLPa2-PC1. In addition, the Bayesian modeling inference is used to explore the uncertainty in extreme flood risks. Comparing the three optimal models, namely, $LN_S$, LNt-2 and $LN_C$-3 in each case, the time trend in AMFP is so minor that the LNt-2 model parameter coefficients are very close to zero. However, the flood quantiles estimated by the $LN_C$-3 model, and that used JJA AO and SLPa2-PC1, oscillate over time along with the variation trend of true observed flood. It should be pointed out that the uncertainty boundary of flood quantiles for the $LN_C$-3 model is relatively high especially for large floods. However, the climate-based model $LN_C$-3 proved reasonable and improves the understanding and interpretation of changing properties of AMFP frequency, which is apparently influenced by the same season AO and SLPa in the Northwest Pacific Ocean. The linkage between the flood extremes and the climatic factors would be useful to provide reliable and valid information under a changing environment.

(3) It is interesting to discover that, based on the nonstationary extreme flood analysis, the return periods associated with extreme flood events, computed by the ENE method restricted to the associated timespan of the covariates, are obviously enlarged compared to the stationary approach; the difference is gradually increasing according to the existing trends. In addition, assigning a return period, although the change rates between the $LN_S$ and $LN_C$-3 model are not registering as high design flood values under current conditions, the larger discrepancies would be found once the climate covariates are located in future uncommon conditions. Nevertheless, the high levels of 50% confidence interval uncertainty boundaries for nonstationary return periods are indeed crossing the 1:1 line in contrast to traditional return levels, which is the major disadvantage of the ENE method. Actually, in order to reduce the uncertainties, the model structures of nonstationary lognormal distribution are set to be simple with a linear trend in distribution parameters. Nevertheless, the uncertainty boundary is still evidently large so that, in the future, new approaches should be pursued to manage or balance the uncertainties of the nonstationary modeling, e.g., by combing more extreme flood events from surrounding stations to collect regional information.

**Author Contributions:** Conceptualization, H.Z. (Hang Zeng) and J.H.; methodology, H.Z. (Hang Zeng) and J.H.; validation, H.Z. (Hang Zeng) and J.H.; formal analysis, H.Z. (Hang Zeng) and J.H.; investigation, Z.L., W.Y. and H.Z. (Hui Zhou); resources, Z.L., W.Y. and H.Z. (Hui Zhou); data curation, Z.L., W.Y. and H.Z. (Hui Zhou); writing—original draft preparation, H.Z. (Hang Zeng) and J.H.; writing—review and editing, H.Z. (Hang Zeng); visualization, J.H.; supervision, H.Z. (Hang Zeng); project administration, H.Z. (Hang Zeng); and funding acquisition, H.Z. (Hang Zeng) and Z.L. All authors have read and agreed to the published version of the manuscript.

**Funding:** This research was funded by the National Natural Science Foundation of China (grant number 51809018), the Natural Science Foundation of Hunan Province, China (grant number 2019JJ50643 and 2018JJ3543) and the Scientific Research Project of Hunan Provincial Education Department, China (grant number 18C0198).

**Institutional Review Board Statement:** Not applicable.

**Informed Consent Statement:** Not applicable.

**Data Availability Statement:** The data used to support the findings of this study are available from the corresponding author upon request.

**Acknowledgments:** The authors gratefully acknowledge the Royal Netherlands Meteorological Institute (KNMI) Climate Explorer and the Hydrology and Water Resources Survey Bureau of Hunan Province in China for providing data used in this study.

**Conflicts of Interest:** The authors declare no conflict of interest.

## References


1. Zisopoulou, K.; Panagoulia, D. An In-Depth Analysis of Physical Blue and Green Water Scarcity in Agriculture in Terms of Causes and Events and Perceived Amenability to Economic Interpretation. *Water* **2021**, *13*, 1693. [CrossRef]
2. Panagoulia, D.; Mamassis, N.; Gkiokas, A. Deciphering the Floodplain Inundation Maps in Greece. In Proceedings of the 8th International Conference Water Resources Management in an Interdisciplinary and Changing Context, Porto, Portugal, 26–29 June 2013; pp. 323–330.
3. Barichivich, J.; Gloor, E.; Peylin, P.; Brienen, R.J.W.; Schöngart, J.; Espinoza, J.C.; Pattnayak, K.C. Recent intensification of Amazon flooding extremes driven by strengthened Walker circulation. *Sci. Adv.* **2018**, *4*, eaat8785. [CrossRef]
4. Mangini, W.; Viglione, A.; Hall, J.; Hundecha, Y.; Ceola, S.; Montanari, A.; Rogger, M.; Salinas, J.L.; Borzì, I.; Parajka, J. Detection of trends in magnitude and frequency of flood peaks across Europe. *Hydrol. Sci. J.* **2018**, *63*, 493–512. [CrossRef]
5. Willner, S.N.; Otto, C.; Levermann, A. Global economic response to river floods. *Nat. Clim. Change* **2018**, *8*, 594–598. [CrossRef]
6. He, C.; Chen, F.; Long, A.; Luo, C.; Qiao, C. Frequency Analysis of Snowmelt Flood Based on GAMLSS Model in Manas River Basin, China. *Water* **2021**, *13*, 2007. [CrossRef]
7. López, J.; Francés, F. Non-stationary flood frequency analysis in continental Spanish rivers, using climate and reservoir indices as external covariates. *Hydrol. Earth Syst. Sci.* **2013**, *17*, 3189–3203. [CrossRef]
8. Ficchì, A.; Cloke, H.; Neves, C.; Woolnough, S.; Coughlan de Perez, E.; Zsoter, E.; Pinto, I.; Meque, A.; Stephens, E. Beyond El Nio: Unsung climate modes drive African floods. *Weather Clim. Extrem.* **2021**, *33*, 100345. [CrossRef]
9. Zhou, Y. Exploring multidecadal changes in climate and reservoir storage for assessing nonstationarity in flood peaks and risks worldwide by an integrated frequency analysis approach. *Water Res.* **2020**, *185*, 116265. [CrossRef]
10. Zhang, Q.; Gu, X.; Singh, V.P.; Xiao, M.; Chen, X. Evaluation of flood frequency under non-stationarity resulting from climate indices and reservoir indices in the East River basin, China. *J. Hydrol.* **2015**, *527*, 565–575. [CrossRef]
11. Kundzewicz, Z.W.; Szwed, M.; Pińskwar, I. Climate Variability and Floods—A global Review. *Water* **2019**, *11*, 1399. [CrossRef]
12. Kundzewicz, Z.W.; Huang, J.; Pinskwar, I.; Su, B.; Szwed, M.; Jiang, T. Climate variability and floods in China—A review. *Earth-Sci. Rev.* **2020**, *211*, 103434. [CrossRef]
13. Zeng, H.; Sun, X.; Lall, U.; Feng, P. Nonstationary extreme flood/rainfall frequency analysis informed by large-scale oceanic fields for Xidayang Reservoir in North China. *Int. J. Clim.* **2017**, *37*, 3810–3820. [CrossRef]
14. Renard, B.; Lall, U. Regional frequency analysis conditioned on large-scale atmospheric or oceanic fields. *Water Res. Res.* **2014**, *50*, 9536–9554. [CrossRef]
15. Yan, L.; Xiong, L.; Guo, S.; Xu, C.Y.; Xia, J.; Du, T. Comparison of four nonstationary hydrologic design methods for changing environment. *J. Hydrol.* **2017**, *551*, 132–150. [CrossRef]
16. Olsen, R.; Lambert, J.H.; Haimes, Y.Y. Risk of extreme events under nonstationary conditions. *Risk Anal.* **1998**, *18*, 497–510. [CrossRef]
17. Salas, J.D.; Obeysekera, J. Revisiting the concepts of return period and risk for nonstationary hydrologic extreme events. *J. Hydrol. Eng.* **2014**, *19*, 554–568. [CrossRef]
18. Parey, S.; Hoang, T.T.H.; Dacunha Castelle, D. Different ways to compute temperature return levels in the climate change context. *Environmetrics* **2010**, *21*, 698–718. [CrossRef]
19. Parey, S.; Malek, F.; Laurent, C.; Dacunha-Castelle, D. Trends and climate evolution: Statistical approach for very high temperatures in France. *Clim. Change* **2007**, *81*, 331–352. [CrossRef]
20. Rootzén, H.; Katz, R.W. Design life level: Quantifying risk in a changing climate. *Water Resour. Res.* **2013**, *49*, 5964–5972. [CrossRef]
21. Liang, Z.; Hu, Y.; Huang, H.; Wang, J.; Li, B. Study on the estimation of design value under non-stationary environment. *South-to-North Water Transf. Water Sci. Technol.* **2016**, *14*, 50–53, (In Chinese with English abstract).
22. Yan, L.; Xiong, L.; Luan, Q.; Jiang, C.; Yu, K.; Xu, C.Y. On the Applicability of the Expected Waiting Time Method in Nonstationary Flood Design. *Water Resour. Manag.* **2020**, *34*, 2585–2601. [CrossRef]
23. Hu, Y.; Liang, Z.; Singh, V.P.; Zhang, X.; Wang, J.; Li, B.; Wang, H. Concept of equivalent reliability for estimating the design flood under non-stationary conditions. *Water Resour. Manag.* **2018**, *32*, 997–1011. [CrossRef]
24. Gu, X.; Zhang, Q.; Singh, V.P.; Xiao, M.; Cheng, J. Nonstationarity-based evaluation of flood risk in the Pearl River basin: Changing patterns, causes and implications. *Hydrol. Sci. J.* **2017**, *62*, 246–258. [CrossRef]
25. Mao, D.; Li, J.; Gong, C.; Peng, J. *Study on the Flood-Waterlogging Disaster in Hunan Province*; Hunan Normal University Press: Changsha, China, 2000. (In Chinese)
26. Du, J.; He, F.; Shi, P.J. Integrated flood risk assessment of Xiangjiang River Basin in China. *J. Nat. Dis.* **2006**, *15*, 8–44. (In Chinese with English abstract)
27. Rayner, N.A.; Parker, D.E.; Horton, E.B.; Folland, C.K.; Alexander, L.V.; Rowell, D.P.; Kent, E.C.; Kaplan, A. Global analyses of sea surface temperature, sea ice, and night marine air temperature since the late nineteenth century. *J. Geophys. Res.-Atmos.* **2003**, *108*, 1063–1082. [CrossRef]
28. Allan, R.; Ansell, T. A New Globally Complete Monthly Historical Gridded Mean Sea Level Pressure Dataset (HadSLP2): 1850–2004. *J. Clim.* **2006**, *19*, 5816–5842. [CrossRef]
29. Song, Z.; Xia, J.; She, D.; Zhang, L. The development of a Nonstationary Standardized Precipitation Index using climate covariates: A case study in the middle and lower reaches of Yangtze River Basin, China. *J. Hydrol.* **2020**, *588*, 125115. [CrossRef]

30. Li, S.; Feng, G.; Wei, H. Summer drought patterns in the middle-lower reaches of the yangtze river basin and their connections with atmospheric circulation before and after 1980. *Adv. Meteorol.* **2016**, *2016*, 8126852. [CrossRef]
31. Qian, C.; Yu, J.Y.; Chen, G. Decadal summer drought frequency in China: The increasing influence of the Atlantic Multi-decadal Oscillation. *Environ. Res. Lett.* **2014**, *9*, 124004. [CrossRef]
32. Gong, D.; Zhu, J.; Wang, S. Significant relationship between spring AO and the summer rainfall along the Yangtze River. *Chin. Sci. Bull.* **2002**, *47*, 948–952. [CrossRef]
33. Wei, F. Relationships between precipitation anomaly over the middle and lower reaches of the Changjiang River in summer and several forcing factors. *Chin. J. Atmos. Sci.* **2006**, *30*, 202–211. (In Chinese with English abstract)
34. Yang, H. The significant relationship between the Arctic Oscillation (AO) in December and the January climate over South China. *Adv. Atmos. Sci.* **2011**, *28*, 398–407. [CrossRef]
35. Gong, D.; Wang, S. Influence of Arctic Oscillation on winter climate over China. *J. Geogr. Sci.* **2003**, *13*, 208–216. [CrossRef]
36. Su, C.; Chen, X. Covariates for nonstationary modeling of extreme precipitation in the Pearl River Basin, China. *Atmos. Res.* **2019**, *229*, 224–239. [CrossRef]
37. Thompson, D.W.J.; Wallace, J.M. The arctic oscillation signature in the wintertime geopotential height and temperature fields. *Geophys. Res. Lett.* **1998**, *25*, 1297–1300. [CrossRef]
38. McLeod, A.I.; Xu, C.; Yanhao, L. Package 'Bestglm'. Available online: http://cran.r-project.org/web/packages/bestglm/bestglm.pdf (accessed on 5 July 2021).
39. Zhao, P.; Zhou, Z. An East Asian subtripical summer monsoon index and its relationship to summer rainfall in China. *Acta Meteor. Sin.* **2009**, *23*, 18–28.
40. Yunyun, L.; Ping, L.; Ying, S. Basic features of the Asian summer monsoon system. In *The Asian Summer Monsoon: Characteristics, Variability, Teleconnections and Projection, Part I*; Elsevier: Amsterdam, The Netherlands, 2019; pp. 3–22.
41. Vogel, R.M.; Wilson, I. Probability distribution of annual maximum, mean, and minimum streamflows in the United States. *J. Hydrol. Eng.* **1996**, *1*, 69–76. [CrossRef]
42. Serago, J.M.; Vogel, R.M. Parsimonious Nonstationary Flood Frequency Analysis. *Adv. Water Res.* **2018**, *112*, 1–16. [CrossRef]
43. Interagency Advisory Committee on Water Data. *Guidelines for Determining Flood Flow Frequency: Bulletin 17b (Revised and Corrected)*; Interagency Committee on Water Data: Washington, DC, USA, 1982; p. 28.
44. Aziz, R.; Yucel, I. Assessing nonstationarity impacts for historical and projected extreme precipitation in Turkey. *Theor. Appl. Clim.* **2021**, *143*, 1213–1226. [CrossRef]
45. Hoffman, M.D.; Gelman, A. The No-U-Turn Sampler: Adaptively Setting Path Lengths in Hamiltonian Monte Carlo. *J. Mach. Learn. Res.* **2014**, *15*, 1593–1623. [CrossRef]
46. Betancourt, M. A Conceptual Introduction to Hamiltonian Monte Carlo. *arXiv* **2017**, arXiv:1701.02434. Available online: https://arxiv.org/pdf/1701.02434.pdf (accessed on 16 September 2021).
47. Stan Development Team. RStan: The R Interface to Stan, Version 2.21.2. Available online: http://mc-stan.org/rstan.html (accessed on 2 July 2021).
48. Vehtari, A.; Gelman, A.; Gabry, J. Practical Bayesian Model Evaluation Using Leave-One-Out Cross-Validation and WAIC. *Stat. Comput.* **2017**, *27*, 1413–1432. [CrossRef]
49. Akaike, H. New look at statistical-model identification. *IEEE Trans. Autom. Control* **1974**, *19*, 716–723. [CrossRef]
50. Schwarz, G. Estimating the dimension of a model. *Ann. Stat.* **1978**, *6*, 461–464. [CrossRef]
51. Spiegelhalter, D.J.; Linde, A.V.D. Bayesian measures of model complexity and fit. *J. R. Stat. Soc. B* **2002**, *64*, 583–616. [CrossRef]
52. Li, Y.; Yu, J.; Zeng, T. Deviance Information Criterion for Bayesian Model Selection: Justification and Variation. *Econ. Stat. Work. Pap.* **2017**, *10*, 1–25. [CrossRef]
53. Cooley, D. Return periods and return levels under climate change. In *Extremes in a Changing Climate*; AghaKouchak, A., Easterling, D., Hou, K., Schubert, S., Sorooshian, S., Eds.; Springer: Dordrecht, The Netherlands, 2013; pp. 97–114.
54. Lilliefors, H.W. On the Kolmogorov-Smirnov test for normality with mean and variance unknown. *J. Am. Stat. Assoc.* **1967**, *62*, 399–402. [CrossRef]

*Article*

# Bivariate Nonstationary Extreme Flood Risk Estimation Using Mixture Distribution and Copula Function for the Longmen Reservoir, North China

**Quan Li** [1,2]**, Hang Zeng** [1,2,*]**, Pei Liu** [3]**, Zhengzui Li** [4]**, Weihou Yu** [4] **and Hui Zhou** [4]

[1] School of Hydraulic and Environmental Engineering, Changsha University of Science & Technology, Changsha 410114, China; liquan9751@163.com
[2] Key Laboratory of Dongting Lake Aquatic Eco-Environmental Control and Restoration of Hunan Province, Changsha 410114, China
[3] Power China Zhongnan Engineering Corporation Limited, Changsha 410014, China; liupei6838840@126.com
[4] Hydrology and Water Resources Survey Center of Hunan Province, Changsha 410008, China; hnlzz@139.com (Z.L.); yweihou@163.com (W.Y.); 2527092@163.com (H.Z.)
* Correspondence: hzeng1989@csust.edu.cn; Tel.: +86-188-9002-8903

**Abstract:** Recently, the homogenous flood generating mechanism assumption has become questionable due to changes in the underlying surface. In addition, flood is a multifaced natural phenomenon and should be characterized by both peak discharge and flood volume, especially for flood protection structures. Hence, in this study, data relating to the 55-year reservoir inflow, annual maximum flood peak (AMFP), and annual maximum flood volume (AMFV) for the Longmen Reservoir in North China have been utilized. The 1-day AMFV exhibits a significant correlation with AMFP. The extreme flood peak-volume pairs are then used to detect the heterogeneity and to perform nonstationary flood risk assessment using mixture distribution as the univariate marginal distribution. Moreover, a copula-based bivariate nonstationary flood frequency analysis is developed to investigate environmental effects on the dependence of flood peak and volume. The results indicate that the univariate nonstationary return period is between the joint OR and the AND return periods. The conditional probabilities of 1-day AMFV, when AMFP exceeds a certain threshold, are likely to be high, and the design flood values estimated by joint distribution are larger than the ones in the univariate nonstationary context. This study can provide useful information for engineers and decision-makers to improve reservoir flood control operations.

**Keywords:** extreme flood risk; mixture distribution; G–H copula; bivariate nonstationary flood frequency analysis; nonstationary return period

**Citation:** Li, Q.; Zeng, H.; Liu, P.; Li, Z.; Yu, W.; Zhou, H. Bivariate Nonstationary Extreme Flood Risk Estimation Using Mixture Distribution and Copula Function for the Longmen Reservoir, North China. *Water* **2022**, *14*, 604. https://doi.org/10.3390/w14040604

Academic Editors: Yuanfang Chen, Dong Wang and Dedi Liu

Received: 1 January 2022
Accepted: 15 February 2022
Published: 16 February 2022

**Publisher's Note:** MDPI stays neutral with regard to jurisdictional claims in published maps and institutional affiliations.

## 1. Introduction

Design flood estimation is necessary for the design of adequate flood control structures such as reservoirs and dams, in order to improve flood preparedness. Flood frequency analysis is the fundamental method for quantifying the design flood and is usually conducted within a univariate flood frequency analysis framework [1–5]. However, an extreme flood events is a multifaced natural phenomenon and is characterized not only by peak discharge but also by flood volume. Moreover, in practice, flood peak discharge and volume are both highly correlated with flood management. Therefore, traditional univariate flood frequency analysis is unable to model the occurrence probability of an extreme flood event [6]. A bivariate frequency analysis has been demonstrated as being desirable and indispensable and is proposed to better understand and capture multiple flood characteristics [7–9]. In recent decades, numerous studies have been conducted to implement bivariate flood frequency analysis. Among them, joint distribution is the most useful tool for capturing flood peak and volume dependence. Copula-based joint distributions have proven to be an effective method in the bivariate framework for flood coincidence risk analysis and for measuring the dependences between flood variables [10–14]. Hu et al. (2022) conducted copula-based bivariate flood frequency analysis and proposed a nonstationary bivariate design flood estimation approach [15]. Brunner et al. (2018) developed a bivariate copula function model to model dependence between flood peak and volume and demonstrated that climate changes not only affect flood peaks but also have an effect on flood characteristics [16]. Duan et al. (2016) evaluated the variations of flood frequency in the Huai River basin by fitting a copula function [17]. Parent et al. (2014) selected copula functions and analyzed the corresponding parameters under a Bayesian framework [18].

In addition, due to the combined actions of different basin properties (e.g., land use change, hydraulic construction, etc.) and meteorological conditions (e.g., thunderstorm, typhoon, etc.), the extreme flood generating mechanisms would be changed and no longer be homogeneous [19,20]. The heterogeneity of extreme flood series resulting from environmental changes would result in changes, both in distribution parameters and in the type of distributions. As emphasized by Alila and Mtiraoui (2002) [21] and Villarini and Smith (2010) [22], prior scientific evidence of mixture populations should be provided to strengthen the physical understanding of the mixed nature of flooding. However, due to the limitation of long-term underlying surface data used for separating the flood population and the complexities of flood generation mechanisms, it is not always feasible to identify the distinct flood populations. Generally, the mixture distribution model does not require flood population separation and is widely utilized in nonstationary flood frequency analysis with various distribution types [21]. Zeng et al. (2014) and Feng and Li (2013) applied mixture distribution on extreme flood series, divided by prior change point detection, and suggested that nonstationary mixture distribution performed much better than stationary single-type distribution [23,24]. Li et al. (2018) proposed the improved mixture distribution for fitting the two subset flood samples, which both consider historical extraordinary floods [25]. Yan et al. (2017) investigated the mixture distribution application from the perspective of the temporal variation of separating the distributions' parameters [20]. Yan et al. (2019) improved the mixture distribution using the flood timescale method to separate it into two flood generation mechanisms [26].

In recent decades, a number of researchers have focused on simultaneously considering the non-stationarity of flood series and modeling flood characteristics by multiple flood variables. Zhang et al. (2019) gave a rigorous comparison of several bivariate nonstationary flood frequency calculation models using different explanatory variables in time-varying marginal distributions [14]. Jiang et al. (2015) applied a time-varying copula function that considered elements of the changing reservoir environment as covariates [27]. Wen et al. (2019) presented a process of employing a time-varying copula model to model the nonstationary dependence structures between two highly correlated flood variables [28]. Generally, because the copula function relaxes the restriction of the marginal distributions' form, most of the above studies adopted time-varying marginal distributions to construct the nonstationary models for multivariate frequency analysis. However, corresponding studies using the nonstationary mixture distributions as marginal distributions and estimating bivariate nonstationary design flood are limited. In this study, the inflow extreme flood series of the Longmen Reservoir, which is located on the southern branch of the Daqing River Basin, are selected as the target flood variables. Because the Longmen Reservoir catchment has undergone extensive measures of returning farmland to forests, along with the construction of soil and water conservation engineering around 1980, the flood generating mechanisms would be heterogeneous, and the traditional flood frequency analysis should not be made available. Because forest cover and hydraulic engineering are the main drivers of controlling runoff processes, this study develops a bivariate nonstationary flood frequency analysis on flood peak and volume variables using a mixture distribution describing the non-stationarity of reservoir inflow annual maximum flood series. Furthermore, the comparison of univariate and bivariate nonstationary flood frequency analysis is extended to investigate and explore the mathematical rules of corresponding design flood for reservoir flood risk management. The findings provide scientific guidance for engineers to manage extreme flood events and improve the theoretical system of bivariate nonstationary flood frequency analysis.

The paper is organized as follows. In Section 2, we describe the study area, the data set, and the statistical–physical analysis for extreme flood characteristics. Section 3 presents the methodologies of the univariate and bivariate nonstationary models, along with a brief description of the bivariate nonstationary return period. The results and discussions are presented in Section 4. Finally, the article concludes in Section 5.

## 2. Study Area and Flood Data

The Longmen Reservoir (Figure 1) is located $39°7'$ N and $115°16'$ E in northern China and has a drainage area of 470 km$^2$ and a total storage capacity of $1.27 \times 10^8$ m$^3$. Due to the temperate semi-arid continental monsoon climate, the rainfall distribution over one year is extremely uneven, with 80% of rainstorms occurring during the flood season in the basin. The uneven water allocation contributes to large flood or drought hazards, which frequently appear in the Longmen Reservoir basin. Thus, the Longmen Reservoir, as one of the four large-scale reservoirs for the Daqing River basin in northern China, was constructed to control flooding, and also to provide an irrigation function. For the reservoir, which was built in February 1958, expanded in 1977, and reinforced in 2002, the design flood control standard reaches a 100-year return period, and the flood check standard is a 2000-year return period.

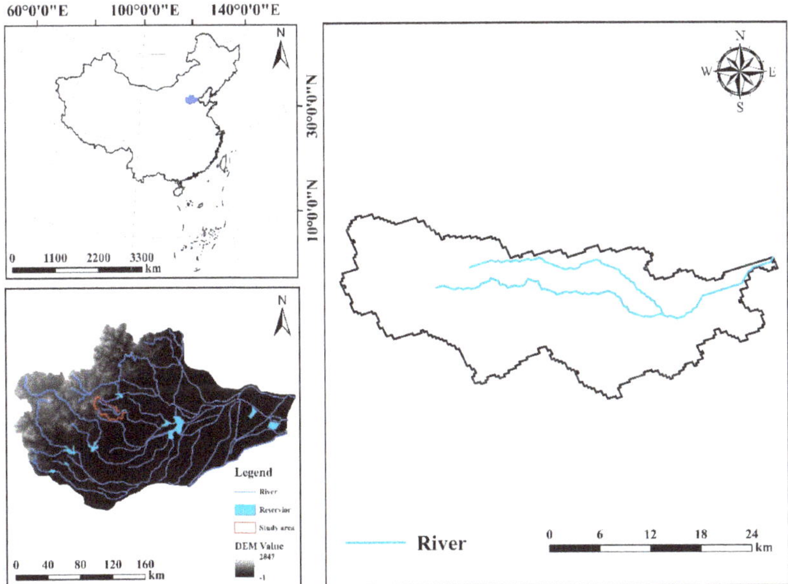


**Figure 1.** Schematic diagram of the Longmen Reservoir catchment and its location in the Daqing River Basin.

### 2.1. Extreme Flood

The annual maximum flood peak series (AMFP) and annual maximum different periods (e.g., 1-day, 3-day, and 6-day) flood volume (AMFV) for the period 1951–2005 are the target variables and were collected from the Hydraulic and Hydropower Design Institute of Hebei Province in China. In addition, the flood in 1963 was the largest recorded flood, with a peak discharge of 4250 m$^3$/s, which is almost 15 times the median annual maximum flood. Moreover, the historically extraordinary flood in 1939, with a peak flow of 4180 m$^3$/s, had the same magnitude as the flood in 1963. Hence, in terms of AMFP and AMFV, there have been two previous extraordinary floods.

In general, in most cases, flood event risks for a reservoir should take into consideration both the flood peak and the volume, which can comprehensively describe the characteristics of a flood event. Traditionally, flood frequency analysis has been aimed at a single flood variable. Thus, in this study, we apply bivariate flood frequency analysis by investigating the correlational relationship between flood peak and volume. Kendall and Spearman correlation tests are used to estimate the dependence between flood peak; $Q$; and annual maximum 1-day, 3-day, and 6-day flood volume series, and their estimated values are listed in Table 1. The results indicate that the 1-day AMFV series has the most significant correlation with AMFP and is selected to display the clearly visible mutually correlated nature, which supports the necessity for bivariate flood frequency analysis.

**Table 1.** The Kendall and Spearman correlation test results between AMFP and AMFVs.

| Flood Series | Kendall Correlation Test | Spearman Correlation Test |
| --- | --- | --- |
| 1-day AMFV | 0.84 | 0.96 |
| 3-day AMFV | 0.79 | 0.93 |
| 6-day AMFV | 0.77 | 0.92 |

*2.2. Statistical-Physical Heterogeneity Analysis*

Whether or not the extreme flood variables series is stationary is a prerequisite and crucial step for implementing flood frequency analysis. In this study, AMFP and 1-day AMFV are utilized in combination with the stationary or nonstationary characteristics in terms of both statistical and physical aspects to establish a bivariate joint distribution model.

Mathematically, as suggested by Zeng et al. (2014) and Xie et al. (2009), the heterogeneity investigation of AMFP and 1-day AMFV records using a two-step diagnose process indicates that the candidate significant change points are presented in the years 1964 and 1979 [23,29]. The two-step diagnose process (for method details, refer to Zeng et al., 2014) includes: (1) First, the Hurst exponent method [30,31] is adopted to identify the long-term memory of flood series, as proposed by Xie et al. (2009) [29], to manifest the diagnosis variation. (2) Second, based on the first diagnosis result, several change point detection tests, including the Mann–Whitney–Pettitt (MWP) test [32], the Brown–Forsythe method [33], and the Moving rank test [34] are applied to ascertain the significant change points. The diagnosis process results, which are shown in Table 2, demonstrate that the AMFP has no variation, the 1-day AMFV exhibits medium variation, and the two leading significant change points appear in 1964 and 1979.

**Table 2.** The statistical heterogeneity results of AMFP and 1-day AMFV.

| Methods | AMFP | 1-Day AMFV |
| --- | --- | --- |
| Hurst exponent value | 0.67 (no variation) | 0.73 (medium variation) |
| MWP | — | 1959–1971, 1974, 1977–1983 |
| Brown–Forsythe | — | 1996, 1964 |
| Moving rank test | — | 1964, 1979, 1998 |
| Change points | — | 1964, 1979 |

Physically, natural hydrologic phenomena that include extreme flood events are investigated by utilizing statistical tools, but the causes are ultimately related to physical factors. Climate and underlying surface causes are both involved to acquire the most significant change point. The annual maximum 30-day precipitation series over the same period, which has significant correlation with 1-day AMFV (Spearman and Kendall correlations test $p$-values are $2.20 \times 10^{-16}$ and $3.52 \times 10^{-15}$, respectively), is chosen to represent the climate effect driver because the annual maximum events generally occur in August during the flood period (June–July–August–September (JJAS)). The annual maximum 30-day precipitation has no significant change point, which reveals that the influence of climate is rather small on the Longmen Reservoir, which has a relatively small area. On the other hand, the underlying surface change is caused by both land-use change and anthropogenic activities. The land-cover types in the Longmen Reservoir, including forest, cultivated land, and grassland, in the years 1970, 1980, and 2000, are showed in Figure 2. The area percentage variation of the three land cover types demonstrates that the forest exhibits an increase, but grassland and cultivated land both have a small decrease between 1970 and 1980. Relatively speaking, there was almost no change between 1980 and 2000. The increase of forest area in the years around 1980 reduced flood generation, to a certain extent. Moreover, in the Daqinghe river basin, as well as in the Longmen Reservoir catchment, since the 1980s, numerous soil and water conservation projects have emerged, such as level trenches, check dams, and riverbank protection engineering. Additionally, dozens of non-engineering measures, such as closing hills for afforestation, greening bare mountains, and forest planting, have been carried out on a large scale, partially due to the 'Soil and Water Conservation Management in Small Basin' document declared by the Ministry of Water and Electricity in China in 1980.

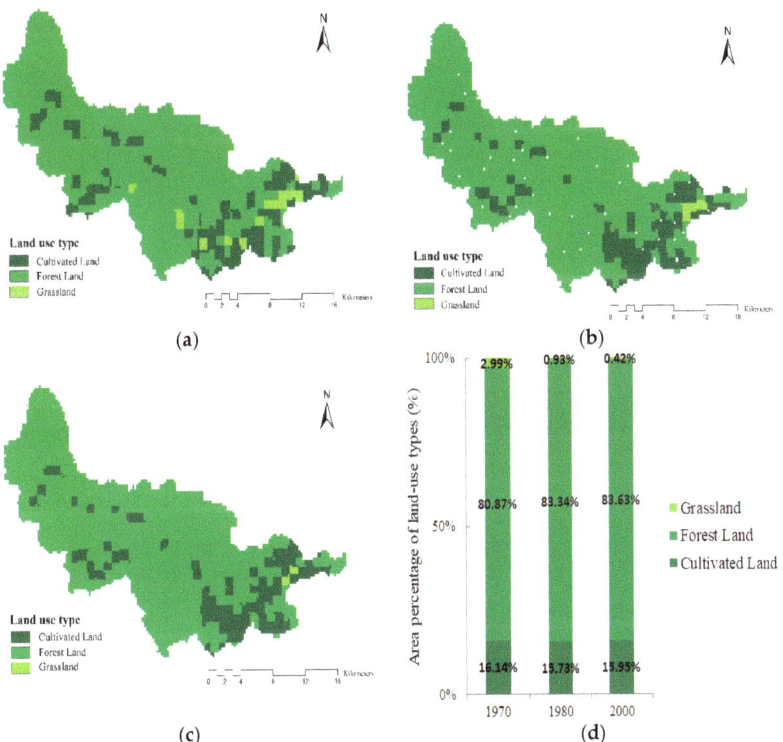


**Figure 2.** The land use and land cover of the Longmen Reservoir in the years (**a**) 1970, (**b**) 1980, and (**c**) 2000; (**d**)The area percentages of land use types in three years.

The change point of 1-day AMFV, which appeared in 1964 and is the lag change resulting from the catastrophic flood event in 1963, is close to the beginning of the flood samples and is therefore discarded. The above land-use changes and frequent anthropogenic activities around 1980 likely destroyed the homogeneity of the flood generating mechanism and contributed to inducing a shift in annual maximum flood series [35]. Consequently, the most significant change point is identified to be in the year 1979 for 1-day AMFV, from both a statistical and a physical viewpoint.

## 3. Methodology

### 3.1. Mixture Distribution as Marginal Distribution

Due to the interaction of underlying surface conditions (or meteorological variations) and flood generation mechanisms, an extreme flood series is generated by distinct complex sources, such as cyclonic rainfall, convective rainfall, land cover situations, and channel characteristics. Once the single flood series are heterogeneous, the conventional stationary physical basis is destroyed. Consequently, the mixture distribution (MD), which is defined in the mathematical statistics field as the probability distribution of a random variable combined with several other random variables, is proposed to address the multiple flood population frequency calculation issue [36]. Theoretically, in this study, we use the basic additive form of finite density mixture distribution, which is described by:

$$\begin{cases} f(x) = \sum\limits_{i=1}^{n} \omega_i f_i(x|\theta_i) \\ \sum\limits_{i=1}^{n} \omega_i = 1 \end{cases} \tag{1}$$

where $f_i(x|\theta_i)$ is the $i$th component probability density distribution with corresponding parameters set, $\theta_i$, $\omega_i$ is the relative weight ($0 \leq \omega_i \leq 1$) denoting the probability of belonging to the $i$th flood component, and $n$ is the number of flood components.

In the application, the $n$ value should be confirmed by flood classification based on the generating mechanism. Nevertheless, it is worth emphasizing that the flood physically-based genesis with underlying surface interaction is complicated, so that the prior subdivision may not be feasible [37–39]. In addition, Alila and Mtiraoui (2002) stressed that an increase in the $n$ value needs a large sample size and makes the parameter estimation less robust, less parsimonious, and less accurate [21]. To keep it to a minimum, in this study, the two single Pearson III type (P-III) probability density distributions are summed up for the mixture model. The selection of P-III distribution is widely applied and recommended in the Regulation for Calculating Design Flood of Water Resources and Hydropower Projects in China. Then the two-component mixture model is given by:

$$f(x) = \omega f_1(x|\theta_1) + (1 - \omega) f_2(x|\theta_2) \tag{2}$$

$$f_i\left(x\middle|\theta_i\right) = \frac{\beta_i^{\alpha_i}}{\Gamma(\alpha_i)}(x - a_{0i})^{\alpha_i - 1} e^{-\beta_i(x - a_{0i})} \tag{3}$$

where all the parameters, namely $\omega$ and $\theta_i(\alpha_i, \beta_i, a_{0i})$, pproximately seven parameters, are jointly estimated from the overall extreme flood series, including historical extraordinary floods by the Simulated Annealing Algorithm (SAA); detailed in Zeng et al., 2014 [23], minimizing the differences between empirical and theoretical cumulative probabilities. It should be noted that the parameters $\theta_i(\alpha_i, \beta_i, a_{0i})$ can be represented by the commonly used statistical parameters mean, $EX_i$, coefficient of deviation, $Cv_i$, and coefficient of skewness, $Cs_i$, which are convenient and visibly manifest flood sample statistical characteristics. The original parameters of P-III distribution and statistical parameter conversion formulas are illustrated by:

$$\begin{cases} EX_i = a_{0i} + \alpha_i / \beta_i \\ Cv_i = \sqrt{\alpha_i} / (\beta_i a_{0i} + \alpha_i) \\ Cs_i = 2 / \sqrt{\alpha_i} \end{cases} \tag{4}$$

### 3.2. Bivariate Copula Functions

Traditional flood frequency analysis usually focuses on individual flood series, and a bivariate assessment of peak discharge and flood volume is not commonly included. A joint consideration of peak discharges and flood volume is, however, crucial when assessing the flood event risks for flood control reservoirs. Moreover, the joint distribution construction is quite difficult, especially for two non-independent random variables. Thus, the copula function as a powerful tool is frequently used by hydrologists for modelling, while jointly considering peak discharge and flood volume without any restrictions on marginal distributions.

The theory of the copula function was proposed by Saklar (1959) [40], and the copula function is found to be multifunctional for constructing joint distribution functions because it allows a variety of independent marginal distributions [41–43]. Based on Sklar's theorem [40], in this study there are two dependent variables, $Q$ and $W$, representing flood peak and flood volume, respectively, and they can be characterized by the associated dependence function copula, which can be expressed as:

$$F(q, w) = C_\theta(F_Q(q), F_W(w)) = C_\theta(u, v) \tag{5}$$

where $u = F_Q(q) = P(Q \leq q)$ and $v = F_W(w) = P(W \leq w)$ are the marginal cumulative distribution functions of univariate random variables $X$ and $Y$, respectively; The bivariate joint probability distribution function, $F$, is expressed with the univariate marginal distributions and the dependence copula function, $C_\theta$, where $\theta$ is the parameter of copula. Moreover, if $F_Q(q)$ and $F_W(w)$ are continuous, then copula, $C$, is unique [43] and captures the dependencies among the random variables. For an extended mathematical introduction and practical approach and details of the copula functions, readers can follow Nelsen (2006) [42], Durante and Sempi (2015) [44], and Salvadori et al. (2007) [45].

Many copula families are frequently employed by hydrologists for modeling extreme flood events, including Archimedean, elliptical, Plackett, and extreme value [46]. The Archimedean family is quite popular due to its massive variety of families, and it is well-adapted for establishing the bivariate joint dependency constructures of the extreme flood characteristics. It is noteworthy that considering the tail dependence in selecting the optimal copula function is of great importance for providing the best fit to flood samples [47]. Thus, in this work, we introduce and test three Archimedean families, i.e., Gumbel–Hougaard (G–H), Clayton, and Frank, for constructing the joint distribution of annual flood characteristics, flood peak discharge, and flood volume series. The three copula functions describe different types of features of dependence structures. For instance, the G–H copula displays a strong capability to model upper-tail dependency, and the Clayton copula is more suitable for modelling lower-tail dependency. On the contrary, the Frank copula exhibits higher versatility and has no tail dependency [48]. In this study, we focus on modeling the extreme flood events and the exceedance probabilities of large flood events, which are of more interest for reservoir flood management. Thus, the G–H copula is selected to model the dependence of the Q-V pair, and the copula dependence parameter $\theta$ is estimated using the relationship between Kendall's tau and $\theta$. The mathematical expression for the bivariate G–H copulas function is illustrated below:

$$\begin{cases} C_\theta(u, v) = \exp\left\{-\left[(-\ln u)^\theta + (-\ln v)^\theta\right]^{1/\theta}\right\}, \theta \in [1, \infty) \\ \tau = 1 - \theta^{-1} \end{cases} \tag{6}$$

### 3.3. Goodness of Fit for Models

For the selected copula to be admissible and capable of depicting the dependency modeling of two extreme flood series, the copula functions are needed to conduct the goodness-of-fit for evaluating the validity. In this study, the Kolmogorov–Smirnov (K–S) test [49] is adopted for the goodness-of-fit test. Thus, the definition of the K–S test is illustrated in the following.

The K–S test statistic $D$ is defined as:

$$D = \max_{1 \leq k \leq n}\left\{\left|C_k - \frac{m_k}{n}\right|, \left|C_k - \frac{m_k - 1}{n}\right|\right\} \tag{7}$$

where $C_k$ is the copula value of the joint observed flood peak and the flood volume pairs samples $(q, w)$, $m_k$ is the number of the $(q, w)$ pairs samples, which simultaneously satisfy the conditions of $Q \leq q$ and $W \leq w$, and $n$ is the sample size. Then, if the statistic $D$ exceeds the critical value at 5% confidence level, it rejects the null hypothesis and reveals that the distribution cannot model the extreme flood variables well.

*3.4. Bivariate Nonstatioarny Return Period*

In the univariate case, extreme flood events for a specific return period are extremely important for the reservoir's design, and the return period is usually defined as a mean inter-arrival time estimation of events exceeding a dangerous flood threshold. Thus, the univariate return period of two variables, flood peak, $Q$, and flood volume, $W$, with thresholds $q$ and $w$, respectively, are given by:

$$T_Q(q) = \frac{1}{1 - F_Q(q)}, \quad T_W(w) = \frac{1}{1 - F_W(w)} \tag{8}$$

In a bivariate domain, in contrast to the univariate case, an extreme flood event can be defined as critical if either flood peak or flood volume exceeds a design flood threshold, or if both flood variables are larger than the prescribed values. Hence, as eight types of possible bivariate joint flood events are proposed by Salvadori and De Michele (2004) [50], the joint "OR" and "AND" return periods (represented by OR-RP and AND-RP, respectively) are two widely used approaches in hydrological applications [51]. They can be expressed as follows:

$$T_{OR} = \frac{\mu}{P(Q > q \, or \, W > w)} = \frac{\mu}{1 - C_\theta(F_Q(q), F_W(w))} \tag{9}$$

$$T_{AND} = \frac{\mu}{P(Q > q \, and \, W > w)} = \frac{\mu}{1 - F_Q(q) - F_W(w) + C_\theta(F_Q(q), F_W(w))} \tag{10}$$

where $\mu$ is the average inter-arrival time between two consecutive events (equals 1 for annual extreme events).

As suggested by Feng and Li (2013) [24], the univariate return periods, OR-RP and AND-RP have the following comparison expression, which is given by:

$$T_{OR} \leq \min[T_Q, T_W] \leq \max[T_Q, T_W] \leq T_{AND} \tag{11}$$

In addition to focusing on the probability of both flood peak and flood volume simultaneously exceeding a certain threshold, the conditional probabilities of flood events are also of great importance for reservoir operations obtained from the copula-based bivariate analysis. The probabilities of flood volume, given flood peak exceeding a certain threshold, are illustrated by:

$$P(W \geq w | Q \geq q) = \frac{P(Q \geq q, W \geq w)}{P(Q \geq q)} = \frac{1 - F_Q(q) - F_W(w) + C_\theta(F_Q(q), F_W(w))}{1 - F_Q(q)} \tag{12}$$

As the two return period approaches for bivariate joint distribution, the design flood peak and volume value calculations are confused and ambiguous. Because the computation of design flood hydrographs for reservoirs is carried out under the assumption that the flood peak and volume events share the same return period, as suggested by Xiao et al. (2007) [52], assuming that $u = v$, Equation (13) of the bivariate OR joint return period $T_{OR}$ and joint copula distribution $C_\theta(u, v)$ can determine the $u$ value. Then the inverse functions of $u = F_Q(q)$ and $v = F_W(w)$ can be used to obtain the design flood peak and volume, respectively, corresponding to the joint return period, $T_{OR}$.

## 4. Results and Discussion

### 4.1. Univarite Mixture Distribution Flood Frequency Analysis

Based on the heterogeneous diagnosis results of AMFP and 1-day AMFV series, the 1-day AMFV is modeled on the nonstationary flood frequency analysis using mixture distribution, and the parameters of MD estimated by SAA are given in Table 3. For the comparison of a stationary benchmark, we used the single-type Pearson type III distribution to fit the AMFP and 1-day AMFV, and the parameters were estimated by the linear moment method [53]. The P-III distribution and MD fitting curves of 1-day AMFV are displayed in Figure 3. The curves indicate that the theoretical frequency curve fitted by MD is a little farther from the upper floods than the P-III distribution. However, the P-III distribution neglected most of the empirical flood data in the corner section, which results in greater differences between the theoretical fitting results and the empirical frequencies compared to MD. However, these small deviations may be overlooked by an inexperienced viewer, the results of which could be quite considerable because even a tiny difference may bring out a huge deviation in the design flood values, and hence contribute to different treatments in flood risk management. Furthermore, the K-S test statistic value, with 0.1668 of MD, is less than the value of 0.3409 with P-III distribution. We thereby suggest that the MD applied in a nonstationary extreme flood series have better modeling performance and improve the fitting capability. Thus, it is necessary to establish the nonstationary model to provide scientific support for the flood control operation of the Longmen Reservoir under the land use changes and the increasing construction of numerous soil and water conservation projects.

**Table 3.** The estimated parameters of MD and P-III distribution in the Longmen Reservoir.

| Flood | $\alpha$ | $EX_1$ | $Cv_1$ | $Cs_1$ | $EX_1$ | $Cv_1$ | $Cs_1$ |
|---|---|---|---|---|---|---|---|
| AMFP ($m^3/s$) | | 265.77 | 2.88 | 6.04 | | | |
| 1-day AMFV (P-III) ($10^8$ $m^3$) | | 0.12 | 2.3 | 5.2 | | | |
| 1-day AMFV (MD) ($10^8$ $m^3$) | 0.34 | 0.18 | 1.7 | 5.1 | 0.09 | 1.95 | 4.00 |

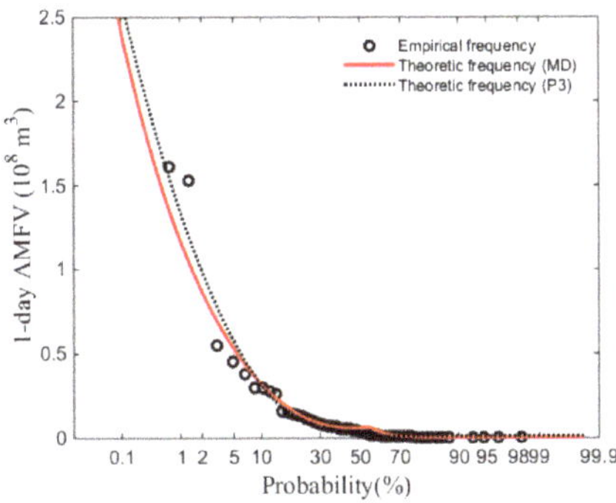


**Figure 3.** The fitting curves of MD and P-III distribution for 1-day AMFV.

Corresponding to the fitting results of MD and P-III distribution, the design flood values with different univariate return periods are provided by graphical information (see Figure 3) as well as by the numerical values summarized in Table 4. The results demonstrate that, given the same return period, the design flood values estimated by MD are smaller than those estimated by P-III distribution. Specifically, the reduced magnitude of 1-day AMFV is approximately 3.1–15.2% between MD and P-III distribution with various return periods. The results indicate that the design flood differences have a great implication for flood control operation, and the nonstationary flood frequency analysis should not be negligible.

**Table 4.** The design flood values of MD and P-III distribution.

| Flood | Distribution | Return Periods (Year) | | | | | |
|---|---|---|---|---|---|---|---|
| | | **2000** | **1000** | **100** | **50** | **20** | **10** |
| 1-day AMFV $(10^8 \text{ m}^3)$ | P-III | 3.02 | 2.61 | 1.34 | 0.99 | 0.58 | 0.32 |
| | MD | 2.79 | 2.36 | 1.14 | 0.84 | 0.51 | 0.31 |
| | Difference (%) | −7.6 | -9.6 | −14.9 | −15.2 | −12.1 | −3.1 |

### 4.2. Fitting Bivariate Joint Distribution

In the above section, MD is used to obtain the nonstationary marginal distribution for the nonstationary 1-day AMFV series, and the significant correlated dependence between AMFP and 1-day AMFV is obviously visible, which supports the necessity for the bivariate flood frequency analysis. Hence, the stationary bivariate copula function is constructed based on the estimated marginal distributions of P-III distribution for both AMFP and 1-day AMFV. In contrast, in the nonstationary context, the copula-based joint distribution is implemented using the P-III distribution for AMFP and MD distribution for 1-day AMFV as the bivariate marginal distributions. The copula parameters, K-S statistical test, OLS, and AIC results under stationary and nonstationary conditions are listed in Table 5. In the nonstationary bivariate context, the applied G–H copula function passes the K-S test, with the statistic $D$ critical value of 0.1817 at the significant level of 0.05, but the copula function model fails the test under the stationary condition. Additionally, the G–H copula under the nonstationary condition is the best-fitted copula function, with smaller $D$, OLS, and AIC values. Hence, the G–H is selected as the most reasonable function for modeling the dependence structure between the AMFP and the nonstationary 1-day AMFV. The Clayton copula and Frank copula functions have also been employed to model the bivariate flood variables, and the fitting results, especially for the extraordinary flood events, are not sufficient. Thus, the Clayton copula and Frank copula parts are not presented in this study.

**Table 5.** The G–H copula function fitting results under stationary and nonstationary conditions.

| Cases | Parameter ($\theta$) | K-S Test ($D$) | OLS | AIC |
|---|---|---|---|---|
| Stationary | 6.26 | 0.3214 | 0.1371 | −220.55 |
| Nonstationary | 6.26 | 0.1419 | 0.0604 | −312.3 |

Figure 4a shows the fitting performance between the theoretical frequency estimated by the optimum G–H copula and the empirical frequency points. Meanwhile, Figure 4b displays the probability–probability plot (PP-plot) of the optimal G–H copula. The good agreement exhibited in Figure 4a,b demonstrates that the selected G–H function has a satisfactory fitting performance.

### 4.3. Estimating Bivariate Nonstationary Return Period and Design Flood

In light of the above mixture marginal distribution and the selected optimal G–H copula function, the copula function joint distribution fitting results and the joint OR-RP and AND-RP for AMFP and 1-day AMFV are illustrated by three-dimensional (3-D) plots in Figure 5. In order to obtaining the bivariate joint return periods intuitively and conveniently, Figure 6a,b displays the isolines of the flood peak-volume pairs for different return periods under the joint OR-RP and AND-RP cases, respectively. Given a flood event in a specific year, the joint return periods in the OR-RP and AND-RP cases are easy to confirm, especially for extraordinary flood events. Taking the largest recorded flood in 1963 as an example, the joint return period of either the flood peak exceeding 4250 m$^3$/s or the flood volume exceeding 1.6 × 10$^8$ m$^3$ is approximately 130 years, and the joint return period of both the flood peak-volume pair exceeding the corresponding threshold is approximately 260 years. The univariate return period for the extreme flood event in 1963 is then 150 years, which lies between the OR-RP and the AND-RP.

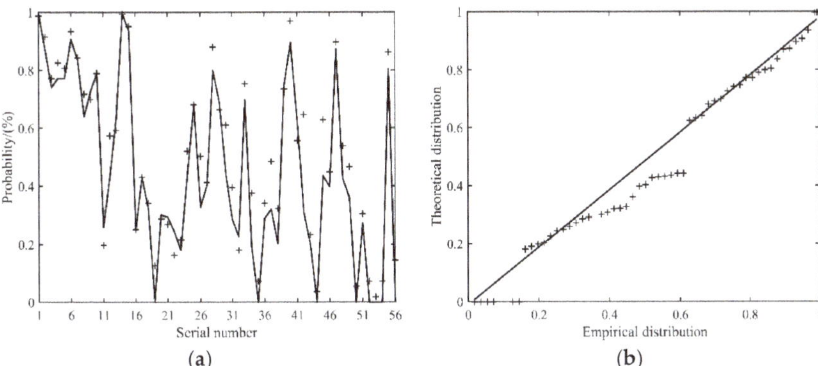

**Figure 4.** (**a**) Fitting performance and (**b**) PP-plot between theoretical and empirical joint distributions.

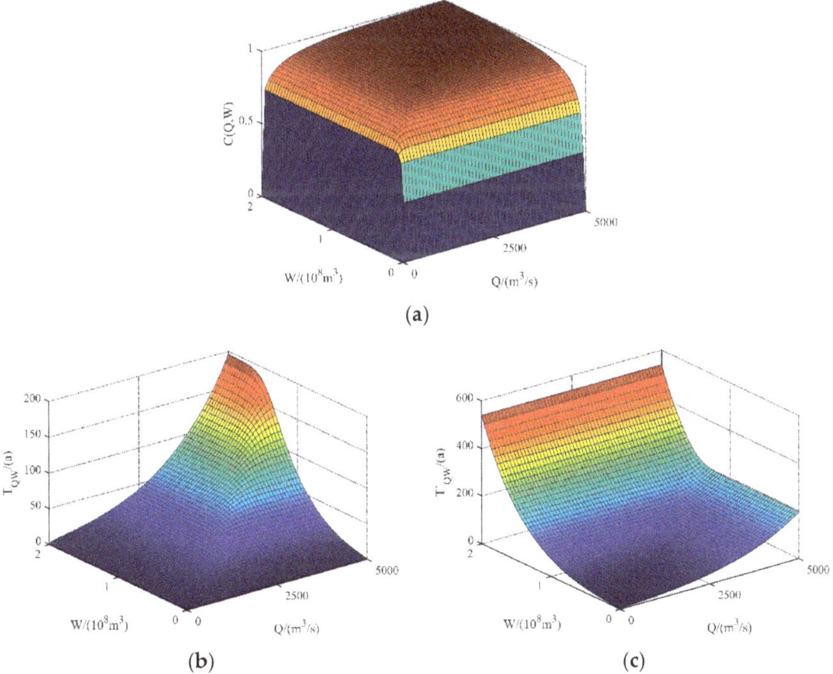

**Figure 5.** The 3D plots of the AMFP and 1-day AMFV pair for (**a**) copula function joint distribution. (**b**) OR-RP, (**c**) AND-RP.

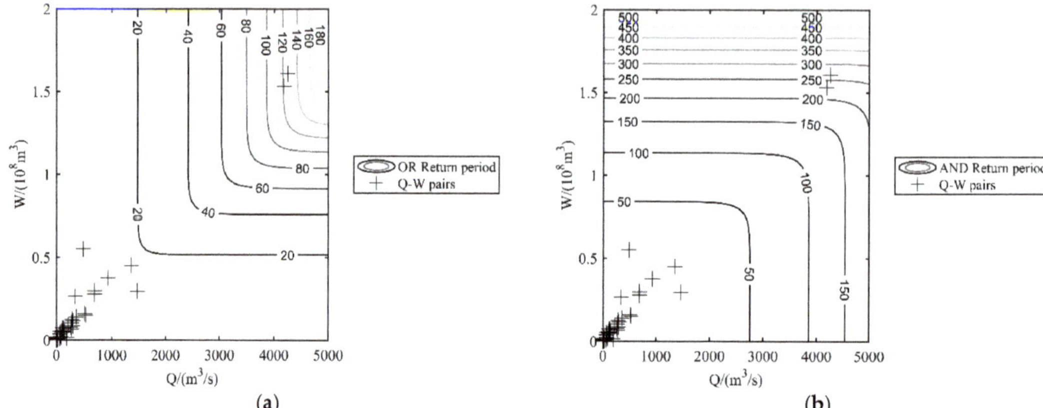

**Figure 6.** The isolines of flood peak-volume pairs under joint (**a**) OR-RP and (**b**) AND-RP conditions, respectively.

According to Equations (9) and (10), through the different univariate return period combinations of AMFP and 1-day AMFV, the bivariate joint OR-RP and AND-RP can be obtained, and their isolines are showed in Figure 7. Additionally, it is obviously visible that, when assigning the same univariate return period, the joint OR-RP is always smaller than the univariate return period; in contrast, the associated AND-RP is greater than the univariate return period. The results are consistent with the mathematical Formula (11).

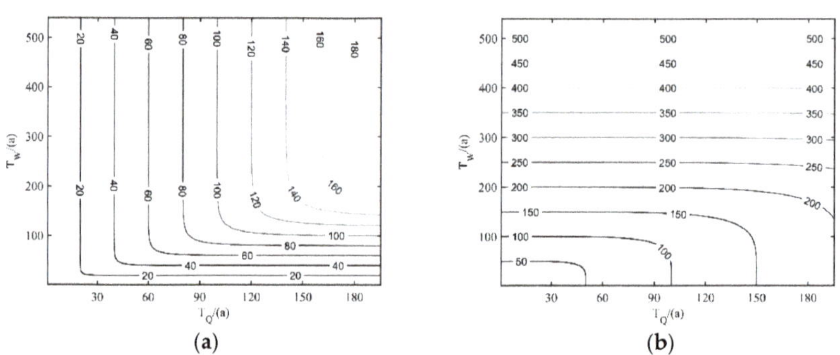

**Figure 7.** The isolines of (**a**) OR-RP and (**b**) AND-RP, respectively, for different univariate return periods of AMFP and 1-day AMFV pairs.

The design flood values for hydraulic engineering are also of utmost important. Because there are two joint return periods, the determination of bivariate design flood values is difficult. As suggested by Li et al. (2013) [22], under the assumption that the flood peak and the volume share the same return period for joint OR-RP, combing Equations (6) and (9), the analytical formulas of bivariate design flood peak and volume are given by:

$$u = v = \left(1 - \frac{1}{T_{OR}}\right)^{2^{-\frac{1}{\theta}}} \tag{13}$$

$$Q = F_Q^{-1}(u), W = F_W^{-1}(v) \tag{14}$$

The design flood values can be calculated for AMFP modeling by P-III distribution and 1-day AMFV modeling by univariate mixture distribution, and the optimal bivariate

G–H copula functions with different return periods ($T$ = 20, 50, 100, 1000, 2000 years) are listed in Table 6. It is worth noting that the bivariate design floods for both flood peak and flood volume are larger than the ones in the univariate P-III distribution and MD conditions, respectively. Compared to the univariate condition, the different percentages of AMFP design flood estimated by bivariate joint distribution increase from 2.29% to 9.3%, in line with the decrease of the return period as well as the 1-day AMFV.

**Table 6.** Design flood results under univariate and bivariate nonstationary conditions.

| Return Period (yr) | Design Flood of Univariate Marginal Distribution | | Design Flood of Bivariate Joint Distribution | | Difference (%) | |
|---|---|---|---|---|---|---|
| | $Q$ (m$^3$/s) | $W_1$ ($10^8$ m$^3$) | $Q$ (m$^3$/s) | $W_1$ ($10^8$ m$^3$) | $Q$ (m$^3$/s) | $W_1$ ($10^8$ m$^3$) |
| 2000 | 9332 | 2.79 | 9546 | 2.86 | 2.29 | 2.51 |
| 1000 | 7996 | 2.36 | 8208 | 2.43 | 2.65 | 2.97 |
| 100 | 3855 | 1.14 | 4038 | 1.19 | 4.75 | 4.39 |
| 50 | 2752 | 0.84 | 2920 | 0.89 | 6.10 | 5.95 |
| 20 | 1473 | 0.51 | 1610 | 0.55 | 9.30 | 7.84 |

An attempt to explore the mathematical rule of design floods between the two cases was made, and it was found that Equation (13) is always smaller than the joint distribution cumulative probability, which is given by:

$$C_\theta(u,v) = 1 - \frac{1}{T_{OR} = T_Q = T_W} \tag{15}$$

which indicates that the smaller cumulative probability directly results in an increase of design flood values under bivariate copula joint distribution.

As the traditional univariate stationary P-III distribution is beyond the above constraint and its calculation is independent of the copula joint distribution, it is worth pointing out that the design flood values of 1-day AMFV under univariate stationary P-III distribution (see Table 4) are larger than the ones estimated by bivariate nonstationary joint distribution (see Table 6). On the other hand, the design flood values of AMFP modelled by bivariate joint distribution are greater than the ones in a stationary context.

### 4.4. Estimating Joint and Condtional Probabilities

The estimation of joint and conditional probabilities for extreme flood events plays a vital role in reservoir flood control operation management. Simultaneous consideration of the probability of flood peak and flood volume exceeding a certain threshold can be invaluable. The joint exceedance probability is the reciprocal value of the joint OR return period. Thus, for the extraordinary flood event in 1963, the joint OR probability of both the flood peak exceeding 4250 m$^3$/s and 1-day AMFV exceeding $1.6 \times 10^8$ m$^3$ is approximately 0.77%.

Considering reservoir flood control, the flood frequency analysis not only focuses on considering the joint probabilities of Q-V pairs exceeding a certain threshold, but also aims to estimate the conditional probabilities of extreme flood events. The outcomes of conditional probabilities are shown in Figure 8. The conditional probability curves of 1-day AMFV when AMFP exceeds a certain threshold are exhibited in Figure 8a. We focus exemplarily on the AMFP exceeding the design flood values with a 100-year (design standard) return period. The conditional probabilities of 1-day AMFV with 100-year and 2000-year return periods are 88% and 5.02%, respectively. The conditional probability results of 1-day AMFV with different return periods when the AMFP exceeds the design flood values with 100-year (design standard) and 2000-year (check standard) return periods, respectively, are summarized in Table 7. The results indicate that, with the high correlation between AMFP and 1-day AMFV, the probability of large flood volume values with the same return period would be high if an extraordinary flood peak occurred.

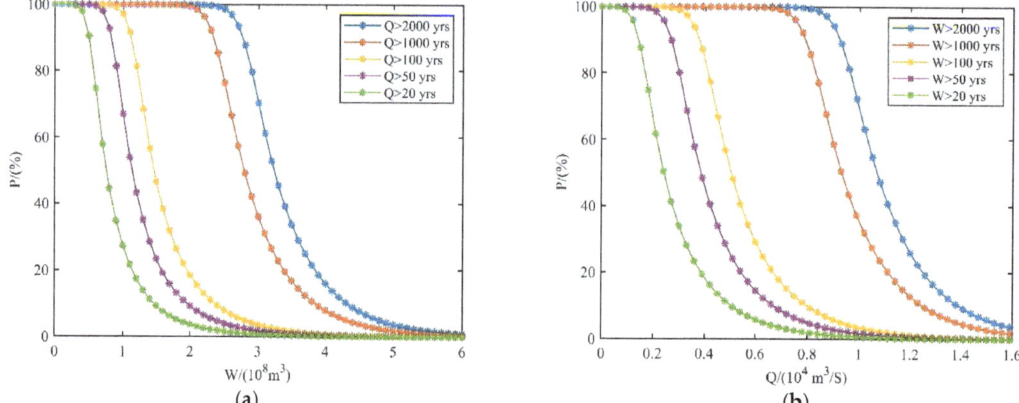

**Figure 8.** The conditional probability curves of (**a**) 1-day AMFV and (**b**) AMFP.

**Table 7.** The conditional probabilities of 1-day AMFV when the AMFP exceeds the design flood values with 100-year and 2000-year return periods.

| Return Period (Year) | | 2000 | 1000 | 100 | 50 | 20 |
|---|---|---|---|---|---|---|
| Design flood Wp ($10^8$ m$^3$) | | 2.79 | 2.36 | 1.14 | 0.84 | 0.51 |
| Conditional probability (%) | 100-year AMFP | 5.02 | 10.05 | 88.00 | 99.63 | 99.99 |
| | 2000-year AMFP | 88.51 | 99.60 | 99.99 | 99.99 | 99.99 |

## 5. Conclusions

This study aimed to investigate the influence of non-stationarity on flood characteristics, considering the dependence between flood peak and flood volume under a changing underlying surface, using nonstationary univariate and bivariate flood frequency analysis models in the Longmen Reservoir in North China. The following main conclusions can be drawn from this study.

(1) The 1-day AMFV exhibits the highest significant correlation with AMFP, which demonstrates the desirability and indispensability of bivariate flood frequency analysis. In addition, the underlying surface changes in the Longmen Reservoir contribute to the heterogeneity of flood generation identified by the statistical methods and physical basis analysis. A significant change point is detected in the year 1979 for 1-day AMFV, but the AMFP is shown to be homogenous.

(2) From univariate nonstationary flood frequency analysis of 1-day AMFV, the fitting performance of mixture distribution is superior to the traditional stationary P-III distribution. Due to the increase of forest land area and some hydraulic engineering construction, the design floods of 1-day AMFV with different return periods estimated by MD are generally smaller than the ones estimated by P-III distribution.

(3) In the case of bivariate analysis, copula-based joint distribution was developed and performed using the stationary P-III distribution for AMFP and nonstationary MD for 1-day AMFV as marginal distributions. There is a relatively large increase for the design floods estimated by bivariate nonstationary joint distribution compared with the ones estimated in a univariate nonstationary context, which can be concluded and proved by rigorous mathematical formula derivation. Furthermore, the results of joint and conditional probabilities demonstrate that, assuming the flood peak and volume share the same return period, the conditional probability of 1-day AMFV exceeding the threshold is likely to be high when the AMFP exceeds the design flood associated with the return period.

**Author Contributions:** Conceptualization, H.Z. (Hang Zeng) and Q.L.; methodology, H.Z. (Hang Zeng) and Q.L.; validation, H.Z. (Hang Zeng) and Q.L.; formal analysis, H.Z. (Hang Zeng) and Q.L.; investigation, P.L.; resources, Z.L., W.Y. and H.Z. (Hui Zhou); data curation, Z.L., W.Y. and H.Z. (Hui Zhou); writing—original draft preparation, H.Z. (Hang Zeng), Q.L. and P.L.; writing—review and editing, H.Z. (Hang Zeng); visualization, Q.L.; supervision, H.Z. (Hang Zeng) and P.L.; project administration, H.Z. (Hang Zeng) and P.L.; funding acquisition, H.Z. (Hang Zeng) and Z.L. All authors have read and agreed to the published version of the manuscript.

**Funding:** This research was funded by the National Natural Science Foundation of China (grant number 51809018), the Natural Science Foundation of Hunan Province, China (grant number 2019JJ50643), and the Scientific Research Project of Hunan Provincial Education Department, China (grant number 18C0198).

**Data Availability Statement:** The data used to support the findings of this study are available from the corresponding author upon request.

**Conflicts of Interest:** The authors declare no conflict of interest.

## References


1. Meaurio, M.; Zabaleta, A.; Boithias, L.; Epelde, A.M.; Sauvage, S.; Sánchez-Pérez, J.-M.; Srinivasan, R.; Antiguedad, I. Assessing the hydrological response from an ensemble of CMIP5 climate projections in the transition zone of the Atlantic region (Bay of Biscay). *J. Hydrol.* **2017**, *548*, 46–62. [CrossRef]
2. Tofiq, F.A.; Guven, A. Prediction of design flood discharge by statistical downscaling and General Circulation Models. *J. Hydrol.* **2014**, *517*, 1145–1153. [CrossRef]
3. Tofiq, F.A.; Güven, A. Potential changes in inflow design flood under future climate projections for Darbandikhan Dam. *J. Hydrol.* **2015**, *528*, 45–51. [CrossRef]
4. Huziy, O.; Sushama, L.; Khaliq, M.N.; Laprise, R.; Lehner, B.; Roy, R. Analysis of streamflow characteristics over Northeastern Canada in a changing climate. *Clim. Dyn.* **2013**, *40*, 1879–1901. [CrossRef]
5. Yin, J.; Guo, S.; He, S.; Guo, J.; Hong, X.; Liu, Z. A copula-based analysis of projected climate changes to bivariate flood quantiles. *J. Hydrol.* **2018**, *566*, 23–42. [CrossRef]
6. Yin, J.; Guo, S.; Liu, Z.; Yang, G.; Zhong, Y.; Liu, D. Uncertainty Analysis of Bivariate Design Flood Estimation and its Impacts on Reservoir Routing. *Water Resour. Manag.* **2018**, *32*, 1795–1809. [CrossRef]
7. Guo, A.J.; Chang, J.X.; Wang, Y.M.; Huang, Q.; Li, Y.Y. Uncertainty quantification and propagation in bivariate design flood estimation using a Bayesian information-theoretic approach. *J. Hydrol.* **2020**, *584*, 124677. [CrossRef]
8. Das, J.; Jha, S.; Goyal, M.K. Non-stationary and copula-based approach to assess the drought characteristics encompassing climate indices over the Himalayan states in India. *J. Hydrol.* **2019**, *580*, 124356. [CrossRef]
9. Jha, S.; Das, J.; Goyal, M.K. Low frequency global-scale modes and its influence on rainfall extremes over India: Nonstationary and uncertainty analysis. *Int. J. Clim.* **2020**, *41*, 1873–1888. [CrossRef]
10. Huang, K.; Chen, L.; Zhou, J.; Zhang, J.; Singh, V.P. Flood hydrograph coincidence analysis for mainstream and its tributaries. *J. Hydrol.* **2018**, *565*, 341–353. [CrossRef]
11. Requena, A.I.; Mediero, L.; Garrote, L. A bivariate return period based on copulas for hydrologic dam design: Accounting for reservoir routing in risk estimation. *Hydrol. Earth Syst. Sci.* **2013**, *17*, 3023–3038. [CrossRef]
12. Salvadori, G.; De Michele, C. Multivariate multiparameter extreme value models and return periods: A copula approach. *Water Resour. Res.* **2010**, *46*, 219–233. [CrossRef]
13. Zhang, L.; Singh, V.P. Bivariate Flood Frequency Analysis Using the Copula Method. *J. Hydrol. Eng.* **2006**, *11*, 150–164. [CrossRef]
14. Zhang, X.; Duan, K.; Dong, Q. Comparison of nonstationary models in analyzing bivariate flood frequency at the Three Gorges Dam. *J. Hydrol.* **2019**, *579*, 124208. [CrossRef]
15. Hu, Y.; Liang, Z.; Huang, Y.; Yao, Y.; Wang, J.; Li, B. A nonstationary bivariate design flood estimation approach coupled with the most likely and expectation combination strategies. *J. Hydrol.* **2021**, *605*, 127325. [CrossRef]
16. Brunner, M.I.; Sikorska, A.E.; Seibert, J. Bivariate analysis of floods in climate impact assessments. *Sci. Total Environ.* **2017**, *616–617*, 1392–1403. [CrossRef]
17. Duan, K.; Mei, Y.; Zhang, L. Copula-based bivariate flood frequency analysis in a changing climate—A case study in the Huai River Basin, China. *J. Earth Sci.* **2016**, *27*, 37–46. [CrossRef]
18. Parent, E.; Favre, A.-C.; Bernier, J.; Perreault, L. Copula models for frequency analysis what can be learned from a Bayesian perspective? *Adv. Water Resour.* **2014**, *63*, 91–103. [CrossRef]
19. Villarini, G. On the seasonality of flooding across the continental United States. *Adv. Water Resour.* **2016**, *87*, 80–91. [CrossRef]
20. Yan, L.; Xiong, L.; Liu, D.; Hu, T.; Xu, C.-Y. Frequency analysis of nonstationary annual maximum flood series using the time-varying two-component mixture distributions. *Hydrol. Process.* **2016**, *31*, 69–89. [CrossRef]
21. Alila, Y.; Mtiraoui, A. Implications of heterogeneous flood-frequency distributions on traditional stream-discharge prediction techniques. *Hydrol. Process.* **2002**, *16*, 1065–1084. [CrossRef]

22. Villarini, G.; Smith, J.A. Flood peak distributions for the eastern United States. *Water Resour. Res.* **2010**, *46*, W06504. [CrossRef]
23. Zeng, H.; Feng, P.; Li, X. Reservoir Flood Routing Considering the Non-Stationarity of Flood Series in North China. *Water Resour. Manag.* **2014**, *28*, 4273–4287. [CrossRef]
24. Ping, F.; Xin, L. Bivariate frequency analysis of non-stationary flood timeseries based on Copula methods. *J. Hydraul. Eng.* **2013**, *44*, 1137–1147, (In Chinese with English abstract). [CrossRef]
25. Li, J.; Zheng, Y.; Wang, Y.; Zhang, T.; Feng, P.; Engel, B.A. Improved Mixed Distribution Model Considering Historical Extraordinary Floods under Changing Environment. *Water* **2018**, *10*, 1016. [CrossRef]
26. Yan, L.; Xiong, L.; Ruan, G.; Xu, C.-Y.; Yan, P.; Liu, P. Reducing uncertainty of designfloods of two-component mixture distributions by utilizing flood timescale to classify flood types in seasonally snow covered region. *J. Hydrol.* **2019**, *574*, 588–608. [CrossRef]
27. Jiang, C.; Xiong, L.; Xu, C.-Y.; Guo, S. Bivariate frequency analysis of nonstationary low-flow series based on the time-varying copula. *Hydrol. Process.* **2014**, *29*, 1521–1534. [CrossRef]
28. Wen, T.; Jiang, C.; Xu, X. Nonstationary Analysis for Bivariate Distribution of Flood Variables in the Ganjiang River Using Time-Varying Copula. *Water* **2019**, *11*, 746. [CrossRef]
29. Xie, P.; Chen, G.; Lei, H. Hydrological alteration analysis method based on Hurst coefficient. *J. Basic Sci. Eng.* **2009**, *17*, 32–39. (In Chinese)
30. Hurst, H.E.; Black, R.P.; Simaika, Y.M. Long-Term Storage: An Experimental Study. *J. R. Stat. Soc. Ser. A (Gen.)* **1966**, *129*, 591–593. [CrossRef]
31. Bărbulescu, A.; Serban, C.; Mafteiv, C. Statistical analysis and evaluation of Hurst coefficient for annual and monthly precipitation time series. *WSEAS Trans. Math.* **2010**, *9*, 791–800. [CrossRef]
32. Pettitt, A.N. A Non-Parametric Approach to the Change-Point Problem. *J. R. Stat. Soc. Ser. C (Appl. Stat.)* **1979**, *28*, 126–135. [CrossRef]
33. Brown, M.B.; Forsythe, A.B. Robust Tests for the Equality of Variances. *J. Am. Stat. Assoc.* **1974**, *69*, 364–367. [CrossRef]
34. Fraedrich, K.; Jiang, J.; Gerstengarbe, F.-W.; Werner, P.C. Multiscale detection of abrupt climate changes: Application to River Nile flood levels. *Int. J. Climatol.* **1997**, *17*, 1301–1315. [CrossRef]
35. Salvadori, N. Evaluation of Non-Stationarity in Annual Maximum Flood Series of Moderately Impaired Watersheds in the Upper Midwest and Northeastern United States. Master's Thesis, Michigan Technological University, Houghton, MI, USA, 2013.
36. Singh, K.P.; Sinclair, R.A. Two-Distribution Method for Flood Frequency Analysis. *J. Hydraul. Div.* **1972**, *98*, 29–44. [CrossRef]
37. Evin, G.; Merleau, J.; Perreault, L. Two-component mixtures of normal, gamma, and Gumbel distributions for hydrological applications. *Water Resour. Res.* **2011**, *47*, W08525. [CrossRef]
38. Villarini, G.; Smith, J.A.; Baeck, M.L.; Krajewski, W.F. Examining Flood Frequency Distributions in the Midwest U.S. *JAWRA J. Am. Water Resour. Assoc.* **2011**, *47*, 447–463. [CrossRef]
39. Grego, J.M.; Yates, P.A. Point and standard error estimation for quantiles of mixed flood distributions. *J. Hydrol.* **2010**, *391*, 289–301. [CrossRef]
40. Sklar, A. Fonctions de répartition à n dimensions et leurs marges. *Publ. del'Institut Statistique L'Université Paris* **1959**, *8*, 229–231.
41. Nelsen, R.B. *An Introduction to Copulas*; Springer: New York, NY, USA, 1999.
42. Nelsen, R.B. *An Introduction to Copulas*, 2nd ed.; Springer: New York, NY, USA, 2006.
43. Shiau, J.-T.; Feng, S.; Nadarajah, S. Assessment of hydrological droughts for the Yellow River, China, using copulas. *Hydrol. Process.* **2007**, *21*, 2157–2163. [CrossRef]
44. Durante, F.; Sempi, C. *Principles of Copula Theory*; Chapman and Hall/CRC: Boca Raton, FL, USA, 2015. [CrossRef]
45. Salvadori, G.; De Michele, C.; Kottegoda, N.T.; Rosso, R. *Extremes in Nature: An Approach Using Copulas*; Water Science and Technology Library; Springer: Berlin, Germany, 2007. [CrossRef]
46. Qi, W.; Liu, J. A non-stationary cost-benefit based bivariate extreme flood estimation approach. *J. Hydrol.* **2018**, *557*, 589–599. [CrossRef]
47. Poulin, A.; Huard, D.; Favre, A.-C.; Pugin, S. Importance of Tail Dependence in Bivariate Frequency Analysis. *J. Hydrol. Eng.* **2007**, *12*, 394–403. [CrossRef]
48. Latif, S.; Mustafa, F. Bivariate joint distribution analysis of the flood characteristics under semiparametric copula distribution framework for the Kelantan River basin in Malaysia. *J. Ocean Eng. Sci.* **2020**, *6*, 128–145. [CrossRef]
49. Lilliefors, H.W. On the Kolmogorov-Smirnov Test for Normality with Mean and Variance Unknown. *J. Am. Stat. Assoc.* **1967**, *62*, 399–402. [CrossRef]
50. Salvadori, G.; De Michele, C. Frequency analysis via copulas: Theoretical aspects and applications to hydrological events. *Water Resour. Res.* **2004**, *40*, W12511. [CrossRef]
51. Shiau, J.T. Return period of bivariate distributed extreme hydrological events. *Stoch. Environ. Res. Risk Assess.* **2003**, *17*, 42–57. [CrossRef]
52. Xiao, Y.; Guo, S.L.; Liu, P.; Fang, B. Derivation of design flood hydrograph based on Copula function. *Eng. J. Wuhan Univ.* **2007**, *4*, 13–17, (In Chinese with English abstract).
53. Hosking, J.R.M. L-Moments: Analysis and Estimation of Distributions Using Linear Combinations of Order Statistics. *J. R. Stat. Soc. Ser. B Stat. Methodol.* **1990**, *52*, 105–124. [CrossRef]

*Article*

# Frequency Analysis of the Nonstationary Annual Runoff Series Using the Mechanism-Based Reconstruction Method

Shi Li [1] and Yi Qin [2],*

[1]  PowerChina Northwest Engineering Corporation Limited, Xi'an 710065, China; lishi0@163.com
[2]  State Key Laboratory of Eco-Hydraulics in Northwest Arid Region of China, Xi'an University of Technology, Xi'an 710048, China
*  Correspondence: qinyi@xaut.edu.cn; Tel.: +86-135-7199-1500

**Abstract:** Due to climate change and human activities, the statistical characteristics of annual runoff series of many rivers around the world exhibit complex nonstationary changes, which seriously impact the frequency analysis of annual runoff and are thus becoming a hotspot of research. A variety of nonstationary frequency analysis methods has been proposed by many scholars, but their reliability and accuracy in practical application are still controversial. The recently proposed Mechanism-based Reconstruction (Me-RS) method is a method to deal with nonstationary changes in hydrological series, which solves the frequency analysis problem of the nonstationary hydrological series by transforming nonstationary series into stationary Me-RS series. Based on the Me-RS method, a calculation method of design annual runoff under the nonstationary conditions is proposed in this paper and applied to the Jialu River Basin (JRB) in northern Shaanxi, China. From the aspects of rationality and uncertainty, the calculated design value of annual runoff is analyzed and evaluated. Then, compared with the design values calculated by traditional frequency analysis method regardless of whether the sample series is stationary, the correctness of the Me-RS theory and its application reliability is demonstrated. The results show that calculation of design annual runoff based on the Me-RS method is not only scientific in theory, but also the obtained design values are relatively consistent with the characteristics of the river basin, and the uncertainty is obviously smaller. Therefore, the Me-RS provides an effective tool for annual runoff frequency analysis under nonstationary conditions.

**Keywords:** frequency analysis; annual runoff; nonstationary; mechanism-based reconstruction

**Citation:** Li, S.; Qin, Y. Frequency Analysis of the Nonstationary Annual Runoff Series Using the Mechanism-Based Reconstruction Method. *Water* **2022**, *14*, 76. https://doi.org/10.3390/w14010076

Academic Editor: Renato Morbidelli

Received: 28 November 2021
Accepted: 29 December 2021
Published: 2 January 2022

**Publisher's Note:** MDPI stays neutral with regard to jurisdictional claims in published maps and institutional affiliations.



## 1. Introduction

River runoff, as the most important form and component of water resources, has changed significantly in a number of rivers worldwide due to the impact of climate change and human activities. The statistical characteristics of annual runoff series exhibit complex, nonstationary changes. This change not only poses a serious threat to regional water resources security [1–4] but also leads to the inability to analyze, predict, and manage water resources effectively, which is because the analysis method of the traditional design annual runoff based on the stationary assumption is no longer applicable. If the traditional frequency analysis method is forcibly used to calculate the design annual runoff and taken as the basis of hydraulic engineering design and water resources planning and management, the rationality and safety of design or planning will be questioned. In China, for example, the annual runoff of many rivers shows a decreasing trend. If this reduction is ignored, the calculated design value will be significantly larger. The larger design annual runoff is bound to lead to misjudgment of water resources shortage, which will further aggravate the current serious water safety problem.

Many scholars have realized the nonstationary problems and carried out corresponding research work. The most representative is the time-varying moments method [5–7],

---

whose basic idea is to assume that the distribution type of hydrological variable is unchanged, but the statistical parameters of the distribution change over time or other covariates. Vogel et al. [8] analyzed the time-varying trend of flood peak discharge series and design flood in the United States using a time-varying moments model established by the exponential model combined with the two-parameter lognormal distribution. They concluded that the flood magnitude in many areas is increasing, and the 100-year flood may become more common. Zeng et al. [9] constructed the time-varying moments model of flood series of Xidayang reservoir based on P-III distribution and verified that the time-varying P-III distribution model fits better to the flood series than the traditional P-III curve. The generalized additive models for location, scale, and shape (GAMLSS) [10] provide a way for the time-varying moments method to link the physical causes, that is, to establish the relationship between statistical parameters and physical factor covariates, including the climate change and human activity factors [11,12].

Although the time-varying moments method can describe the nonstationary changes of hydrological series well, it is difficult to apply in practice because there might be different design values every year for the same design standard [13,14]. For example, Villarini et al. [15] found that the 100-year design flood in the Little Sugar Creek in the United States could range from the minimum of 2.1 $m^3 s^{-1}$ $km^{-2}$ (1957) to the maximum of 5.1 $m^3 s^{-1}$ $km^{-2}$ (2007). In order to solve this problem, many scholars have proposed the methods of expected waiting time (EWT) [16,17] and expected number of exceedance (ENE) [18,19] to redefine the return period concept. Some studies have also proposed equivalent reliability (ER) [20], design life level (DLL) [21], and average design life level (ADLL) [22] methods based on the concept of reliability. These methods effectively solve the multi-value problem of hydrological design, but some controversy remains. Some studies believed that the trend exhibited in an observed hydrological series, which is often regarded as a type of nonstationarity, may actually be periodic frequency swings in a stationary process [23]; even the word "trend" is not well defined [24]. In practice, the design quantile obtained for a given reliability over the design lifetime varies with the choice of initial time and the curve type used for fitting the relationship between the statistical parameters and the covariates. This means that the reliability of the future design values depends heavily on the time-varying characteristics of statistical parameters; however, the uncertainty about the prediction of statistical parameters is greatly increased due to the lack of ergodicity of the time series. As Serinaldi and Kilsby [25] pointed out, when the model structure cannot be inferred in a deductive manner, and nonstationary models are fitted by inductive inference, the model structure introduces an additional source of uncertainty so that the resulting nonstationary models can provide no practical enhancement of the credibility and accuracy of the predicted extreme quantiles, whereas possible model misspecification can easily lead to physically inconsistent results.

Obviously, the core problem of the nonstationary frequency analysis problem is the non-simplicity of the sample series. If the nonstationary sample series can be converted into a stationary one, all the aforementioned problems will no longer exist because there are already mature analytical theories and technical methods for this simple series. To retain the advantages of traditional frequency analysis method and avoid the weaknesses of current nonstationary frequency analysis methods, Qin and Li [26] proposed a Mechanism-based Reconstruction (Me-RS) method to reconstruct nonstationary series into stationary series according to the physical mechanism. In this paper, we propose a complete nonstationary frequency analysis method for annual runoff series based on the theory of Me-RS. We then took the nonstationary annual runoff series in the Jialu River Basin (JRB) in northern Shaanxi as an example and analyzed the uncertainty of the deduced design annual runoff by Bootstrap method to verify the practicability and reliability of the nonstationary frequency analysis method proposed in this paper.

## 2. Methodology

### 2.1. Me-RS Method

The core thought of the Me-RS method is based on the assumption that "the nonstationary changes of hydrological variable $Y$ is caused by the nonstationary changes of its influencing factors." For example, the trend variations in runoff may be due to the trends in some of its meteorological factors, such as precipitation, temperature, etc., while the abrupt changes in some specific stages may be caused by the intensification of some human activities, such as the sudden increase of water consumption due to the construction of irrigation projects, and the adjustment and utilization of runoff by reservoir projects. From the perspective of causality, these meteorological factors or human activity factors are the root of runoff change, and these influence factors always act on runoff in their specific ways. In the Me-RS method, the action function describes the mechanism of the influence factor on the research variable is defined as the Mechanism function.

In general, the change in a hydrological variable $Y$ is the result of the influence of multiple factors $X_1$, $X_2$, ... , $X_i$, as shown in Figure 1. Under the condition that the influence of other factors remains unchanged, the effect of single influence factor $X_i$ on the hydrological variable $Y$ is described as a Mechanism function $f_i(X_i)$. The Mechanism function, which represents the physical mechanism of hydrological phenomena, will never change. For instance, in the flow discharge $Q = AV$, $A$ is the cross-sectional area, and $V$ is the flow velocity. When $V$ or $A$ or both change with time, $Q(t)$ also changes with the values of $A(t)$ or $V(t)$ or both, but the Mechanism function $f_A(\cdot)$ or $f_V(\cdot)$ remains unchanged. In other words, $V$ and $A$ may change with time in the specific environment, but the mechanism that the flow discharge equal to the multiplication of these two Mechanism functions will never change. Therefore, the physical mechanism will remain unchanged no matter how the state of $X_i$ changes with time.

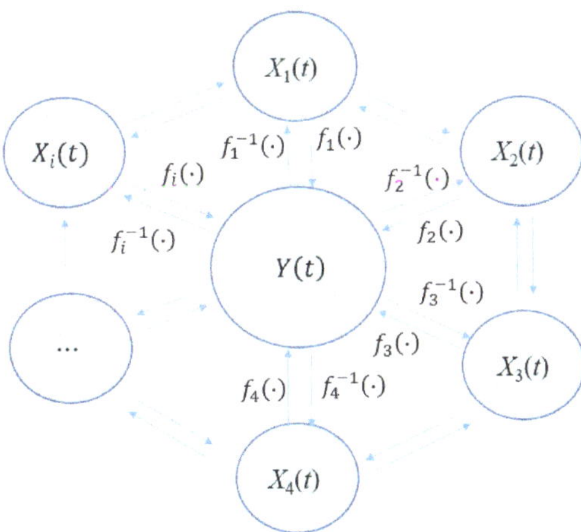


**Figure 1.** The relationship between hydrological variable $Y$ and its influencing factors.

According to the above concept of Mechanism function, the relationship model between $Y$ and the Mechanism function of multiple influence factors can be established. Due to the complexity of hydrological systems and the analytical capability, statistical models are usually used to describe the relationship between $Y$ and its explanatory variables. There are two general forms of statistical models: the superposition model and the multiplication model.

$$Y(t) = \sum_{1}^{N} f_i(X_i(t)),$$
(1)

$$Y(t) = \prod_{1}^{N} f_i(X_i(t)), \tag{2}$$

where $N$ is the number of influencing factors of hydrological variable $Y$, and $f_i(X_i(t))$ is the Mechanism function of influencing factor $X_i$. Assuming that $n$ of $N$ influence factors are known, the effect of the remaining $N\text{-}n$ unrecognized influence factors shows randomness and is denoted by $\xi$. Then, the Equations (1) and (2) can be written as:

$$Y(t) = \sum_{1}^{n} f_i(X_i(t)) + \xi, \tag{3}$$

$$Y(t) = \prod_{1}^{n} f_i(X_i(t))\xi. \tag{4}$$

Considering the universal interaction among the hydrological elements in the hydrological system, there is no absolute independence among the explanatory variables, and their interactions make the hydrological system exhibit a highly complex nonlinear state. Therefore, the multiplication model is more applicable to describe the hydrological system than the superposition model. Since the nonstationary change of $Y(t)$ is caused by the nonstationary changes in some influencing factors, if the effects of all the nonstationary factors are removed from $Y(t)$, as shown in Equation (5), the remainder will show stationarity.

$$\frac{Y(t)}{\prod_1^m f_i(X_i(t))} = \prod_{m+1}^{n} f_{m+1}(x_{m+1}(t)) \cdot \xi = \delta, \tag{5}$$

where $m$ is the number of nonstationary influencing factors. The right side of Equation (5) is the multiplication of the Mechanism functions of the remaining stationary factors and other unrecognized influence factors, which is denoted as $\delta$, and is usually considered as a natural random variable with a probability distribution characterized by the mean $\mu$ and variance $\sigma^2$. If only a subset of the nonstationary factor $X_1, X_2, \ldots, X_l$ $(l < m)$ is considered, it is still a nonstationary series. However, when the most important nonstationary factors are removed, the remaining series can achieve the statistical stationary state. According to the Me-RS idea, the Me-RS function of $Y(t)$ is defined as

$$RS(t) = \frac{Y(t)}{\prod_1^m f_i(X_i(t))}. \tag{6}$$

The new stationary series reconstructed by the Me-RS function is called the Me-RS series, denoted as $RS_t$. Theoretically, when all the influencing factors causing nonstationary changes in $Y$ are identified (i.e., $l = m$), and the Mechanism functions are constructed accurately, the Me-RS series $\{RS_t\}$ $(t = 1, 2, \ldots)$ calculated by the Me-RS function is stationary and can be used in any case where a stationary series is required. Although it is impossible to obtain an absolute stationary process due to the limitations of our understandings and existing methods, it is practical to achieve the statistical stationary state. As the correct explanatory variable Mechanism functions are continuously added to the Me-RS function, the Me-RS series will gradually tend to be statistically stationary and closer to a random noise.

Due to the causal relationship between the research variable $Y$ and its influence factors, the nonstationarity (linear, nonlinear, or abrupt change) of the research variable $Y$ is consistent with the corresponding nonstationarity of the influence factors. After removing the influence of the nonstationary factors according to the Me-RS method, the numerical characteristic of the Me-RS series is a constant, i.e.,

$$E[RS(t)] = E\left[\frac{Y(t)}{\prod_1^m f_i(X_i(t))}\right] = E(\delta) = \mu, \tag{7}$$

$$Var[RS(t)] = Var\left[\frac{Y(t)}{\prod_1^m f_i(X_i(t))}\right] = Var(\delta) = \sigma^2. \tag{8}$$

Therefore, the Me-RS method is not limited to a specific type of nonstationarity and is effective for any kind of nonstationary change.

### 2.2. Frequency Analysis of Nonstationary Annual Runoff Series

After obtaining the stationary Me-RS series, the traditional method can be used to conduct the frequency analysis, including the distribution and parameter estimation, and then obtain the design value $RS_p$ of the Me-RS series under a certain standard (frequency or return period). Based on the assumption that "the nonstationary change of the hydrological variable $Y$ is caused by the nonstationary change of its influence factor," as long as the value of the influence factor $X_i$ at a certain time is determined, the design value of the nonstationary annual runoff series $Y(t)$ can be obtained according to the definition of the Me-RS function as follows:

$$Y_p = RS_p \cdot \prod_{i=1}^{m} f_i\left(X_{i,design}\right),\tag{9}$$

in which the $X_{i,design}$ is the value of the influence factor at the design stage, which can be the current value or the predicted value.

### 2.3. Uncertainty Analysis Using the Bootstrap Method

In order to provide relatively robust design results to engineering design and water resources management, it is necessary to evaluate the rationality of the Me-RS method from the perspective of uncertainty of design value. We used the Bootstrap method [27] to quantitatively analyze the uncertainty of the design value, that is, resample the Me-RS series $RS_t$ $N$ times to obtain $N$ sample series, calculate the design value $RS_{p,j}$ ($j = 1, 2, \ldots, N$) of each sample series, obtain the design annual runoff $Y_{p,j}$ ($j = 1, 2, \ldots, N$) according to Equation (9), and then deduce the uncertainty confidence interval $\left(Y^*_{p,N\alpha/2}, Y^*_{p,N(1-\alpha/2)}\right)$ of design value under the significance level $\alpha$.

### 2.4. Nonstationary Analysis

Strictly speaking, the Mechanism function should be determined by theoretical derivation or experimental analysis. However, due to the complexity of hydrological behaviors and our limited understandings, it is difficult to obtain the absolutely accurate mathematical expressions of the Mechanism function that represents the physical mechanism of hydrological behaviors. Since the influence law between hydrological elements can be implied in the statistical law, the statistical relationship between $Y$ and $X_i$ can be used to estimate the Mechanism function. It is clear that the estimated Mechanism function has certain uncertainty, so the nonstationary test of $RS_t$ is necessary. The tests used in this paper include the Mann-Kendall (M-K) test [28,29] for the trend analysis of the first moment, the Pettitt test [30] for the change-point analysis, and the Breusch-Panan (B-P) test [31] for the trend analysis of the second moment. The significance level $\alpha$ of each test is 0.05.

## 3. Study Area and Data

### 3.1. Study Area

In this paper, we took the JRB in the Yulin region of Shaanxi Province in China as the study area and conducted nonstationary frequency analysis of the annual runoff series by the Me-RS method. The Jialu River is located along the Yellow River between Hekou and Longmen Station and the southern edge of the Mu Us Desert. The river originates from Duanqiao Village, Yulin City, Shaanxi Province, and flows from northwest to southeast and joins the Yellow River at Muchangwan in Jia County. The JRB has an approximate river length of 93 km and a drainage area of 1134 km$^2$ and lies between the geographical coordinates of 37°58′–38°29′ N and 109°56′–110°32′ E (Figure 2). Shenjiawan Hydrological station is the control station for this area. The average annual precipitation in this basin is about 402.3 mm, 75% of which falls during the flood season. Most of the rainfall is in the form of short, intense rainstorms.

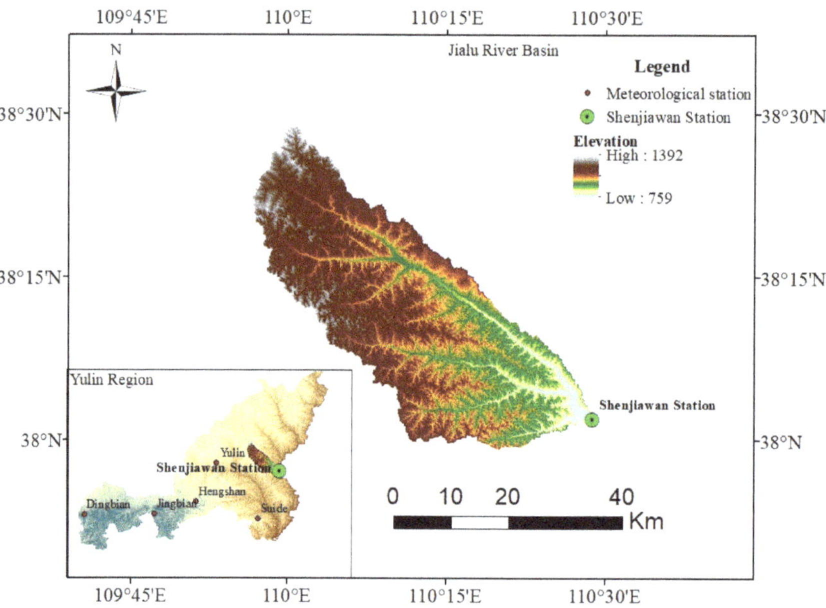



**Figure 2.** Geographic location of the JRB and meteorological and hydrological data.

### 3.2. Data

In the Loess Plateau area where the JRB is located, the serious water and soil loss are common. To control this phenomenon, dozens of check dams have been built in JRB. As an engineering measure of water and soil conservation, the check dams can intercept and deposit the sediment in front of the check dams, and the upstream water is slowly drained out by the horizontal pipe. However, in fact, the upstream water is often stored in front of the dam for daily use of residents; that is, the check dam is used as a small reservoir. The number of check dams is very large, with more than 20,000 in Yulin, Shaanxi Province and more than 700 in JRB alone. The continuous construction of the check dam projects results in a continuous downtrend of annual runoff series in JRB, as shown in Figure 3. The nonstationarity test methods in Section 2.4 were adopted to analyze the nonstationarity of the annual runoff series in JRB, as shown in Table 1.

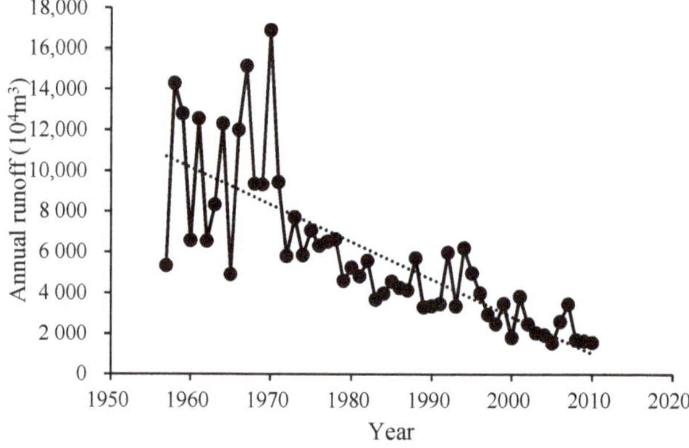



**Figure 3.** The annual runoff series of the JRB from 1959 to 2010.

**Table 1.** The nonstationary tests of the annual runoff series in JRB.

| Object | M-K Test for Mean | | B-P Test for Variance | | Pettitt Test for Change Point | |
|---|---|---|---|---|---|---|
| | $Z$ | $Z_\alpha$ | $\chi$ | $\chi$ | Change-Point | $p$-Value |
| Annual runoff series | −7.49 | 1.96 | 23.27 | 3.84 | 1982 | $6.80 \times 10^{-8}$ |
| Annual precipitation series | 0.50 | 1.96 | 0.01 | 3.84 | - | - |

According to Figure 3 and Table 1, the annual runoff series of the JRB exhibits a strong nonstationary change. The first and second moments both show a significant trend, and an abrupt change occurred in 1982. Before the Me-RS analysis of annual runoff series, it is necessary to identify the main influence factors and determine the Mechanism functions. The influencing factors of runoff mainly include climatic factors and underlying surface factors. Among the climatic factors, precipitation is a direct factor affecting the runoff. The annual precipitation series from 1969 to 2010 was collected in this study and is shown in Figure 4. The same nonstationary analysis is shown in Table 1.

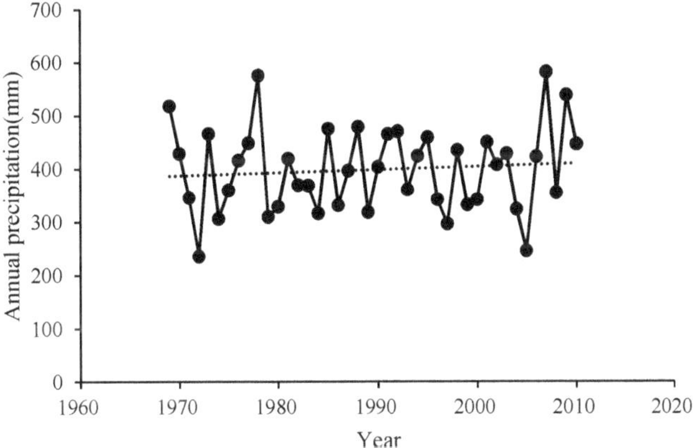

**Figure 4.** The annual precipitation series of the JRB from 1969 to 2010.

According to the test results in Table 1, there is no significant nonstationary change in annual precipitation in JRB, so precipitation is not the main influencing factor causing the nonstationary change in runoff in JRB. Therefore, we turned our attention to the underlying surface factor. Through the field survey, it is concluded that human activities, such as the soil and water conservation engineering measures represented by check dams constructed in JRB in recent years, are the cause of the nonstationarity in annual runoff. The influence of the check dams on the annual runoff is believed to be controlled by the storage capacity and the basin area of the check dams. Therefore, the reservoir index (*RI*) proposed by López and Francés [11] is used to quantify the impact of check dams.

$$I = \sum_{i=1}^{N} \left( \frac{A_i}{A_T} \right) \left( \frac{C_i}{C_T} \right), \tag{10}$$

where $A_i$ is the control area of each reservoir, $A_T$ is the basin area, $C_i$ is the capacity of each reservoir, $C_T$ is the average annual runoff of the basin, and $N$ is the number of reservoir in the basin. The *RI* series (Figure 5) exhibits a monotonic upward trend.

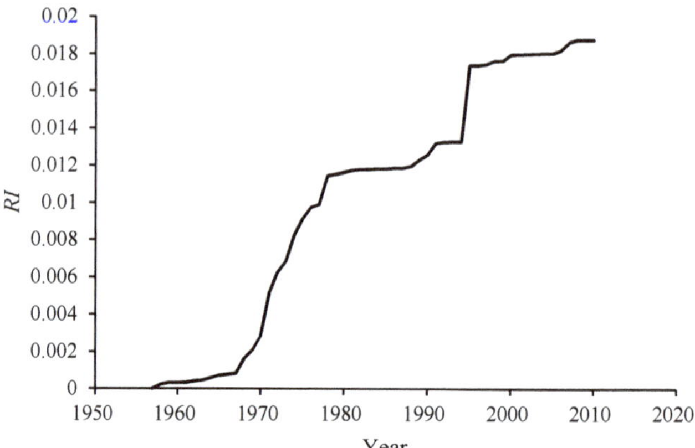

**Figure 5.** Time-varying characteristic of the *RI* series in the JRB.

## 4. Reconstruction of the Annual Runoff Series

### 4.1. The Mechanism Function of RI

As pointed out in Section 2.4, there are almost no absolutely accurate Mechanism function determined by theoretical derivation or experimental analysis at present. Still, we can estimate the approximation of the Mechanism function one by one through regression approach. We first established the relationship between $Y$ and $X_1$, $f_1(X_1(t))$, as the Mechanism function of $X_1$ and then removed the influence of the first factor, i.e., $\frac{Y(t)}{f_1(X_1(t))}$; then, the Mechanism function $f_2(X_2(t))$ of the second factor $X_2$ was established according to the relationship between the remained series $\frac{Y(t)}{f_1(X_1(t))}$ and the second factor $X_2$; this process was repeated iteratively until the reconstructed series achieved the stationarity. In the case of this study, *RI* was taken as the main factor causing the nonstationary change in annual runoff in JRB, and the Mechanism function of *RI* was estimated by regression analysis, as shown in Figure 6 and Equation (11).

$$f(RI) = e^{-78.79RI},\tag{11}$$

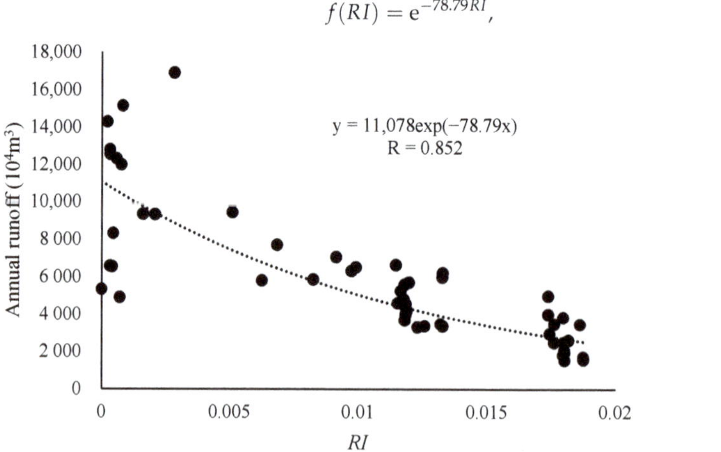


**Figure 6.** The estimated Mechanism function of the *RI* on the annual runoff.

### 4.2. The Me-RS Function and the Me-RS Series

After obtaining the Mechanism function $f(RI)$, the Me-RS function of the annual runoff in JRB was determined according to Equation (6) as follows:

$$RS_{RI}(t) = \frac{R(t)}{e^{-78.79\,RI(t)}}.$$

(12)

We input the values of annual runoff and *RI* into the Me-RS function to obtain the Me-RS series of annual runoff, as shown in Figure 7. Since the influence of nonstationary factor *RI* is removed, the Me-RS series of annual runoff should be stationary; otherwise, other nonstationary factors should continue to be considered. Therefore, the three nonstationary test methods introduced in Section 2.4 were used to analyze the stationarity characteristics of the Me-RS series. The test results are list in Table 2.

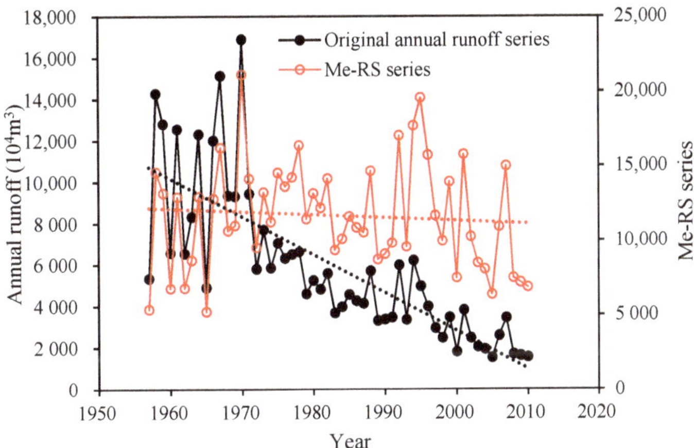


**Figure 7.** Comparison of the Me-RS series and the original annual runoff series in JRB.

**Table 2.** The nonstationary tests of the Me-RS series of the annual runoff in JRB.

| Object | M-K Test for Mean | | B-P Test for Variance | | Pettitt Test for Change Point | |
|---|---|---|---|---|---|---|
| | $Z$ | $Z_\alpha$ | $\chi$ | $\chi$ | Change-Point | $p$-Value |
| Annual runoff series | −7.49 | 1.96 | 23.27 | 3.84 | 1982 | $6.80 \times 10^{-8}$ |
| Me-RS series | −0.97 | 1.96 | 0.12 | 3.84 | - | - |

Compared with Table 1, after the reconstruction with *RI*, the original annual runoff series with significant first and second moment trends and significant change-point achieved the stationarity in all aspects, which verified that the Me-RS method is effective for any kind of nonstationary change. In this case, the Me-RS series reconstructed by single-factor *RI* has excellent stationarity, so no additional factors are added.

## 5. Frequency Analysis of the Annual Runoff Series

*5.1. Calculation of the Design Value of the Me-RS Series*

Once the Me-RS series was tested to be stationary, the design value of the Me-RS series could be calculated according to the traditional frequency analysis method. We selected four distributions, the Pearson type III (P-III), Weibull (WEI), Log-normal (LNO), and Gumbel (GU) distributions, as the alternative distributions of the Me-RS series (Table 3).

The distribution parameters were estimated by the L-moments method [32]. To evaluate the fitting accuracy of the four alternative distributions, the Kolmogorov-Smirnov test [33], the Nash-Sutcliffe efficiency [34], and the root mean square error were used to determine the optimal distribution. Based on our analysis, the optimal distribution of the Me-RS series of annual runoff in JRB is the WEI distribution (Table 4), and the Q-Q plot in Figure 8 also shows the good fitting effect of WEI distribution.

**Table 3.** Probability density functions of the alternative distributions.

| Distribution | Probability Density Function |
|---|---|
| P-III | $f(x) = \frac{1}{\sigma\lvert\mu\kappa\rvert\Gamma(1/\kappa^2)}\left(\frac{x-\mu}{\mu\sigma\kappa} + \frac{1}{\kappa^2}\right)^{\frac{1}{\kappa^2}-1} exp\left[-\left(\frac{z-\mu}{\mu\sigma\kappa} + \frac{1}{\kappa^2}\right)\right], -\infty < x < \infty, \sigma > 0, \kappa \neq 0, \frac{x-\mu}{\mu\sigma\kappa} + \frac{1}{\kappa^2} \geq 0$ |
| WEI | $f(x) = \frac{\kappa}{\sigma}\left(\frac{x-\mu}{\sigma}\right)^{\kappa-1} exp\left[-\left(\frac{x-\mu}{\sigma}\right)^{\kappa}\right], x > 0, \mu > 0, \sigma > 0, -\infty < \kappa < \infty$ |
| LNO | $f(x) = \frac{\beta^{\mu}}{(x-\mu)\sigma\sqrt{2\pi}} exp\left[\frac{-ln[(x-\mu)-\kappa]^2}{2\sigma^2}\right], x > \mu, \sigma > 0$ |
| GU | $f(x) = \frac{1}{\sigma} exp\left\{-\left(\frac{x-\mu}{\sigma}\right) - exp\left[-\left(\frac{x-\mu}{\sigma}\right)\right]\right\}, -\infty < x < \infty, -\infty < \mu < \infty, \sigma > 0$ |

**Table 4.** Parameter estimation of the optimal distribution for the Me-RS series of the JRB annual runoff.

| Object | Optimal Distribution | Estimated Parameters |
|---|---|---|
| Me-RS series $RS_{RI,t}$ | WEI | $\mu = 2889.237, \sigma = 9846.289, \kappa = 2.598$ |

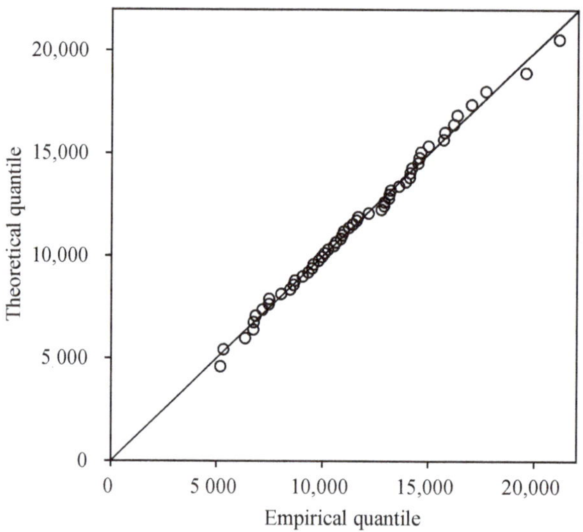


**Figure 8.** Q-Q plot of the theoretical and empirical quantiles for the Me-RS series of the JRB annual runoff.

According to the optimal distribution and the estimated parameters, we can determine the design quantiles for various return periods in the Me-RS series (Figure 9).

*5.2. Calculation of the Design Annual Runoff*

Once the design value of the stationary Me-RS series was calculated, and we could then determine the corresponding quantiles for the original nonstationary annual runoff series according to Equation (9). Therefore, it was necessary to determine the value of the *RI* at the design stage. Considering that a large number of check dams has been constructed in JRB, and the construction has been saturated in recent years, the *RI* calculated based on the control area and the storage capacity of the check dams should be basically maintained at the level of 2010, so the *RI* data in 2010, as shown in Figure 5, were taken as the value at the design stage. Then, according to Equation (9), the design annual runoff is shown in Figure 10.

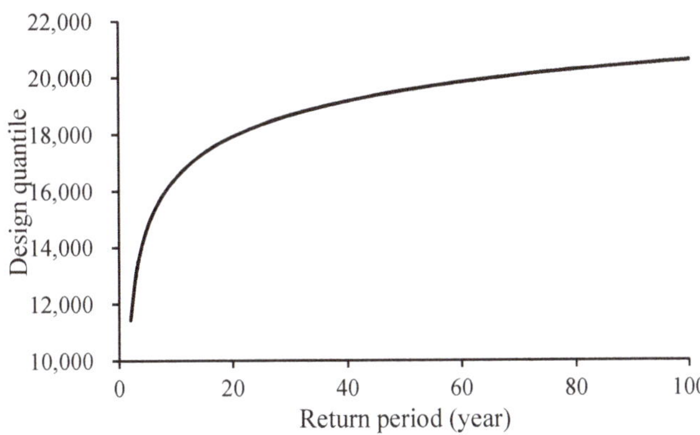

**Figure 9.** The design quantiles of different return periods of the Me-RS series in JRB.

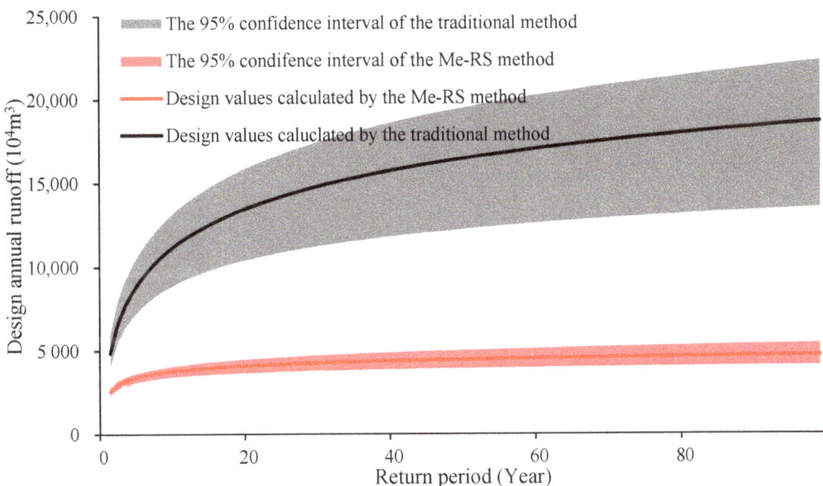


**Figure 10.** The design values of different return periods of the Me-RS series in JRB with the 95% bootstrapped confidence intervals.

## 6. Discussion

### 6.1. Rationality Analysis of the Design Annual Runoff

The rationality of the design value obtained by the Me-RS method can be analyzed from two aspects. One is from the theoretical perspective: compared with the traditional frequency analysis method, the Me-RS method only conducts the de-nonstationarity transformation on the sample series to ensure that the sample series used for frequency analysis is the simple sample. This method does not change the approaches of estimating the population and calculating the design values. Moreover, the physical meaning of the Me-RS function $RS(t)$ is the research variable under the influence of unit Mechanism function value. Since the design value $RS_p$ is only a sample of the $RS_t$ population, by using the Equation (9), the $RS_p$ is expanded by the Mechanism function value of the design stage, and the obtained design value of the nonstationary annual runoff series is the result of the action of the influence factors under the design conditions. The second method is from the perspective of the design value: as shown in Figures 4 and 5, the precipitation has stayed the same since 1982, but with the continuous construction of the check dams, the water storage volume and the water surface area has gradually increased, resulting in the increase of the total evaporation and the decrease of annual runoff. The average annual runoff measured in recent 30 years and 20 years are 33.7 million m$^3$ and 30.9 million m$^3$, respectively, which is basically equivalent to the 50%-frequency design value 26.1 million m$^3$, indicating that the calculation results are basically consistent with the reality.

*6.2. Comparison with the Traditional Frequency Analysis Method*

As the current nonstationary frequency analysis theories are far from mature, many hydraulic engineering methods are still designed using the traditional frequency analysis method regardless of whether the sample series is stationary, which is called here the direct traditional frequency analysis method (DTFAM). In order to compare the difference between the Me-RS method and the DTFAM, we also used the DTFAM to calculate the design values of the annual runoff of JRB. The optimal distribution and distribution parameters of the annual runoff series are listed in Table 5. The design values of different return period are shown by the black line in Figure 10. In practice, it is usually necessary to analyze the design values of wet, medium, and dry years, so the results at the frequencies of 20%, 50%, and 80% are list in Table 6. The results of the two methods are greatly different, and the reasons mainly include two aspects. On the one hand, for the Me-RS method, the current *RI* value is used to calculate the annual runoff design value according to Equation (9). Therefore, the obtained design values reflect the state after the annual runoff series has been reduced. However, the sample series used by the traditional method covers the whole downtrend process of the annual runoff series, which cannot reflect the state of the present stage nor the state of any moment but the average state over the years. On the other hand, the ranking of each sample point has changed after the conversion of the original series to the Me-RS series according to Figure 7. For example, in 1967, from the second place in the original series to the fifth place in the Me-RS series, the change in ranking will also lead to a change in the design value. According to the measured annual runoff data, the design annual runoff at the 50% frequency obtained by the DTFAM is 48.5 million m$^3$, which is 44% or 57% larger than the measure average annual runoff values over recent 30 or 20 years, respectively, and far away from the reality.

**Table 5.** Parameter estimation of the optimal distribution for the original annual runoff series of JRB.

| Object | Optimal Distribution | Estimated Parameters |
|---|---|---|
| Annual runoff series $R_t$ | WEI | $\mu = 1449.781, \sigma = 4649.454, \kappa = 1.166$ |

**Table 6.** Nonstationary design values and their uncertainties at different frequencies.

| Method | Design Value (Width of 95% Confidence Intervals)/10$^4$ m$^3$ | | |
|---|---|---|---|
| | Frequency of 20% | Frequency of 50% | Frequency of 80% |
| Me-RS method | 3357.52 (571.35) | 2610.38 (514.76) | 1920.62 (450.62) |
| Traditional method | 8442.30 (3124.45) | 4845.28 (1866.65) | 2734.43 (1133.84) |

According to the Bootstrap method described in Section 2.3, we further analyzed and calculated the uncertainty interval of the design annual runoff deduced by the Me-RS method and the DTFAM, as shown in the shadow in Figure 10. As the return period increases, the uncertainty of the design value increases. However, the uncertainty change rate of the Me-RS method is significantly smaller than that of the DTFAM, and the uncertainty interval width is also far smaller. For the frequency of 50%, the 95% confidence intervals of design value calculated by the DTFAM vary from 40.3 million m$^3$ to 58.9 million m$^3$, and even the lower limit is 19% or 30% larger than the measure average annual runoff values over recent 30 or 20 years, respectively. If the calculation results of DTFAM are used for water resources planning and management, it will have a great impact on water security. It is not consistent with the statistical principle to analyze the nonstationary hydrological series by using the traditional method based on stationary samples, so the obtained design values cannot be guaranteed to conform to the reality. Therefore, the direct use of nonstationary sample series for traditional frequency analysis should be avoided.

*6.3. Application Problem of the Me-RS Method*

According to the above discussion, the Me-RS method is obviously superior to the DTFAM no matter the theory, the coincidence between the design value and the measured data, or the uncertainty of the design value. The main reason is that the Me-RS method considers the physical cause of nonstationary changes of annual runoff series and obtains the stationary sample series needed for the frequency analysis according to this so as to ensure the correct use of the traditional frequency analysis method. However, the design value is related to the state of the influence factor at the design stage, which brings two problems. First, if the state of the influencing factors changes in the future, then the project constructed at the current stage will inevitably encounter unsuitable problems; for example, the check dams may be continuously silted in the future. Second, changes in some new influencing factors, such as significant changes in rainfall, or in other factors may lead to the unsuitable problem of the projects in the future.

For the first problem, the solution is to recalculate the new Me-RS series according to the changed impact factor. Since the new Me-RS series still reflects the unit action of the original factors, the design value can be recalculated combined with the existing Me-RS series and then according to the new design value to analyze the countermeasures of the hydraulic projects. The second problem can be discussed in two ways. If the new nonstationary influencing factor is a hydrological element with observed data, we only need to reconstruct the annual runoff series by the observed data, then recalculate the design value and analyze the countermeasures of the hydraulic projects. However, if the new factor has never been observed in the past, such as the influence of some human activities that has never occurred, the Me-RS method fails because the influence has not been recorded in history.

Based on the above discussion, it can be concluded that under the nonstationary conditions, all the exiting hydraulic projects will encounter unsuitable problems, and the Me-RS method can provide a reasonable basis for solving this problem.

## 7. Conclusions

Human activities and climate change lead to nonstationary changes in the originally stationary hydrological series, which brings a theoretical bottleneck to hydrological frequency analysis based on simple samples. As there is no effective solution at present, engineers are still forced to use the traditional frequency analysis method to conduct frequency analysis on the nonstationary hydrological series. However, the results cannot be evaluated, so the safety and economy of the design scheme cannot be judged. The Me-RS proposed in this paper provides an effective tool for annual runoff frequency analysis under nonstationary conditions. The case study on the calculation of design annual runoff in JRB shows that compared with the directly frequency analysis of the nonstationary hydrological series, the Me-RS method not only has theoretical support, but also the obtained design values are consistent with the actual condition and has much smaller uncertainty. Furthermore, the Me-RS method can consider not only the current design conditions but also the future design conditions

The traditional frequency analysis method is mature in theory and has been tested by engineering practice, while the Me-RS method can achieve good effect because it combines physical cause (Mechanism function) with statistical theory and establishes the Me-RS function according to the Mechanism function of the influence factor, obtains a stationary Me-RS series, and ensures the correct use of the traditional frequency analysis method. It is not consistent with the statistical principle to analyze the nonstationary hydrological series by using the traditional method based on stationary samples, so the obtained design values cannot be guaranteed to conform to the reality. Therefore, the direct use of nonstationary sample series for frequency analysis should be avoided.

**Author Contributions:** Conceptualization, Y.Q.; methodology, Y.Q. and S.L.; analysis and calculation, S.L.; writing—original draft preparation, S.L.; amending manuscripts, Y.Q.; visualization, S.L.; project administration, Y.Q.; funding acquisition, Y.Q. All authors have read and agreed to the published version of the manuscript.

**Funding:** This research was funded by National Natural Science Foundation of China, grant number 51679184.

**Data Availability Statement:** The data presented in this study are available on request from the corresponding author.

**Acknowledgments:** The field survey of this investigation was carried out as an activity of the Yulin Water Authority of Shaanxi Province, and Yulin Hydraulic engineering design Team for the hydrologic manual revision. Moreover, the observed data were obtained from the team. We would like to express our gratitude and thank all who supported us in this research.

**Conflicts of Interest:** The authors declare no conflict of interest.

# References


1. Li, S.; Qin, Y.; Song, X.; Bai, S.; Liu, Y. Nonstationary frequency analysis of the Weihe River annual runoff series using de-nonstationarity method. *J. Hydrol. Eng.* **2021**, *26*, 04021034. [CrossRef]
2. Zhang, Q.; Singh, V.P.; Sun, P.; Chen, X.; Zhang, Z.; Li, J. Precipitation and streamflow changes in China: Changing patterns causes and implications. *J. Hydrol.* **2011**, *410*, 204–216. [CrossRef]
3. Milly, P.C.D.; Dunne, K.A.; Vecchia, A.V. Global pattern of trends in streamflow and water availability in a changing climate. *Nature* **2005**, *438*, 347–350. [CrossRef] [PubMed]
4. Milly, P.C.; Betancourt, J.; Falkenmark, M.; Hirsch, R.M.; Kundzewicz, Z.W.; Lettenmaier, D.P.; Stouffer, R.J. Stationarity is dead: Whither water management? *Science* **2008**, *319*, 573–574. [CrossRef]
5. Strupczewski, W.G.; Singh, V.P.; Feluch, W. Non-stationary approach to at-site flood frequency modelling I. Maximum likelihood estimation. *J. Hydrol.* **2001**, *248*, 123–142. [CrossRef]
6. Strupczewski, W.G.; Kaczmarek, Z.; Feluch, W. Non-stationary approach to at-site flood frequency modelling II. Weighted least squares estimation. *J. Hydrol.* **2001**, *248*, 143–151. [CrossRef]
7. Strupczewski, W.G.; Singh, V.P.; Mitosek, H.T. Non-stationary approach to at-site flood frequency modelling III. Flood analysis of Polish rivers. *J. Hydrol.* **2001**, *248*, 152–167. [CrossRef]
8. Vogel, R.M.; Yaindl, C.; Walter, M. Nonstationarity: Flood magnification and recurrence reduction factors in the United States. *J. Am. Water Resour. Assoc.* **2011**, *47*, 464–474. [CrossRef]
9. Zeng, H.; Feng, P.; Li, X. Reservoir flood routing considering the non-stationarity of flood series in north China. *Water Resour. Manag.* **2014**, *28*, 4273–4287. [CrossRef]
10. Rigby, R.A.; Stasinopoulos, D.M. Generalized additive models for location, scale and shape. *J. R. Stat. Soc. Ser. C* **2005**, *54*, 507–544. [CrossRef]
11. López, J.; Francés, F. Non-stationary flood frequency analysis in continental Spanish rivers, using climate and reservoir indices as external covariates. *Hydrol. Earth Syst. Sci.* **2013**, *17*, 3189–3203. [CrossRef]
12. Xiong, L.; Jiang, C.; Du, T. Statistical attribution analysis of the nonstationarity of the annual runoff series of the Weihe River. *Water Sci. Technol.* **2014**, *70*, 939–946. [CrossRef] [PubMed]
13. Salas, J.D.; Obeyskekera, J. Revisiting the concepts of return period and risk for non-stationary hydrologic extreme events. *J. Hydrol. Eng.* **2014**, *19*, 554–568. [CrossRef]
14. Yan, L.; Xiong, L.; Guo, S.; Xu, C.Y.; Xia, J.; Du, T. Comparison of four nonstationary hydrologic design methods for changing environment. *J. Hydrol.* **2017**, *551*, 132–150. [CrossRef]
15. Villarini, G.; Smith, J.A.; Serinaldi, F.; Bales, J.; Bates, P.D.; Krajewski, W.F. Flood frequency analysis for nonstationary annual peak records in an urban drainage basin. *Adv. Water Resour.* **2009**, *32*, 1255–1266. [CrossRef]
16. Olsen, J.R.; Lambert, J.H.; Haimes, Y.Y. Risk of extreme events under nonstationary conditions. *Risk Anal.* **1998**, *18*, 497–510. [CrossRef]
17. Wigley, T.M.L. The effect of changing climate on the frequency of absolute extreme events. *Clim. Chang.* **2009**, *97*, 67–76. [CrossRef]
18. Parey, S.; Malek, F.; Laurent, C.; Dacunha-Castelle, D. Trends and climate evolution: Statistical approach for very high temperatures in France. *Clim. Chang.* **2007**, *81*, 331–352. [CrossRef]
19. Parey, S.; Hoang, T.T.H.; Dacunha-Castelle, D. Different ways to compute temperature return levels in the climate change context. *Environmetrics* **2010**, *21*, 698–718. [CrossRef]
20. Hu, Y.; Liang, Z.; Singh, V.P.; Zhang, X.; Wang, J.; Li, B.; Wang, H. Concept of equivalent reliability for estimating the design flood under non-stationary conditions. *Water Resour. Manag.* **2018**, *32*, 997–1011. [CrossRef]

21. Rootzén, H.; Katz, R.W. Design Life Level: Quantifying risk in a changing climate. *Water Resour. Res.* **2013**, *49*, 5964–5972. [CrossRef]
22. Read, L.K.; Vogel, R.M. Reliability, return periods, and risk under nonstationarity. *Water Resour. Res.* **2015**, *51*, 6381–6398. [CrossRef]
23. Mandelbrot, B.B.; Wallis, J.R. Computer experiments with fractional gaussian noises: Part 1, averages and variances. *Water Resour. Res.* **1969**, *5*, 228–241. [CrossRef]
24. Lins, H.F.; Cohn, T.A. Stationarity: Wanted dead or alive. *J. Am. Water Resour. Assoc.* **2011**, *47*, 475–480. [CrossRef]
25. Serinaldi, F.; Kilsby, C.G. Stationarity is undead: Uncertainty dominates the distribution of extremes. *Adv. Water Resour.* **2015**, *77*, 17–36. [CrossRef]
26. Qin, Y.; Li, S. Study on mechanism-based reconstruction method to cope with the nonstationary change of hydrological series. *J. Hydraul. Eng.* **2021**, *52*, 807–818. (In Chinese) [CrossRef]
27. Efron, B. Bootstrap methods: Another look at the jackknife. *Ann. Stat.* **1979**, *7*, 1–26. [CrossRef]
28. Mann, H.B. Nonparametric tests against trend. *Econom. J. Econom. Soc.* **1945**, *13*, 245–259. [CrossRef]
29. Kendall, M.G. *Rank Correlation Methods*; Charles Griffin: London, UK, 1975.
30. Pettitt, A.N. A non-parametric approach to the change-point problem. *Appl. Stat.* **1979**, *28*, 126–135. [CrossRef]
31. Breusch, T.S.; Pagan, A.R. A simple test for heteroscedasticity and random coefficient variation. *Econometrica* **1979**, *47*, 1287–1294. [CrossRef]
32. Hosking, J.R.M. L-moments: Analysis and estimation of distributions using linear combinations of order statistics. *J. R. Stat. Soc. Ser. B Methodol.* **1990**, *52*, 105–124. [CrossRef]
33. Kolmogorov-Smirnov, A.N.; Kolmogorov, A.; Kolmogorov, M. Sulla determinazione empírica di una legge di distribuzione. *G. Institto Ital. Attuari* **1933**, *4*, 83–91.
34. Nash, J.E.; Sutcliffe, J.V. River flow forecasting through conceptual models part I-A discussion of principles. *J. Hydrol.* **1970**, *10*, 282–290. [CrossRef]

MDPI

St. Alban-Anlage 66

4052 Basel

Switzerland

Tel. +41 61 683 77 34

Fax +41 61 302 89 18

www.mdpi.com

*Water* Editorial Office

E-mail: water@mdpi.com

www.mdpi.com/journal/water